Computer Architecture and Design Methodologies

Twilight zone of Moore's law is affecting computer architecture design like never before. The strongest impact on computer architecture is perhaps the move from unicore to multicore architectures, represented by commodity architectures like general purpose graphics processing units (gpgpus). Besides that, deep impact of application-specific constraints from emerging embedded applications is presenting designers with new, energy-efficient architectures like heterogeneous multi-core, accelerator-rich System-on-Chip (SoC). These effects together with the security, reliability, thermal and manufacturability challenges of nanoscale technologies are forcing computing platforms to move towards innovative solutions. Finally, the emergence of technologies beyond conventional charge-based computing has led to a series of radical new architectures and design methodologies.

The aim of this book series is to capture these diverse, emerging architectural innovations as well as the corresponding design methodologies. The scope covers the following.

- Heterogeneous multi-core SoC and their design methodology
- Domain-specific architectures and their design methodology
- Novel technology constraints, such as security, fault-tolerance and their impact on architecture design
- Novel technologies, such as resistive memory, and their impact on architecture design
- Extremely parallel architectures

More information about this series at https://link.springer.com/bookseries/15213

Anubhab Baksi

Classical and Physical Security of Symmetric Key Cryptographic Algorithms

Springer

Anubhab Baksi
Temasek Laboratories
Nanyang Technological University
Singapore, Singapore

ISSN 2367-3478 ISSN 2367-3486 (electronic)
Computer Architecture and Design Methodologies
ISBN 978-981-16-6521-9 ISBN 978-981-16-6522-6 (eBook)
https://doi.org/10.1007/978-981-16-6522-6

This Springer imprint is published by the registered company Springer Nature Singapore Pte Ltd.
The registered company address is: 152 Beach Road, #21-01/04 Gateway East, Singapore 189721, Singapore

About This Book

Symmetric key cryptography is one of the cornerstones of security in the modern era of electronic communication. The symmetric key algorithms, known as the ciphers, are to satisfy certain requirements in order to be considered secure, which are broadly classified as classical attack and physical attack. We show new results in context of both the classical and physical attacks to advance the state of the art.

In classical attack, we first show an issue related to a common modelling using Mixed Integer Linear Programming (MILP). We provide a new MILP modelling that overcomes this issue and explores heuristic options to reduce the solution time taken by the MILP solver. Our analysis shows that the solution time can be improved nearly ten-folds by using a proper heuristic. Second, we show how Machine Learning (ML) can be used as a generic tool in the analysis of the symmetric key ciphers. In the process, we demonstrate how the existing security notions (that do not use ML) underestimate the vulnerability of the ciphers. To the best of our knowledge, this is the first generic application of ML in this field.

In physical attack, we start with new mathematical results related to the Differential Fault Attack (DFA) from the point of view of the cipher designer. Next, we make use of these results to propose a cipher named DeFault, which has an inbuilt resistance against DFA. While all other methods to thwart DFA rely on some form of duplication, DeFault has an inherent protection against DFA that does not use duplication and hence is the first of its kind. Third, we analyse the so-called infective countermeasure that is used as a duplication-based DFA countermeasure in more depth. We construct new schemes, show weakness of an existing scheme, and propose a simple patch to another scheme to fix its weakness, among other results. Last, we propose a low-cost countermeasure to a newly proposed fault model, named Statistical Ineffective Fault analysis (SIFA). In contrast to the existing SIFA countermeasures that rely on some form of triplication at its core, our countermeasure uses duplication and thus is the most cost-effective.

Contents

1 Introduction 1
1.1 Context and Motivation 1
1.2 Research Directions 3
1.2.1 Cipher Design and Classical Cryptanalysis 3
1.2.2 Realization/Mapping 5
1.2.3 Physical Attack and Countermeasure 5
1.3 Standardization of Ciphers 6
1.4 Organization 7
References 10

2 Fundamentals of Symmetric Key Cryptography 13
2.1 Building Blocks 13
2.1.1 Boolean Function 13
2.1.2 Substitution Box (SBox) 14
2.1.3 Linear Layer 19
2.2 Primitives 19
2.2.1 (Un-keyed) Permutation 19
2.2.2 Block Cipher 20
2.2.3 Stream Cipher 21
2.2.4 Hash Function 22
2.2.5 Message Authentication Code (MAC) 24
2.2.6 Authenticated Encryption with Associated Data (AEAD) 25
2.3 Cipher Families 26
2.3.1 Substitution Permutation Network (SPN) 26
2.3.2 Feistel Network 27
2.3.3 Add–Rotation–XOR (ARX) Construction 29
2.4 Description of Exemplary Ciphers 30
2.4.1 ADVANCED ENCRYPTION STANDARD (AES) 30
2.4.2 PRESENT-80 33
2.4.3 GIFT-128 35

2.4.4 CHASKEY ... 36
2.5 Formidability of the Attacker ... 39
2.5.1 Conventional Notions of Security ... 39
2.5.2 Power of the Attacker ... 39
2.5.3 Objective of the Attacker ... 40
2.6 Major Classical Attacks ... 42
2.6.1 Differential Attack ... 42
2.6.2 Linear Attack ... 44
2.6.3 Algebraic Attack ... 44
2.6.4 Integral/Cube Attack ... 46
2.6.5 Impossible Differential Attack ... 47
2.7 Device Implementation ... 47
2.8 Additional Topics ... 49
2.8.1 Black Box–Grey Box–White Box Models ... 49
2.8.2 Mixed Integer Linear Programming (MILP) ... 50
2.8.3 Machine Learning (ML) ... 51
2.8.4 Competitions ... 54
References ... 55

3 Fault Attack ... 59
3.1 Introduction ... 59
3.2 Fault Models ... 60
3.2.1 Precise Bit Flip ... 60
3.2.2 Single/Multiple Fault Adversary ... 60
3.2.3 Random/Deterministic Fault Model ... 61
3.2.4 Information Theoretic View ... 61
3.2.5 Other Aspects ... 61
3.3 Data Alteration Methods ... 62
3.3.1 Volatility ... 62
3.3.2 Modification of Operation ... 62
3.3.3 Modification of Operand ... 63
3.4 Sources of Fault Injection ... 65
3.5 Analysis Methods ... 68
3.5.1 Difference-Based Fault Analysis ... 69
3.5.2 Collision-Based Fault Analysis ... 76
3.5.3 Statistics-Based Fault Analysis ... 77
3.5.4 Others ... 80
3.6 Generalized Fault Attack Automation Frameworks ... 81
3.6.1 Cipher Level Approaches ... 81
3.6.2 Implementation Level Approaches ... 82
3.7 Countermeasures ... 82
3.7.1 Detection ... 84
3.7.2 Infection ... 85
3.7.3 Prevention ... 86

3.7.4 Re-keying, Tweak and Tweak-in-Plaintext, Masking Plaintext ... 86
3.7.5 Attacks on Countermeasures ... 87
3.7.6 Specialized Countermeasures Against Statistical Ineffective Fault Attack ... 88
References ... 89

4 Side Channel Attack ... 99
4.1 Introduction and Background ... 99
4.2 Power Analysis ... 101
4.2.1 Simple Power Analysis ... 102
4.2.2 Differential Power Analysis (DPA) ... 102
4.2.3 Template Attack ... 103
4.2.4 Correlation Power Analysis (CPA) ... 103
4.2.5 Countermeasures ... 103
4.3 Case Study: Side Channel Analysis of CHASKEY ... 104
4.3.1 Practical Attack Setups ... 104
4.3.2 Experimental Results ... 105
References ... 107

5 New Insights on Differential and Linear Bounds Using Mixed Integer Linear Programming ... 109
5.1 Introduction ... 110
5.2 Background ... 111
5.2.1 Branch Number to Model SBox (Inscrypt'11) ... 112
5.2.2 Convex Hull to Model SBox—Active SBox Count (Eprint'13) ... 113
5.2.3 Convex Hull to Model SBox—Exact Bound (Eprint'14) ... 115
5.2.4 Redundant Constraints to Reduce Solution Time (Eprint'19) ... 116
5.3 Problem with Convex Hull Modelling ... 116
5.4 Automated Bounds with MILP: Our Proposal ... 121
5.4.1 Modelling ... 122
5.4.2 Optimizations ... 123
5.4.3 Results ... 126
5.5 Conclusion ... 131
5.6 Supplementary Discussion ... 132
5.6.1 Detailed Description on MILP Modelling of XOR ... 132
5.6.2 Illustration with MILP Model for 1-Round Differential Bound for GIFT-128 ... 133
5.6.3 Illustration with MILP Model of Previous Constraints for 4-Round Differential Bound for GIFT-128 ... 133
References ... 139

6 Machine Learning-Assisted Differential Distinguishers for Lightweight Ciphers 141
6.1 Introduction 142
6.2 Background 144
6.2.1 Markov Ciphers 144
6.2.2 Gohr's Work on SPECK (CRYPTO'19) 146
6.3 Basic Description of the Ciphers 147
6.3.1 GIMLI 147
6.3.2 ASCON 149
6.3.3 KNOT 149
6.4 Machine Learning-Based Distinguishers 150
6.4.1 Model 1: Multiple Input Differences 150
6.4.2 Model 2: One Input Difference 153
6.4.3 Comparison with Existing Models 154
6.5 Results on Round-Reduced Ciphers 154
6.5.1 GIMLI (Model 1) 154
6.5.2 ASCON and KNOT (Model 1) 156
6.5.3 CHASKEY (Model 2) 157
6.6 Choice of Machine Learning Model 157
6.7 Conclusion and Follow-Up Problems 159
References 161

7 Differential Paradox: How an SBox Plays Against Differential Fault Analysis 163
7.1 Introduction 163
7.2 Difference Distribution Table-Related Properties 164
7.3 Characterizing SBoxes in View of DFA 167
7.4 Implication of Our Analysis and Future Work 174
References 175

8 DEFAULT: Cipher-Level Resistance Against Differential Fault Attack 177
8.1 Introduction 177
8.2 Background 180
8.2.1 DFA Models 180
8.2.2 DFA Protection 181
8.2.3 Feasibility of Cipher-Level Protection Against Faults 182
8.2.4 Working Principle for DFA 184
8.3 Construction of DFA-Resistant Layer/Cipher 185
8.3.1 Ad-hoc DFA Protection to Any Cipher (DEFAULT-LAYER) 185
8.3.2 Extension to a Full-Fledged Cipher (DEFAULT) 186
8.3.3 Construction of DEFAULT-LAYER 187
8.3.4 Construction of DEFAULT-CORE (and DEFAULT) 189

8.4 Design Rationale 191
8.4.1 Design Philosophy 191
8.4.2 Structure of the DEFAULT PermBits 193
8.4.3 Selection of the DEFAULT SBoxes 194
8.4.4 Unbiased Linear Structures 197
8.5 Security Analysis 198
8.5.1 Protection Against Differential Fault Attack 199
8.5.2 Protection Against Classical Cryptanalysis 201
8.5.3 Protection Against Side Channels Attacks 203
8.6 Automated Bounds for Differential and Linear Attacks 204
8.7 Performance 206
8.7.1 Hardware Benchmark 206
8.7.2 Software Benchmark 207
8.8 Conclusion 208
8.9 Supplementary Discussion 210
References 214

9 To Infect or Not to Infect: A Critical Analysis of Infective Countermeasures in Fault Attacks 217
9.1 Introduction 217
9.2 Background 220
9.2.1 Context of Differential Fault Analysis 220
9.2.2 Early Countermeasures: Detection-Based 220
9.2.3 Evolution of Infective Countermeasures 221
9.2.4 Notations and Terminologies 222
9.2.5 Necessity and Sufficiency of Randomness 226
9.2.6 Scope and Applicability 226
9.2.7 Connection with Side Channel Countermeasures 229
9.3 Type I Constructions 230
9.3.1 Multiplication-Based Constructions 230
9.3.2 Derivative-Based Constructions 231
9.3.3 New Type I Schemes 231
9.3.4 Benchmarking Results for Type I Schemes 235
9.4 Type II/Cipher-Level Constructions 237
9.4.1 Critical Look at CHES'14 Countermeasure 241
9.4.2 Our Patch for LatinCrypt'12 Countermeasure 243
9.5 Conclusion 249
References 250

10 A Novel Duplication-Based Countermeasure to Statistical Ineffective Fault Analysis 255
10.1 Introduction 255
10.2 Fault Attack Preliminaries 258
10.2.1 Differential Fault Attack (DFA) 258
10.2.2 General Countermeasures Against Fault Attacks 258
10.3 Statistical Ineffective Fault Attack (SIFA) 259

10.3.1 Duplication-Based Countermeasures and Need for Specialization 259
10.3.2 Existing SIFA Countermeasures 261
10.4 Our Proposed Solution 262
10.4.1 Adopting Inverted Logic to Symmetric Key Ciphers 265
10.4.2 Benchmarks 268
10.4.3 Evaluation 269
10.4.4 Comparison with Existing Countermeasures 269
10.4.5 Connection with Side Channel Countermeasures 272
10.5 Conclusion 272
References 273

11 Concluding Remarks 277
11.1 Synopsis 277
11.2 Interesting Problems for Future Research 279
References 281

Index 283

Chapter 1
Introduction

1.1 Context and Motivation

In the present world, electronic communication takes a vital role in our life. We frequently deal with sensitive/confidential information. As such, the security of such information is of utmost importance. *Cryptography* is the branch of computer science (often involves electronics, mathematics, and statistics) which studies the area security mostly from a theoretical perspective. In short, it deals with data *confidentiality* (or *privacy*), *integrity*, *authentication*, and *non-repudiation*.[1] Confidentiality means only the intended recipient will be able to properly get the message (the message will appear as garbage to any third party). Integrity refers to the identification if the message is tampered with. Authentication is the way to ascertain that the sender is verifiable. Finally, non-repudiation means the inability for the sender to deny the dispatch of the message after it has been sent.

Traditionally, the work-flow is described by scenario where Alice wants to send a message to Bob securely, where Eve is eavesdropping. We will refer to Eve as the *attacker* or the *adversary*. Alice and Bob share some secret component (*key*), which is not known to Eve. In symmetric key (also known as *private key*) systems, the key shared between Alice and Bob is identical, which contrasts with *public key* (also known as *asymmetric key*) cryptography, where the secret keys are different. Symmetric key systems are generally simpler (hence faster) compared to the asymmetric key systems.

Alice converts her message (*plaintext*) using some agreed-upon algorithm (*cipher*) with the help of her key. This process is called as *encryption*. Now, to Eve, this encrypted message (*ciphertext*) will look like gibberish. Bob, with his key, will be able to convert the ciphertext back to the plaintext; this process is known as *decryption*. In another direction, Alice may create a *tag* for her message (which may

[1] Notions like proof-of-work and proof-of-stake are now considered within cryptography. Such notions are related to niche applications like crypto-currency or blockchain and are not considered within the scope of this document.

A. Baksi, *Classical and Physical Security of Symmetric Key Cryptographic Algorithms*, Computer Architecture and Design Methodologies,
https://doi.org/10.1007/978-981-16-6522-6_1

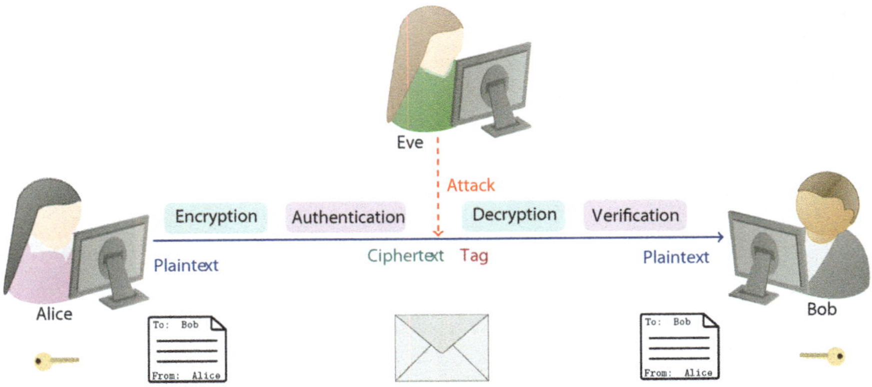

Fig. 1.1 Alice–Bob–Eve model

or may not require the secret key). This tag would be verified by Bob to ensure the authenticity (that the message came from Alice) and/or integrity (the message has not been tampered with). Figure 1.1 gives a quick overview of the system.

Nowadays, the cipher description is almost always public in the academia and industry. The key is the only secret component. Thus, Eve will not be able to recover the plaintext from the ciphertext, or forge an authentication tag if she does not know the key, although she is well aware of the cipher. Now, she will try to get more information on the key and/or plaintext(s) based on her knowledge on the cipher and the ciphertext(s)/tag(s). In other words, the attacker can query (possibly a reduced version of) the cipher and receive the corresponding output. This type of evaluation is referred to as *cryptanalysis* or an *attack*. In general, cryptanalysis refers to evaluating the security and thereby finding any weakness in a cryptographic primitive.

Cryptography, at least commonly, does not deal with other security-related issues like anonymity, availability, deniability, time/duration of communication, amount of data exchange, etc. It also does not deal with cybersecurity/network security/hardware security-related issues like fallback authentication, hacking, biometric authentication, computer virus, etc. In other words, cryptography is mathematical model for analysing the security of the ciphers under some generality assumptions.

It may be stated that, often the term *cryptology* is used as the superset of cryptography (cipher design/implementation part) and cryptanalysis (cipher analysis part). However, as the design and implementation are done by keeping the attacks in mind, cryptography covers the entire cryptanalysis in that sense. Hence, there is little advantage to consider cryptology is different from cryptography.

1.2 Research Directions

In the theoretical model of cipher analysis, it is assumed that the attacker can perform all sorts of analysis. For example, the attacker has access to high-performance computing resources. Any weakness can be utilized by the attacker, so goes the saying: "Security is only as strong as the weakest link" (as shown in Fig. 1.4[2]). Relatively new models further allow the attacker access to the device performing some cryptographic operation. The communication channel or the other devices in the network can also be subject to attack.

Figure 1.2 represents an overall (and non-exhaustive) view of research directions in the context of the book. In addition, Fig. 1.3 shows the overall connection of research topics. It may be noted that the attacker has the access to the input and output information for the classical as well as device-dependent attacks.

1.2.1 Cipher Design and Classical Cryptanalysis

The cryptographic community designs ciphers for different applications. These ciphers are designed carefully in order to satisfy certain security requirement, and generally released in public for peer review. Some standardized analysis methods are used to verify the security claims. These methods attempt to find some inherent

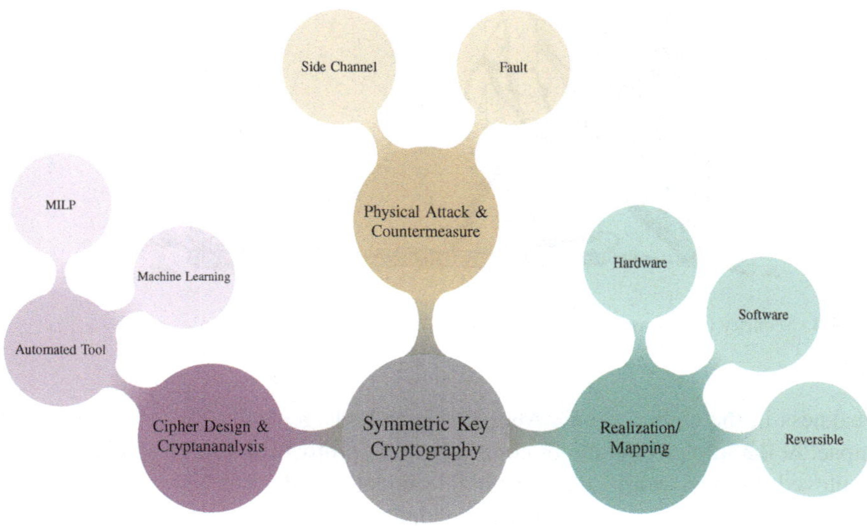

Fig. 1.2 Major research directions (in current context) in symmetric key cryptography

[2] A sketch by Sukumar Ray, adopted from https://commons.wikimedia.org/wiki/File:Chordhara.gif.

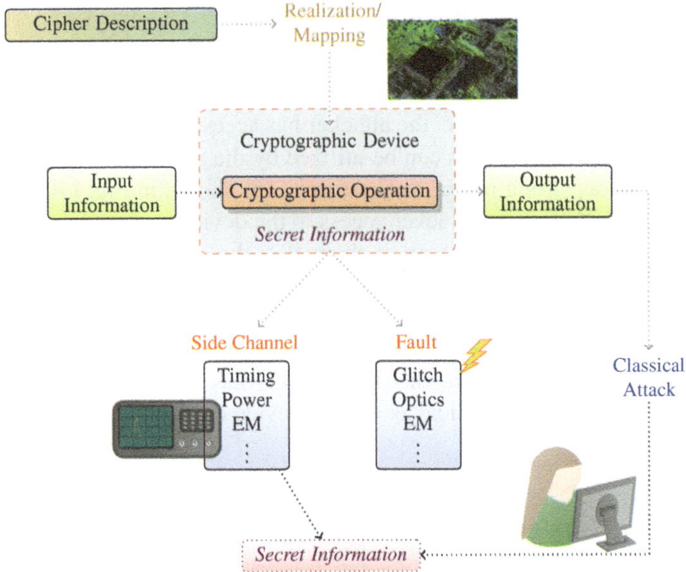

Fig. 1.3 Overview of realization/mapping, classical, and physical attacks

Fig. 1.4 Security is only as strong as the weakest link

weakness in the cipher, if any. Also, those methods generally do not consider the impact of the implementation or communication protocol where this cipher will be employed. In other words, these methods simulate an attacker who can only query the cipher with inputs (can only know the input to the cipher or can choose the input to the cipher, depending on the model), collect the corresponding cipher outputs, and attempts to learn about the secret component of the cipher through rigorous observation. These methods, which we loosely refer to as attacks, have been well studied over the last couple of decades. If no major weakness is reported in a couple of years or so, then the ciphers is generally considered to be secure.

1.2.2 *Realization/Mapping*

Realization/mapping deals with the problem of implementing a cipher to a device. Once a cipher is designed and deemed secure against the classical attack model, it is essential to test the cost of secure implementation of the cipher under different platforms. Broadly, the devices can be classified into two categories, namely, hardware and software. Hardware refers to ASIC (application-specific integrated circuit) and FPGA (field programmable gate array) platforms, whereas software refers to CPU (microprocessor or microcontroller) and GPGPU (general-purpose graphics processing unit) implementations.

1.2.3 *Physical Attack and Countermeasure*

Relative recently, a new way of cryptanalysis has been emerged. As the small-scale devices started becoming more ubiquitous, the attacker may be able to access the devices while those devices are performing some cryptographic operation. These include smart card, Radio Frequency Identification (RFID) chips, sensors, embedded systems, etc. The following types of physical attacks are considered within the scope:

- The attacker can now observe power consumption, electromagnetic radiation, etc. of the device. This type of attack is known as *Side Channel Analysis* or *Side Channel Attack* (SCA).
- The attacker can also tamper the device with, e.g., glitch at the input power line, injecting a LASER shot, etc. This may affect the operation, e.g., flipping a bit from 0 to 1 or vice versa, skipping one instruction, etc. This attack methodology is known as *Fault Analysis* or *Fault Attack* (FA).

In this document, we distinguish between FA and SCA (instead of considering FA and SCA as active and passive variants of the same underlying method) for the following reasons:

- FAs are more closely related to classical attacks (can be thought as simplified versions of the classical attacks), whereas SCAs are typically not.
- FAs are easy to simulate. This is not so easy for SCAs as the environmental noise is to be taken into account.

Normally, one of the two attacks is sufficient to recover the secret component in the design. A cipher, which is secure against the classical analysis, will likely be vulnerable to this type of attacks unless some countermeasure is adopted. As the devices are easily available, the attacker can study it even up to the micro-architecture

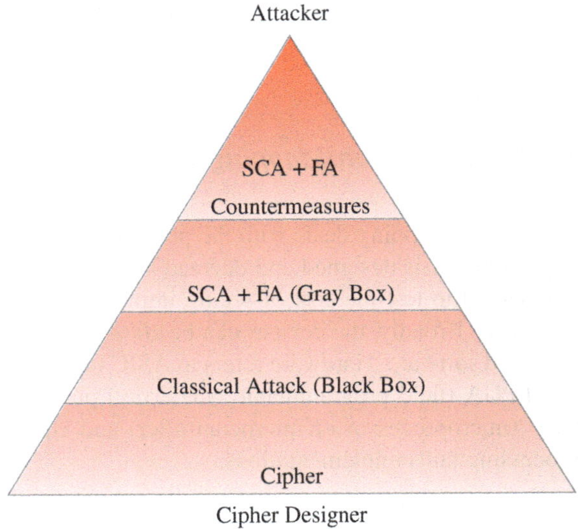

Fig. 1.5 Hierarchy of models based on attacker's power (in current context)

level. This knowledge enables her to carry out attacks on the system without directly attacking the theoretical backbone (that ascertains the classical security of the cipher). A basic overview of the models based on the perceived power of the attacker, mostly tailored towards the contribution of the book, is given in Fig. 1.5. Traditionally, classical attacks on the cipher are considered first. After that the cipher is implemented on a device. Finally, the SCA and FA countermeasures are put in place.

1.3 Standardization of Ciphers

Not long ago, the cipher description used to be kept secret. However, as it has become apparent over time, this practice does not necessarily offer a stronger security. Part of the reason is, it is possible for a dedicated third party to reverse engineer the cipher.[3] On the other hand, there is a possibility that a serious weakness will go unnoticed (since no third-party analysis on the algorithm is done). For example, the KEELOQ block cipher (that was used in cars by a number of manufacturers) is now known to have serious security concern [Bog07]. Apart from that, there is a suspicion that some ciphers may contain a hidden *backdoor* (see [PW20] for more information) that can be exploited by the cipher designer. Thus, now the common practice is to release the cipher description in public and get it analysed by the community to gain trust.

---

[3] This is the case for RC4, description of which is leaked on the Internet anonymously in 1994: https://web.archive.org/web/20010722163902/http://cypherpunks.venona.com/date/1994/09/msg00304.html.

Consequently, the academia and industry work together to design, analyse, standardize, and deploy ciphers. The ciphers (specification, design rationale, security claims, known vulnerabilities, implementation cost, etc.) are available in public. Despite the collaboration though, the industry is generally slow to respond to the threats reported by the academia. For instance, the industry seems to use ciphers like `RC4` or `SHA-1` which are considered weak in the cryptographic community, despite having known vulnerabilities.[4]

1.4 Organization

In this book, results related to classical cryptanalysis as well as device-dependent attack/countermeasure are presented to advance the state of the art, together with related background information. Figure 1.6 shows basic interconnections among various parts of the document. The organization of the rest is given next.

Prerequisite and Postscript

Chapter 2 covers a concise background on symmetric key cryptography. We briefly describe the basic notions, including the building blocks, primitives, and classical attack models. All in all, basic concepts on cipher design, implementation, and cryptanalysis are mentioned. For more details, one may refer to textbooks like [Sti06, PP10, SSL15, Ava16] or dissertations such as [Jea13, Per17, Roh19].

Fault attack and its countermeasure are discussed in **Chap.** 3. This is also available in [BBB+20] with minor modification. We discuss fault attack/countermeasure from a point of view of classical attacks. Also, we provide information related to fault automation and particulars of fault injection equipment, among other related things.

Side channel attack and countermeasure are briefly described in **Chap.** 4. Here we present basic information relating to side channel attack, example of attacking the XOR operation and provide new information related to this. For better clarity, we demonstrate an experiment.[5] One may refer to other references such as [MOP07, Pee13, SSL15, Bil15, RD20] for more discussion.

Chapter 11 wraps-up the book. We discuss a few ideas which span over multiple topics which could potentially culminate in possible follow-up research works in the future.

---

[4] See, for example, https://www3.ntu.edu.sg/home/wuhj/research/msoffice/index.html or https://security.googleblog.com/2017/02/announcing-first-sha1-collision.html.

[5] The experimental results are reported for the first time in the literature. Yet, we treat that as a part of the prerequisite/background.

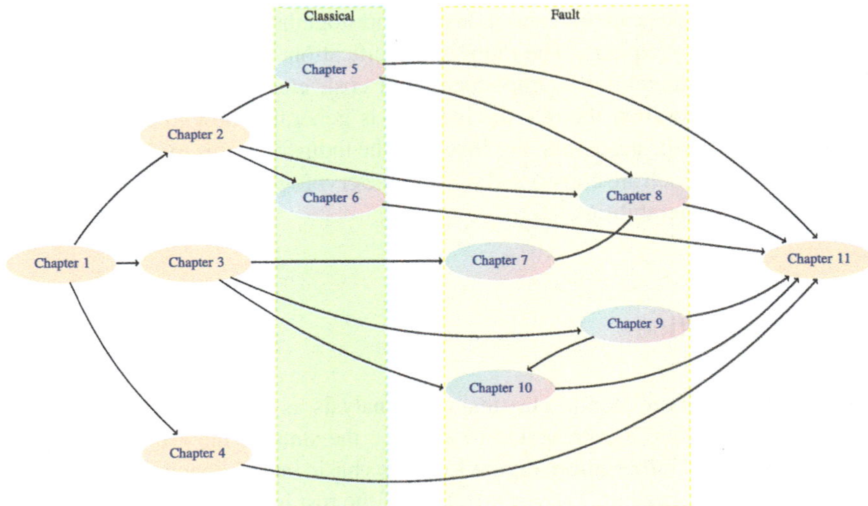

Fig. 1.6 Major interconnections among introduction/background, research contribution, and conclusion

Research Contribution

New ideas and results on Mixed Integer Linear Programming (MILP) modelling for two classical attacks are reported in **Chap.** 5. It is available in [Bak20'1], with major modification (and also in [Bak20'2], with minor modification). We first show an unreported problem related to the common MILP modelling. Next, we propose a new MILP modelling which is free from that problem. We further study the effect of redundant constraints which may reduce the solution time taken by an MILP solver by a few folds, and improved bounds for a popular cipher named `GIFT-128` [BPP+17].

We show how Machine Learning (ML) can be used as a generic tool for classical analysis. This is the first time such analysis is reported in the literature in **Chap.** 6, which is available in [BBC+21] with major modification, and in [BBC+20]) with minor modification. With four ciphers, we demonstrate how an attacker with capability to deploy machine learning tools may be able to significantly reduce the complexity of an attack. We also experiment with multiple options for the choice of the machine learning model.

Theoretical analysis of the most common type of fault attack in **Chap.** 7. To the best of our knowledge, this is the first-and-only analysis of its kind. We observe few interesting and somewhat paradoxical results.

The main contribution of **Chap.** 8 is to propose a cipher with inherent protection against certain type of fault attack, known as the *Differential Fault Attack* (DFA). As far as we know, no such research has been reported so far. In particular, all constructions which are used to protect against this fault attack rely on some form of duplication, followed by error detection. Our cipher does not use duplication, instead the security is achieved by using a completely new paradigm of cipher design.

We revisit a duplication-based fault countermeasure in **Chap.** 9, which is available in [BSS20] and [BSS19], with minor change. The contributions here are manifold. We categorize the countermeasures existing in the literature, which enables us to analyse those with meticulous details. We show a few improved countermeasures in one category. In the other category, we uncover a problem related to a particular countermeasure and propose a patch to an old (and broken) countermeasure.

Our work in **Chap.** 10, which is available in [BKK+20'2]/[BKK+20'1], with minor modification, deals with a recent fault attack model. This model is shown to bypass the duplication-based countermeasures. To protect against it, few countermeasures have been proposed in the literature, all of which triplicate to correct the error. Our solution only uses duplication, thus offering a great reduction of overhead compared to the rest. We show its effectiveness through simulation.

Overall, the contribution covers a diverse set of topics in the realm of symmetric key cryptography, both in classical and device-dependent setting. Our results shed new light in the current modelling for certain classical attacks, which are essential while designing/analysing a cipher. We further show new results in terms of fault attack as well as propose countermeasures . In total, three countermeasures are proposed/analysed here that can thwart two types of fault[6]. One important aspect of our contribution is to interpret the device-dependent attacks in the same way as classical attacks, so that a systematic and robust study of both attacks is possible from a cipher designer's point of view. The machine learning-based analysis (Chap. 6) proposes the first generic application to the symmetric key cryptography. Similarly, the theoretical analysis of a fault attack type (Chap. 7) and the cipher-level countermeasure to it (Chap. 8) offer the first such insight. We are confident, our contribution will be useful to the community and motivate further research and development.

Topics Not Included

This book assumes basic familiarity with linear and abstract algebra, coding and information theory, artificial intelligence/machine learning, and probability. As such, discussion on those topics is omitted for the sake of brevity.

Theoretical notions of cryptography such as provable security or codes (such as non-malleable codes) are not included. Similarly, public key cryptography, including post-quantum cryptography, is not included. Quantum analysis of symmetric key cryptography is not included either.

Since this book mostly deals from a classical point of view, topics related to device (such as physically unclonable function, random number generation) are omitted.

[6] Although two countermeasures protect against the same fault, the modi operandi are significantly separate.

Tools Used

- **Sage.**[7] Sage is an open-source tool, written on top of Python. It natively offers several Application Programming Interfaces (APIs) which are frequently used in cryptography.
- **Gurobi.**[8] Gurobi is an efficient tool to solve MILP and other related problems. Although commercial, the academic license is available free of cost.

Notations and Stylizations

- For better clarity, $\sharp$ is used for shorthand for length/size (typically in bits), whereas # is used shorthand for number.
- *Italics* is used when a new term introduced in that particular chapter.
- The SBoxes are represented in string format (instead of the more common table-based format). Similarly, bit permutations are described as arrays, instead of the more common table-based format.
- `Typewriter` is used to indicate cipher name, cipher component name, mode of application (for a block cipher), and nibble. Names of the ciphers are capitalized for uniformity. An SBox (given in a look-up format) is capitalized, while a nibble is not. Specialized tools (which are only designed/used in the context of symmetric key cryptography) are written as capitalized.
- We use the three terms attack, analysis and cryptanalysis interchangeably in context of classical cryptanalysis. Similarly, the terms plaintext and message, and the terms the attacker, the adversary, and the name Eve are used interchangeably.
- In case where multiple versions of the same research work exist (with practically the same title), only one version (generally the more detailed version or freely available version) is cited.
- Counting (unless mentioned otherwise) starts from 0.
- The shorthand $a(d)a + nd$ is used for $a, a + d, \ldots, a + nd$.
- The notation $\|$ is used to denote append.
- An element from $\mathbb{F}_2^n$ is represented by its corresponding integer value from $[0, 2^n - 1]$ throughout the document for simplicity. In other words, conversion of elements of $\mathbb{F}_2^n$ to-and-from $\mathbb{F}_{2^n}$ is assumed intrinsically.

References

[Bog07] A. Bogdanov, Cryptanalysis of the keeloq block cipher. Cryptology ePrint Archive, Report 2007/055 (2007). https://eprint.iacr.org/2007/055

---

[7] http://www.sagemath.org/.

[8] https://www.gurobi.com/.

[Ava16] R. Avanzi, A salad of block ciphers. Cryptology ePrint Archive, Report 2016/1171 (2016), https://eprint.iacr.org/2016/1171

[Jea13] J. Jean, Cryptanalysis of Symmetric-Key Primitives Based on the AES Block Cipher. Ph.D. thesis, Ecole Normale Supérieure de Paris (2013), https://tel.archives-ouvertes.fr/tel-00911049/document

[Per17] L.P. Perrin, Cryptanalysis, Reverse-Engineering and Design of Symmetric Cryptographic Algorithms. Ph.D. thesis, University of Luxembourg (2017), https://orbilu.uni.lu/handle/10993/31195

[Roh19] R. Rohit, Design and Cryptanalysis of Lightweight Symmetric Key Primitives. Ph.D. thesis, University of Waterloo (2019), https://uwspace.uwaterloo.ca/bitstream/handle/10012/15556/rohit_raghvendra.pdf?sequence=5&isAllowed=y

[Bil15] B. Bilgin, Threshold Implementations As Countermeasure Against Higher-Order Differential Power Analysis. Ph.D. thesis, Katholieke Universiteit Leuven and University of Twente (2015), https://www.esat.kuleuven.be/cosic/publications/thesis-256.pdf

[Bak20'1] A. Baksi, New insights on differential and linear bounds using mixed integer linear programming, in *International Conference on Information Technology and Communications Security (SecITC)* (2020), https://link.springer.com/chapter/10.1007%2F978-3-030-69255-1_4

[BBC+20] A. Baksi, J. Breier, Y. Chen, X. Dong, Machine learning assisted differential distinguishers for lightweight ciphers (extended version) (2020), https://eprint.iacr.org/2020/571

[BSS20] A. Baksi, D. Saha, S. Sarkar, To infect or not to infect: A critical analysis of infective countermeasures in fault attacks. J. Cryptogr. Eng. (2020), https://link.springer.com/article/10.1007%2Fs13389-020-00224-9

[PW20] T. Peyrin, H. Wang, The malicious framework: embedding backdoors into tweakable block ciphers. Cryptology ePrint Archive, Report 2020/986 (2020), https://eprint.iacr.org/2020/986

[Sti06] D.R. Stinson, *Cryptography - Theory and Practice*. Discrete Mathematics and Its Applications Series. CRC Press (2006). ISBN: 978-1-58488-508-5

[PP10] C. Paar, J. Pelzl, *Understanding Cryptography - A Textbook for Students and Practitioners*. Springer (2010). ISBN: 978-3-642-04100-6, https://doi.org/10.1007/978-3-642-04101-3

[SSL15] K. Sakiyama, Y. Sasaki, Y. Li, *Security of Block Ciphers - From Algorithm Design to Hardware Implementation* (Wiley, 2015). ISBN: 978-1-118-66001-0, http://eu.wiley.com/WileyCDA/WileyTitle/productCd-1118660013.html

[MOP07] S. Mangard, E. Oswald, T. Popp, *Power Analysis Attacks - Revealing the Secrets of Smart Cards* (Springer, 2007). ISBN: 978-0-387-30857-9

[Pee13] E. Peeters, *Advanced DPA Theory and Practice: Towards the Security Limits of Secure Embedded Circuits*, 1st edn (Springer, New York, 2013), https://www.springer.com/gp/book/9781461467823

[RD20] M. Randolph, W. Diehl, Power side-channel attack analysis: a review of 20 years of study for the layman. Cryptogr. **4**(2), 15 (2020), https://doi.org/10.3390/cryptography4020015

[Bak20'2] A. Baksi, New insights on differential and linear bounds using mixed integer linear programming (full version) (2020), https://eprint.iacr.org/2020/1414

[BPP+17] S. Banik, S.K. Pandey, T. Peyrin, Y. Sasaki, S.M. Sim, Y. Todo, Gift: a small present. Cryptology ePrint Archive, Report 2017/622 (2017), https://eprint.iacr.org/2017/622

[BBC+21] A. Baksi, J. Breier, Y. Chen, X. Dong, Machine learning assisted differential distinguishers for lightweight ciphers, in *Design, Automation and Test in Europe Conference* (2021), https://eprint.iacr.org/2020/571.pdf

[BSS19] A. Baksi, D. Saha, S. Sarkar, To infect or not to infect: a critical analysis of infective countermeasures in fault attacks (2019), https://eprint.iacr.org/2019/355

[BKK+20'2] A. Baksi, V.B. Kumar, B. Karmakar, S. Bhasin, D. Saha, A. Chattopadhyay, A novel duplication based countermeasure to statistical ineffective fault analysis, in *Australasian Conference on Information Security and Privacy* (2020), https://link.springer.com/chapter/10.1007/978-3-030-55304-3_27

[BKK+20'1] A. Baksi, V.B.Y. Kumar, B. Karmakar, S. Bhasin, D. Saha, A. Chattopadhyay, A novel duplication based countermeasure to statistical ineffective fault analysis (2020), https://eprint.iacr.org/2020/1268

[BBB+20] A. Baksi, S. Bhasin, J. Breier, D. Jap, D. Saha, Fault attacks in symmetric key cryptosystems (2020), https://eprint.iacr.org/2020/1267

Chapter 2
Fundamentals of Symmetric Key Cryptography

2.1 Building Blocks

2.1.1 Boolean Function

Definition 2.1 (*Boolean Function*) An n-bit Boolean function is a map $f : \{0, 1\}^n \rightarrow \{0, 1\}$.

Definition 2.2 (*Vectorial Boolean Function*) An $m \times n$ vectorial Boolean function is a map $f : \{0, 1\}^m \rightarrow \{0, 1\}^n$.

The term "vectorial" is dropped if $n = 1$. Given a natural number n, there are 2^{2^n} such Boolean functions. A Boolean function can be represented in a few canonical ways, such as a truth table, *Algebraic Normal Form* (ANF), Univariate Polynomial Form, *Conjunctive Normal Form* (CNF), Disjunctive Normal Form, and few other non-canonical ways such as Sum of Product, Product of Sum. ANF is most commonly used in symmetric key cryptography. In ANF, a function is expressed in terms of Exclusive OR and AND of the variables. In other words, it expresses an n-bit Boolean function in terms of a multivariate polynomial in $\mathbb{F}_2[x_0, x_1, \ldots, x_{n-1}]/(x_0^2 + x_0, x_1^2 + x_1, \ldots, x_{n-1}^2 + x_{n-1})$. The algebraic degree of a function is the maximum of algebraic degree(s) of its monomials. The Boolean function with algebraic degree at most 1 is known as an *affine* function. In CNF, logical AND operations connect multiple clauses, each of which is composed by logical OR operations of one or more (possibly negated) literals.

Definition 2.3 (*Coordinate Function and Component Function*) Suppose $F : \mathbb{F}_2^n \rightarrow \mathbb{F}_2^n$ is defined as $F(x) = (f_0(x), \ldots, f_{n-1}(x))$ for all $x \in \mathbb{F}_2^n$, where $f_i : \mathbb{F}_2^n \rightarrow \mathbb{F}_2$ for $i = 0, \ldots, n-1$. Then each f_i is called a coordinate function of F. Further, the linear combinations of f_i's are called the component functions of F.

A. Baksi, *Classical and Physical Security of Symmetric Key Cryptographic Algorithms*, Computer Architecture and Design Methodologies,
https://doi.org/10.1007/978-981-16-6522-6_2

Table 2.1 Calculation for non-linearity of $f(x_0, x_1) = x_0x_1$

x_0	x_1	f	f_0	f_1	f_2	f_3	f_4	f_5	f_6	f_7
0	0	0	0	1	0	0	1	1	0	1
0	1	0	0	1	0	1	1	0	1	0
1	0	0	0	1	1	0	0	1	1	0
1	1	1	0	1	1	1	0	0	0	1

Definition 2.4 (*Non-linearity*) The non-linearity of the Boolean function $f : \mathbb{F}_2^n \rightarrow \mathbb{F}_2$ is the minimum Hamming distance of f to the set of all affine functions. Further, the non-linearity of $F : \mathbb{F}_2^n \rightarrow \mathbb{F}_2^n$ is the minimum of the non-linearities of all the component functions of F.

Example 2.1 Consider the 2-bit Boolean functions over the variables x_0, x_1. There are $2^{2^2} = 16$ such functions, of which the following 8 are affine:

$$f_0(x_0, x_1) = 0,\ f_1(x_0, x_1) = 1,\ f_2(x_0, x_1) = x_0,\ f_3(x_0, x_1) = x_1,$$
$$f_4(x_0, x_1) = x_0 \oplus 1,\ f_5(x_0, x_1) = x_1 \oplus 1,\ f_6(x_0, x_1) = x_0 \oplus x_1,\ f_7(x_0, x_1) = x_0 \oplus x_1 \oplus 1.$$

Now consider the AND operation, $f(x_0, x_1) = x_0x_1$. The calculation of its non-linearity is described in Table 2.1. The Hamming distances of f with the affine functions $(f_0, \ldots, f_7)$ are given by the vector $(1, 3, 1, 1, 3, 3, 3, 1)$. Since the minimum of this vector is 1, the non-linearity of f is 1.

2.1.2 *Substitution Box (SBox)*

A *Substitution Box* (*SBox*, for short) is a fundamental component in a lot of symmetric key ciphers. On several designs, it is the only component that inserts non-linearity. Technically, an $m \times n$ SBox is a look-up table, which maps an m-bit input to an n-bit output. In other words, it is an $m \times n$ vectorial Boolean function. Bijective SBoxes, which are a permutation of n-bits, are more commonly used nowadays and are often described as an n-bit SBox. There are $2^n!$ many $n \times n$ SBoxes, for example, there are $16! \approx 2^{44.25014}$ 4-bit SBoxes, and $256! \approx 2^{1683.99629}$ 8-bit SBoxes.

`AES` (see Sect. 2.4.1) uses an 8-bit SBox. Recently, the community is more focused on lightweight designs (see Sect. 2.7). Since an 8-bit SBox is generally considered costly, smaller (generally, 4-bit) SBoxes are predominantly used in recent lightweight cipher like `PRESENT` [BKL+07] or `GIFT` [BPP+17] (see Sect. 2.4.3 for description of `GIFT-128`). For example, $S =$ `1A4C6F392DB7508E` is the SBox used in `GIFT`. In words, the SBox can be described as: The input `0` is mapped to `1`, the input `1` is mapped to output `a`, ..., the input `f` is mapped to `e`. The inverse SBox S^{-1} can be derived from the definition of S.

Algebraic Properties

The algebraic properties of an SBox are given by the its coordinate functions. The Möbius transform [Can16] finds the coordinate functions for a given SBox.

Example 2.2 The component functions $(y_0, \ldots, y_{15})$ of the 4×4 SBox `126CDE39F58BA047` are given here with input variables (x_3, x_2, x_1, x_0) in ANF. The coordinate functions $(\mathbf{y_1}, \mathbf{y_2}, \mathbf{y_4}, \mathbf{y_8})$ are marked in bold.

$$\begin{aligned}
y_0 &= 0,\\
\mathbf{y_1} &= \mathbf{x_0x_1 \oplus x_0x_3 \oplus x_0 \oplus x_1x_2 \oplus x_1 \oplus x_2x_3 \oplus 1},\\
\mathbf{y_2} &= \mathbf{x_0 \oplus x_1 \oplus x_3},\\
y_3 &= x_0x_1 \oplus x_0x_3 \oplus x_1x_2 \oplus x_2x_3 \oplus x_3 \oplus 1,\\
\mathbf{y_4} &= \mathbf{x_1 \oplus x_2 \oplus x_3},\\
y_5 &= x_0x_1 \oplus x_0x_3 \oplus x_0 \oplus x_1x_2 \oplus x_2x_3 \oplus x_2 \oplus x_3 \oplus 1,\\
y_6 &= x_0 \oplus x_2,\\
y_7 &= x_0x_1 \oplus x_0x_3 \oplus x_1x_2 \oplus x_1 \oplus x_2x_3 \oplus x_2 \oplus 1,\\
\mathbf{y_8} &= \mathbf{x_0x_1 \oplus x_0x_3 \oplus x_1x_2 \oplus x_2x_3 \oplus x_2 \oplus x_3},\\
y_9 &= x_0 \oplus x_1 \oplus x_2 \oplus x_3 \oplus 1,\\
y_{10} &= x_0x_1 \oplus x_0x_3 \oplus x_0 \oplus x_1x_2 \oplus x_1 \oplus x_2x_3 \oplus x_2,\\
y_{11} &= x_2 \oplus 1,\\
y_{12} &= x_0x_1 \oplus x_0x_3 \oplus x_1x_2 \oplus x_1 \oplus x_2x_3,\\
y_{13} &= x_0 \oplus 1,\\
y_{14} &= x_0x_1 \oplus x_0x_3 \oplus x_0 \oplus x_1x_2 \oplus x_2x_3 \oplus x_3,\\
y_{15} &= x_1 \oplus x_3 \oplus 1.
\end{aligned}$$

This SBox has 0 non-linearity, as at least one of its non-trivial component functions (i.e., except y_0) is affine.

Differential Properties

A *Difference Distribution Table* (DDT) is a table used in context of cipher design/cryptanalysis. For an $n \times n$ SBox S, it is basically the $2^n \times 2^n$ matrix, where the row δ $(= 0, 1, \ldots, 2^n - 1)$ and column Δ $(= 0, 1, \ldots, 2^n - 1)$, which is denoted as $\mathrm{DDT}_S[\delta, \Delta]$, stores the number of solution(s) x for $S(x) \oplus S(x \oplus \delta) = \Delta$. Notice that $\mathrm{DDT}_S[0, 0] = 2^n$ as $S(x) \oplus S(x \oplus 0) = 0$ holds for all x. The entries in the DDT are even. It follows from the observation that if x is a solution to $S(x) \oplus S(x \oplus \delta) = 0$ $(\delta \neq 0)$, then so is $x \oplus \delta$.

The maximum entry at the DDT of S, except the case $\delta = \Delta = 0$, is called the *Differential Uniformity* (DU), and we denote it by DU_S (the subscript S is dropped if understood from context).

Remark 2.1 The DDT of S^{-1} can be obtained by taking the transpose of the DDT of S. The sum of each row and each column of a DDT is equal to 2^n. The entries in DDT are non-negative and even, so the DU ≥ 2.

Almost Perfect Non-linear (APN) Function. The SBoxes with DU $= 2$ are known as *Almost Perfect Non-linear* (APN) functions. It is known for that for the SBox S with $S(x) = x^{-1} = x^{n-2}$ over GF(2^n), except for 0 which is mapped to 0, gives DU $= 2$ when n is odd (for example, the 3×3 SBox `01567234`) and DU $= 4$ when n is even [Nyb93]. It is known that no such SBox exists for $n = 4$. The problem whether bigger even-sized SBox with DU $= 2$ exists is known as the *Big APN problem.* A major breakthrough is reported in [BDM+10], where the authors show the a 6×6 SBox with DU $= 2$. Later, this method is reverse-engineered to get several other 6×6 SBoxes in [PUB16]. It is proven afterwards in [CDP16] that the method of [PUB16] can only yield SBoxes from the same affine equivalence class of [BDM+10]. To the best of our knowledge, the following two problems are open till date:

- Whether there exists a 6×6 SBox with DU $= 2$ which belongs to a different affine equivalence class than that of [BDM+10];
- Whether there exists any $n \times n$ SBox with DU $= 2$, where $n \geq 8$ is even.

Definition 2.5 (*Linear Structure*) For $F : \mathbb{F}_2^n \rightarrow \mathbb{F}_2^n$, an element $a \in \mathbb{F}_2^n$ is called linear structure (LS) of F, if for some constant $c \in \mathbb{F}_2^n$, $F(x) \oplus F(x \oplus a) = c$ holds $\forall x \in \mathbb{F}_2^n$.

Note that zero is the trivial linear structure. Also, the set of all linear structures of F denoted as $\mathcal{L}(F)$ forms a subspace of $\mathbb{F}_2^n$ and is termed as the *linear space* of F. If $F : \mathbb{F}_2^n \rightarrow \mathbb{F}_2^n$ has a (non-zero) linear structure then 2^n becomes an entry in the DDT.

Definition 2.6 (*Differential Branch Number*) The Differential Branch Number (DBN) of the Boolean function F is defined as $\min_{\forall x, \forall \delta \neq 0}\{\text{HW}(\delta) + \text{HW}(F(x) \oplus F(x \oplus \delta))\}$, where HW$(\cdot)$ is the Hamming weight.

The DBN of an $n \times n$ SBox is lower bounded by 2, and upper bounded by $\lceil \frac{2n}{3} \rceil$ (both inclusive) [SS17, SMS19].

Example 2.3 The DDT for the 4×4 SBox `126CDE39F58BA047` is given in Table 2.2, where entries corresponding to $\delta = \Delta = 0$ and other zero entries are omitted for the sake of better readability. The differential uniformity for this SBox is 16 ($= 2^4$) which corresponds to $\delta =$ `5`, `a`, `f`. Hence, {`5`, `a`, `f`} are the non-trivial linear structures. This SBox has DBN of 3.

Linear Properties

A *Linear Approximation Table* (LAT) is an important concept in cipher design or cryptanalysis. The LAT for an $n \times n$ SBox S is the $2^n \times 2^n$ matrix where the row γ $(= 0, 1, \ldots, 2^n - 1)$ and column Γ $(= 0, 1, \ldots, 2^n - 1)$ store $|\{\gamma \cdot \vec{x} \oplus \Gamma \cdot \vec{y} = 0\}| - 2^{n-1}$, $\vec{x}$ denotes the input variables to the SBox, $\vec{y}$ denotes the output variables, and $\cdot$ denotes dot product. Notice that LAT[0, 0] $= 2^{n-1}$.

Table 2.2 DDT for SBox `126CDE39F58BA047`

δ \ Δ	1	2	3	4	5	6	7	8	9	a	b	c	d	e	f
1			8							8					
2							8							8	
3				8									8		
4					8							8			
5															16
6		8									8				
7	8							8							
8							8							8	
9				8									8		
a									16						
b			8							8					
c		8									8				
d	8							8							
e					8							8			
f						16									

Remark 2.2 The entries in the LAT are even. For a sketch of proof, assume for some (γ, Γ) that (ν, ω) is a solution to $\Psi(\vec{x}, \vec{y}) \equiv \gamma \cdot \vec{x} \oplus \Gamma \cdot \vec{y} = 0$. Now consider $\Theta = \{(\nu, \omega), (\overline{\nu}, \omega), (\nu, \overline{\omega}), (\overline{\nu}, \overline{\omega})\}$, where $\overline{z}$ indicates the complement of z, i.e., $z \oplus \overline{z} = 0$. We claim that there are exactly two or exactly four solutions to Ψ belonging from Θ. Note that, $\Psi(\bar{\nu}, \omega) =$ parity of ν, $\Psi(\nu, \bar{\omega}) =$ parity of ω, and $\Psi(\bar{\nu}, \bar{\omega}) =$ parity of $\nu \oplus$ parity of ω. Either exactly one of the parities is 0 or exactly three are.

Definition 2.7 (*Linear Branch Number*) The Linear Branch Number (LBN) of an $n \times n$ SBox is defined as $\min_{\alpha \neq 0, \beta \neq 0, \mathrm{LAT}[\alpha, \beta] \neq 0}\{\mathrm{HW}(\alpha) + \mathrm{HW}(\beta)\}$; where $\mathrm{HW}(\cdot)$ denotes the Hamming weight.

The lower bound of LBN of an $n \times n$ SBox is 2 and the upper bound is $n - 1$ (both inclusive) [SS17, SMS19].

Example 2.4 The 4×4 SBox `126CDE39F58BA047` has LBN of 3. Its LAT is given in Table 2.3, where the (input mask, output mask) are denoted by (γ, Γ), $\gamma = \{1, 2, \ldots, 15\}$ and $\Gamma = \{1, 2, \ldots, 15\}$ and the zero null entries are suppressed for better readability. Note that, the SBox has seven non-trivial affine relations among its input and output bits (that hold with probability 1), of which three are linearly independent. The seven non-trivial affine relations correspond to the (γ, Γ) pairs: $\{(\mathtt{1}, \mathtt{d}), (\mathtt{4}, \mathtt{b}), (\mathtt{5}, \mathtt{6}), (\mathtt{a}, \mathtt{f}), (\mathtt{b}, \mathtt{2}), (\mathtt{e}, \mathtt{4}), (\mathtt{f}, \mathtt{9})\}$. Denoting the input bits as

Table 2.3 LAT for SBox `126CDE39F58BA047`

γ \ Γ	1	2	3	4	5	6	7	8	9	a	b	c	d	e	f
1													−8		
2			−4		+4					+4		+4			
3	−4						−4	−4						+4	
4											−8				
5						+8									
6	−4						−4	+4						−4	
7			+4		−4					+4		+4			
8			−4		−4					−4		+4			
9	−4						+4	+4						+4	
a															−8
b		+8													
c	+4						−4	+4						+4	
d			−4		−4					+4		−4			
e				+8											
f									−8						

(x_3, x_2, x_1, x_0) and output bits as (y_3, y_2, y_1, y_0), the following three linearly independent affine relations hold with probability 1:

$$
\begin{aligned}
&(\gamma = \mathtt{b}, \Gamma = \mathtt{2}) \qquad && x_3 \oplus x_1 \oplus x_0 \oplus y_1 = 0,\\
&(\gamma = \mathtt{e}, \Gamma = \mathtt{4}) \qquad && x_3 \oplus x_2 \oplus x_1 \oplus y_2 = 0,\\
&(\gamma = \mathtt{4}, \Gamma = \mathtt{b}) \qquad && x_2 \oplus y_3 \oplus y_1 \oplus y_0 = 1.
\end{aligned}
$$

Affine Equivalence

We call SBoxes S_0 and S_1 *Affine Equivalent* (AE) if there exist two non-singular binary matrices A_0, A_1 (each with compatible dimension) such that $S_1 = A_0 \circ S_1 \circ A_1$; i.e., $S_1(x) = A_0 S_0(A_1(x \oplus c_0)) \oplus c_1$, for all inputs x and for some binary vectors c_0, c_1 of compatible dimension.

Affine equivalence SBoxes have the same DDT and same absolute LAT up to a permutation. In particular, the SBox $T(x) = S(x \oplus c_0) \oplus c_1$, where c_0, c_1 are constants, will have same DDT and same absolute LAT of S. Therefore, differential uniformity is invariant under affine equivalence, so are the other cryptographic properties like non-linearity, algebraic degree, etc. However, DBN is not preserved [SS17, SMS19].

It is to be noted that the affine equivalence classification of all 4×4 SBoxes has been done already—there are 302 such classes. We follow the representative SBoxes given in [DeC07, Chap. 5.4.2] throughout.

To find affine equivalent SBoxes of a given $n \times n$ SBox, binary non-singular matrices are required. The General Linear group over $\mathbb{F}_q$ of degree n, denoted by $\mathrm{GL}(n, \mathbb{F}_q)$, consists of $n \times n$ non-singular matrices with each entry is from $\mathbb{F}_q$. It can be shown that the order of $\mathrm{GL}(n, \mathbb{F}_q)$ is $\prod_{k=0}^{n-1} (q^n - q^k)$. Therefore, number of 3×3 and 4×4 binary (so, $q = 2$) non-singular matrices are 168 and 20160, respectively.

2.1.3 Linear Layer

The linear layer of a cipher indicates a binary non-singular matrix. This matrix is multiplied with the cipher state. However, generally the linear layers are described in a variety of ways (from which the binary non-singular matrix representation can be derived). For example, the linear layers of `AES` (see Sect. 2.4.1 for its description) are given by `ShiftRows` and `MixColumns` (the latter is typically described in terms of elements over $\mathrm{GF}(2^8)$).

Loosely, the linear layer can be classified into two categories. In the first category, the size of the basic block of the linear layer is smaller than the state size of the cipher (so it is applied iteratively over the state), like `AES`. In the second category, the linear layer applies to the entire state of the cipher, like `PRESENT` [BKL+07].

Remark 2.3 A linear layer can be equivalently described as a bijective linear SBox (i.e., all its coordinate functions are affine). For example, the `AES` `MixColumns` can be described as a 32-bit linear SBox.

2.2 Primitives

2.2.1 (Un-keyed) Permutation

A permutation with good cryptographic properties is typically the starting point of a cipher design. Indeed, when the (secret) key is introduced in the permutation, it becomes a full-fledged cipher. It can be said that the (secret) key chooses a permutation from the family of all permutations.

A permutation can be built with SBoxes and as typically done with *block ciphers* (Sect. 2.2.2), or using simpler Boolean functions which is common in *stream ciphers* (Sect. 2.2.3). Further, using a permutation with a proper *mode*, primitives like *hash functions* (Sect. 2.2.4), *Message Authentication Code* (Sect. 2.2.5), or *Authenticated Encryption with Associated Data* (Sect. 2.2.6) can also be designed.

2.2.2 Block Cipher

A block cipher takes a fixed length plaintext and returns the ciphertext with the help of a (secret) key during the encryption process, and returns back the plaintext from the ciphertext if the same (secret) key is used. Therefore, when the key is fixed, a block cipher works as a bijection, and chaining the key makes the block cipher choosing a different bijection. All the major block ciphers use a relatively simple update function over its state. The state is thus updated over multiple rounds of iteration (thus making the overall computation complex).

The *key schedule* is a separate subroutine that takes the key and generates the corresponding *round keys*, which are typically XORed with the cipher state. Generally, the key schedule subroutine is a bijection, so the key can be retrieved if any one of the round keys becomes known. Some ciphers like `PRINCE` [BCG+12], `LED-64` [GPP+12] use the identity function as the key schedule.

A *Tweakable Block Cipher* (TBC) takes a public input, named *tweak*, in addition to key and plaintext. Examples include the `SKINNY` family [BJK+16].

Padding and Mode of Operation

Since the size of the plaintext is fixed, two types of situations may arise. If the length of the plaintext is less than what is needed for the cipher, it is needed to pad the plaintext to match the desired length (discussion on padding methods is omitted for the interest of brevity). On the other hand, if the length of the plaintext is longer than what can be processed by the cipher, the plaintext is to be split and managed iteratively. The mode of operation states how to manage plaintexts/ciphertexts in such cases.

Following [BBB+18, Sect. II.D], we classify block cipher modes into two categories: non-symmetric and symmetric modes. A visualization, which is adopted from [BBB+18, Fig. 2], is shown in Fig. 2.1. In non-symmetric modes, the original plaintext is encrypted with the `encrypt()` subroutine, but decrypted with the `decrypt()` subroutine. In contrast to this, in symmetric modes, a counter/*Initialization Vector* (IV)/*nonce*[1] is both encrypted and decrypted with the `encrypt()` subroutine, the result is XORed with the plaintext. The operations in common symmetric modes are shown in Fig. 2.3 and the non-symmetric in Fig. 2.2. The `Electronic Code-book` mode (`ECB`) processes each blocks of plaintexts independently of the other. The `Cipher Block Chaining` mode (`CBC`) XORs an IV with the initial block of the plaintext, and the ciphertext blocks are XORed with the next plaintext blocks for encryption. The `Counter` (`CTR`) mode uses separate IV blocks as the plaintext, and then XORs the outputs with the plaintext blocks to get the ciphertext blocks. The `Cipher Feedback` (`CFB`) mode uses the ciphertext blocks (i.e., after the plaintext blocks are XORed) as the next plaintext block

[1] Nonce: Number used once.

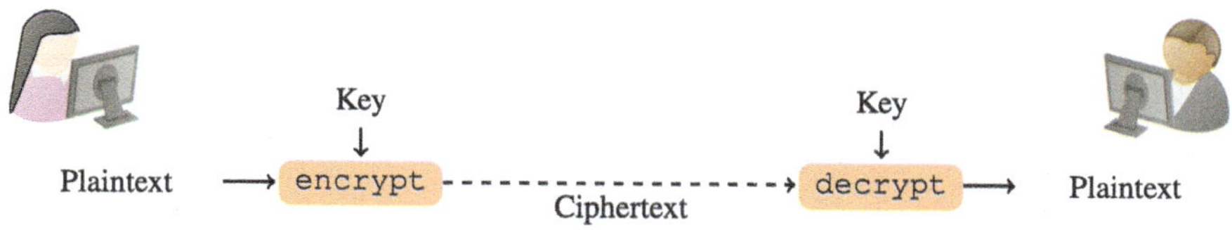

(a) Operation in non-symmetric modes

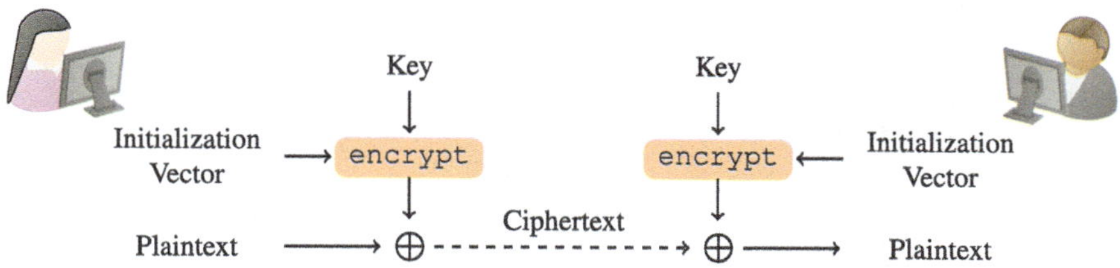

(b) Operation in symmetric modes

Fig. 2.1 Two types of block cipher modes

for encryption. The `Output Feedback` (`OFB`) mode uses the output from the cipher (i.e., before the plaintext blocks are XORed) as the next plaintext block for encryption.

2.2.3 *Stream Cipher*

Stream ciphers uses the (secret) key with an IV to create a *key-stream*. The key-stream is XORed with the plaintext during encryption, and the ciphertext is XORed with the same key-stream to get back the plaintext. The same (key, IV) pair cannot be used more than once, as this would repeat the same key-stream which can be recovered, thus destroying confidentiality.

Remark 2.4 Stream ciphers are similar to the symmetric modes of the block cipher operation.

Typically, a stream cipher is constructed using one or more *Feedback Shift Registers* (FSR). A feedback shift register shifts its contents at each clock while filling up the empty location with a so-called update function (that takes multiple locations of the register as input). See Fig. 2.4 for a basic work-flow for an FSR. If the update function is affine, then the FSR is called a *Linear Feedback Shift Register* (LFSR). Similarly, if the update function is non-linear, then the FSR is called a *Non-linear Feedback Shift Register* (NLFSR).

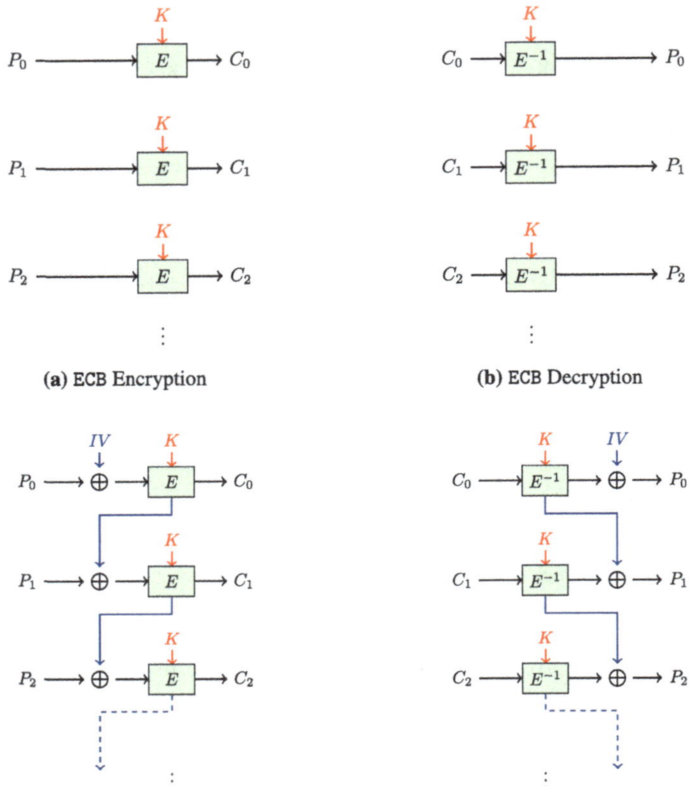

(a) ECB Encryption **(b)** ECB Decryption

(c) CBC Encryption **(d)** CBC Decryption

Fig. 2.2 Common non-symmetric modes

Example 2.5 The stream cipher `GRAIN-v1` [HJM06] uses one 80-bit NLFSR (X) and one 80-bit LFSR (Y), a schematic is given in Fig. 2.5. The update function of X is $g(\cdot)$ which takes 26 tap positions, whereas the update function of Y is f which takes six tap positions. The combiner function, $h(\cdot, \cdot)$, uses tap positions from both FSRs.

2.2.4 *Hash Function*

A hash function takes an arbitrary length input data to convert it to a fixed length output data, which is called *hash* or *digest*. A hash function mimics the theoretical notion of *one way function* (i.e., it is computationally hard to get the input given its output).

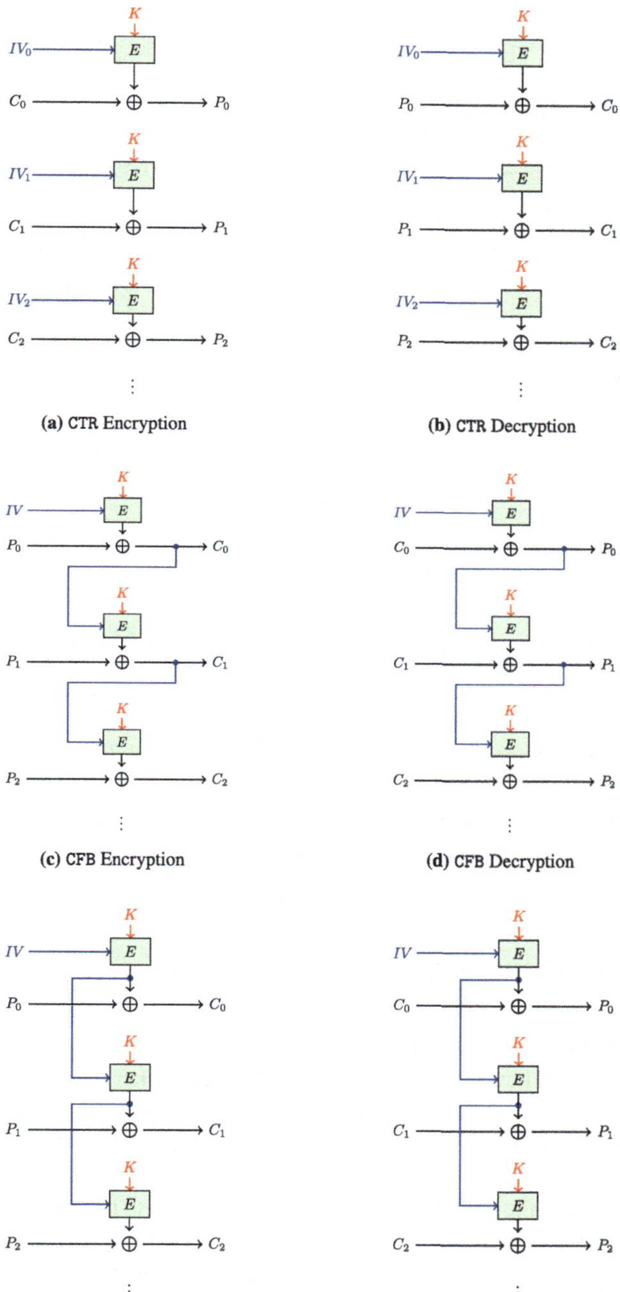

(a) CTR Encryption (b) CTR Decryption

(c) CFB Encryption (d) CFB Decryption

(e) OFB Encryption (f) OFB Decryption

Fig. 2.3 Common symmetric modes

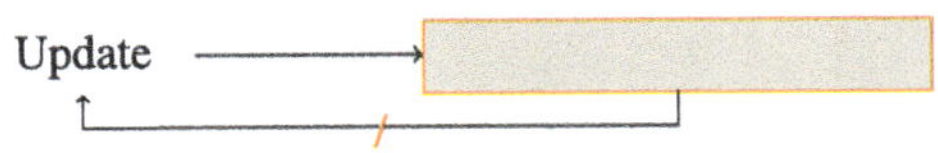

Fig. 2.4 Work-flow of a feedback shift register

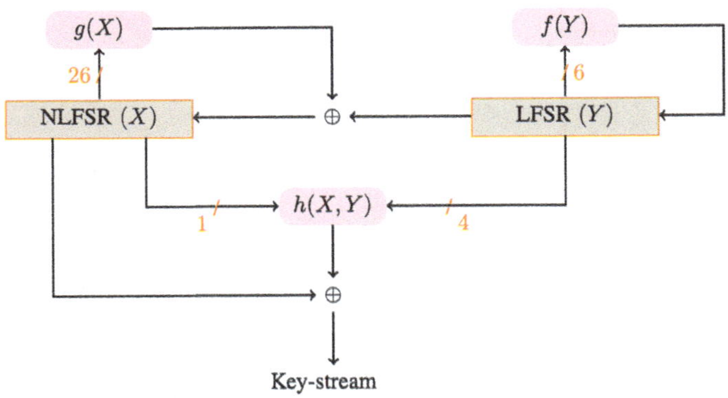

Fig. 2.5 Structure of GRAIN-v1

A secure hash function $H : \{0, 1\}^\star \rightarrow \{0, 1\}^n$ is to satisfy the following properties:

1. **Pre-image Resistance.** Given a hash $y \in \{0, 1\}^n$, it is computationally hard to find x for which that $H(x) = y$.
2. **Collision Resistance.** It is computationally hard to find distinct x_0, x_1 such that $H(x_0) = H(x_1)$.
3. **Second Pre-image Resistance.** Given x_0 (and thus $H(x_0)$), it is infeasible to find any x_1 ($\neq x_0$) such that $H(x_0) = H(x_1)$.

2.2.5 Message Authentication Code (MAC)

A Message Authentication Code (MAC) is used for authentication of a message m using a shared secret key K such that it is hard for an attacker to forge authentication for a new message m without the knowledge of the secret key K. The MAC achieves this security notion through generation of a deterministic tag τ from the combination of message m and secret key K (i.e., $\tau = f(m, K)$). There are a couple of well-known MAC constructions like the HMAC (based on universal hash functions) and the CBC-MAC (based on the CBC mode of block cipher) which are the most commonly used in practical applications [KBC97, PR00]. However, as far as implementation of MAC on embedded devices (16-bit, 32-bit microcontrollers) are concerned, those methods may incur heavy computational cost due to the underlying operations based on hash functions and full-round block ciphers.

2.2.6 *Authenticated Encryption with Associated Data (AEAD)*

The Authenticated Encryption with Associated Data (AEAD) [BN00] is a relatively recent concept in symmetric key cryptography where a cipher aims at providing confidentiality as well as authenticity. Given a message (referred to as plaintext) and associated data, such a scheme encrypts the plaintext (confidentiality) and generates a tag depending on both the plaintext and the associated data (authentication). The sender (Alice) generates the ciphertext and the tag and sends them along with the associated data through an insecure channel (where the attacker, Eve, is active) to the recipient (Bob). Upon receiving the ciphertext and tag, Bob generates the plaintext and the tag. If both the tags match, then the received ciphertext (hence the plaintext retrieved is original) and the associated data are considered not disturbed by Eve. Otherwise, Bob learns there is a disturbance caused by Eve, generates an invalid signal, and discards the entire data received. For the transmission of plaintext, both confidentiality and authenticity are required, whereas only authenticity is required for the associated data. This is practical in scenarios like Internet packets, where the header is not confidential (acts as the associated data) but the body is confidential (acts as the plaintext).

In a loose sense, an AEAD can be constructed, for example, by using a block cipher or by using a feedback shift register, with a suitable update function along with a suitable mode, that allows taking input of the plaintext and the associated data and giving output of the ciphertext and the authentication tag. In practice, more factors are to be considered, such as whether to release ciphertext if verification of the tag fails (on Bob's side).

For illustration, ACORN[Wu16] is an AEAD that uses six feedback shift registers. Figure 2.6 shows a schematic of the cipher—here f is a feedback function (which takes multiple state bits) and through m the key, IV, associated data (if any) and plaintext (if any) are passed. Similarly, the AEAD AEGIS [WP15] uses one round of unkeyed AES permutation as its round function (Fig. 2.7).

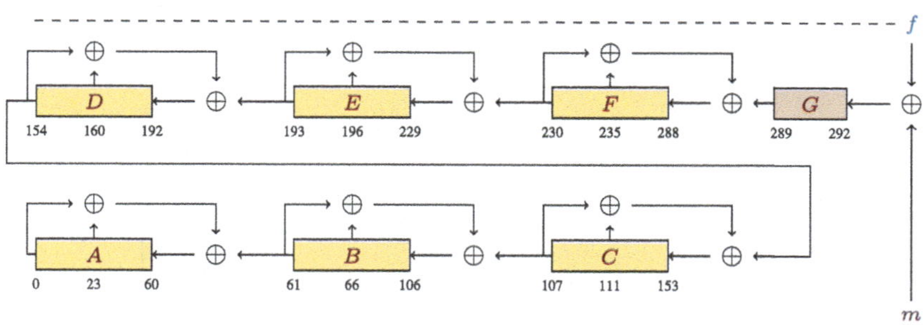

Fig. 2.6 Structure of ACORN

Fig. 2.7 One round of a typical SPN cipher

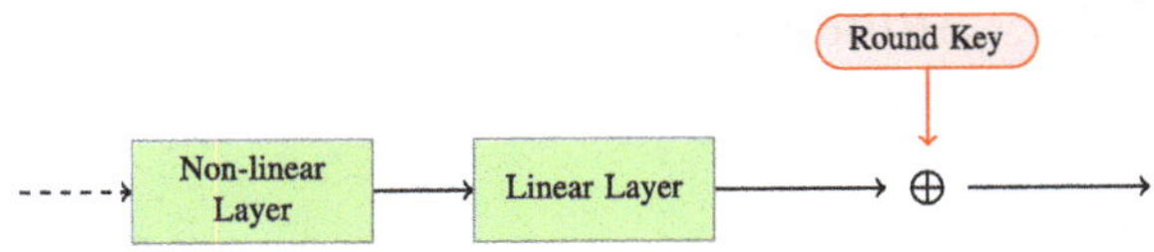

2.3 Cipher Families

2.3.1 Substitution Permutation Network (SPN)

The *Substitution Permutation Network* (SPN) can be loosely described as iteration of the cipher state over multiple rounds. SPN ciphers are very common, a lot of the high-profile block ciphers are based on it, and a few other ciphers like hash functions or AEADs use an SPN structure at its core. Examples include AES (see Sect. 2.4.1 for a short description of AES), PRESENT [BKL+07] (a brief description of which is given in Sect. 2.4.2).

For encryption, a round can be described as a layer of non-linear layer (typically SBoxes), followed by a linear layer, and then the round key addition. Some ciphers like GIFT-128 [BPP+17] additionally use another operation at which predetermined constants are XORed at selected locations of the state. A top level pictorial

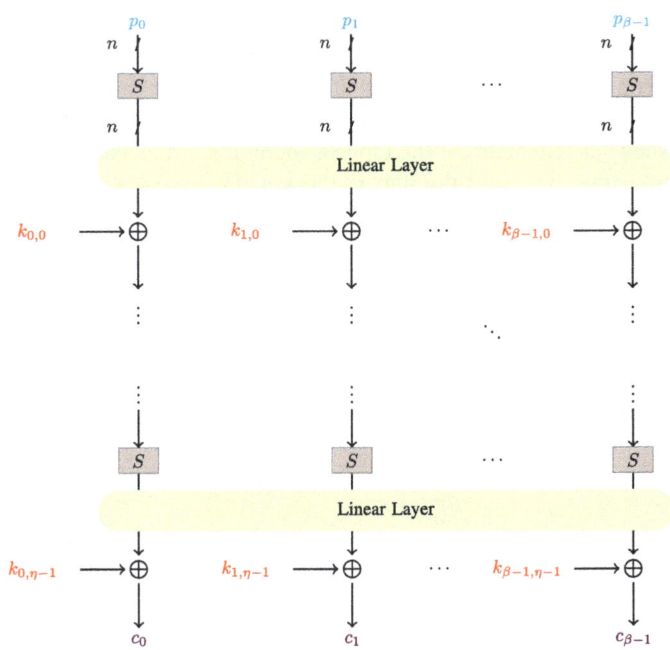

Fig. 2.8 Schematic of a typical SPN cipher

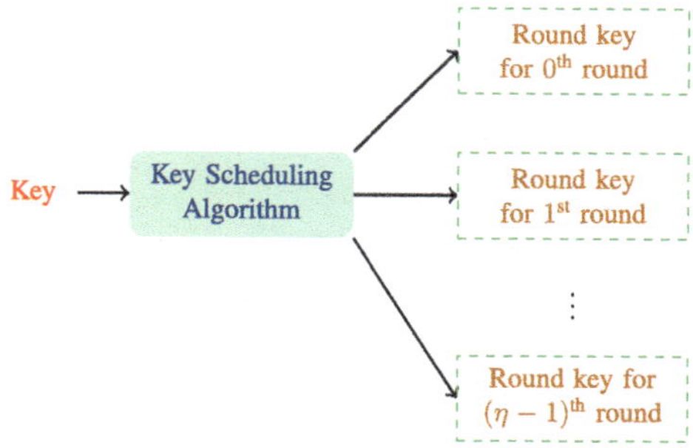

Fig. 2.9 Schematic of a key schedule

description is given in Fig. 2.8. Here, n-bit SBoxes are used repeatedly for β times (so the state size is $n\beta$). The linear layer, which acts as multiplication by a $n\beta \times n\beta$ binary non-singular matrix[2] on the state, is applied after this. The last operation in one round is that the corresponding round keys are XORed. This is iterated η times (so the cipher is said to have η rounds). The plaintext $P = (p_0, p_1, \ldots, p_{\beta-1})$ is split into the β SBoxes at the beginning (round 0), and finally the ciphertext $C = (c_0, c_1, \ldots, c_{\beta-1})$ is obtained from the state. The key schedule takes the key and generates the ith round keys, $K_i = (k_{0,i}, k_{1,i}, \ldots, k_{\beta-1,i})$, for $i = 0, 1, \ldots, \eta - 1$. A schematic of key schedule is given in Fig. 2.9.

The decryption is defined as the inverse of the inverse function. Therefore, with our current example, the ith round, $i = 0, 1, \ldots, \eta - 1$, decryption is defined as follows—XORing the $(\eta - 1 - i)$th round key, applying the inverse of the linear layer (i.e., multiplying with the inverse matrix), and applying the layer with inverse of the SBox.

2.3.2 *Feistel Network*

A *Feistel Network* can be roughly described as a two-branch structure where one branch is passed through the a non-linear function, f, and the output from the f is used to update the other branch. The same f function is used for encryption as well as decryption, hence f not have to be bijective. The round keys may need to be independent to ensure security.

A typical round structure of a Feistel network-based block cipher can be found in Fig. 3.9. The encryption and decryption operations can be described as given in Fig. 2.10. The plaintext is split into two parts $l_0 \parallel r_0$, which are periodically updated (see Fig. 2.10a) for η rounds. The round keys $(k_0, k_1, \ldots, k_{\eta-1})$ are fed to the f function. The ciphertext is obtained at the end $r_{\eta-1} \parallel l_{\eta-1}$. During decryption, the steps are reversed (see Fig. 2.10b). Generalization of the Feistel network has been

[2] The linear layer may not be directly given in the form of a binary matrix, but this form can be derived easily.

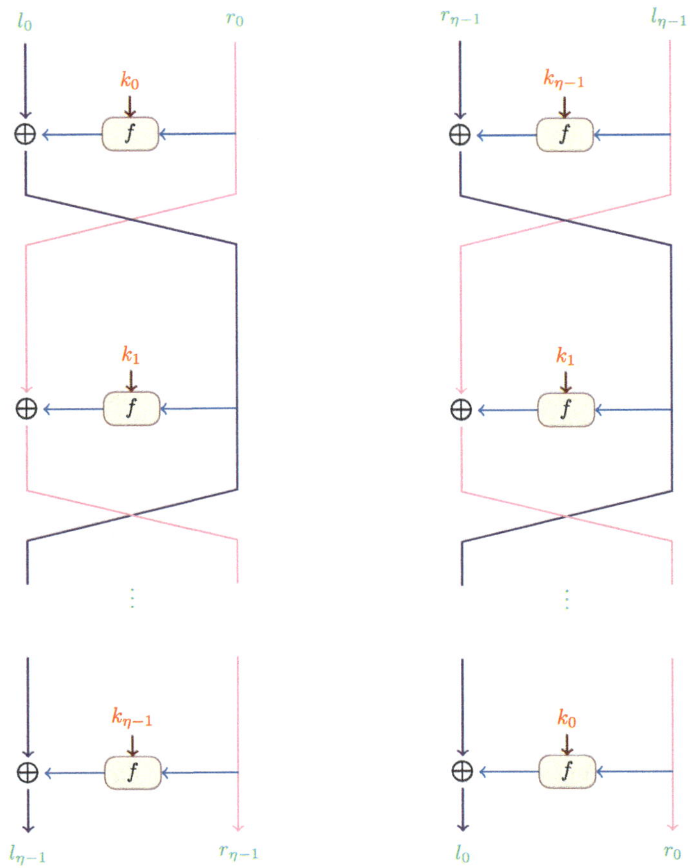

(a) Encryption **(b)** Decryption

Fig. 2.10 Work-flow in a Feistel network-based block cipher

proposed in literature [HR10], of which three common types (referred to as Type I, II, and III) are shown in Fig. 2.11.

Remark 2.5 Preferably, at least one *whitening key* is to be used at the end of encryption ($r_{\eta-1}$ of Fig. 2.10a) to prevent the attacker to trivially reduce one round.

Remark 2.6 Since the f function can be a hash function, it can be said that a Feistel network can convert a hash function to a block cipher .

Remark 2.7 It has been shown in [LR88] that, under some regularity conditions, three rounds of a Feistel network based block cipher are sufficient to meet security requirement.

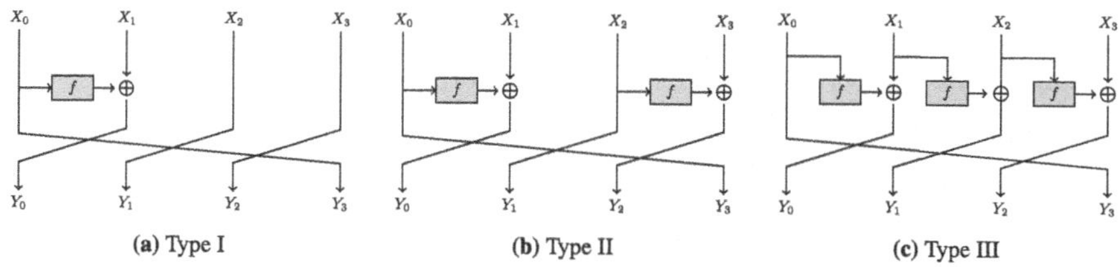

Fig. 2.11 Generalized Feistel networks

Remark 2.8 The f function of a Feistel construction can be an SPN-based unkeyed permutation. In that case, an n-bit SPN permutation will be used to build a block cipher of $2n$-bits.

A Feistel construction can be considered not so efficient compared to its SPN counterpart for the following reasons. A Feistel construction needs at least three rounds since a 1-probability *differential trail* (stated in Sect. 2.6.1) will pass through a 2-round Feistel construction, whereas a 2-round of a SPN construction can achieve full security in theory. An SPN construction can use identical round keys (which can be equal to the master key) thus saving cost for key scheduling if round keys are not precomputed, but identical round keys will likely weaken the security of a Feistel construction. Nevertheless, the Feistel construction is an interesting concept because it gives the unique prospect of designing an invertible cipher from a non-invertible function.

Remark 2.9 Though a 2-round Feistel construction trivially fails the distinguisher game (see Sect. 2.5.3) due to 1-probability differential transition, it is not trivial to recover the secret key.

2.3.3 Add–Rotation–XOR (ARX) Construction

The *Add–Rotation–XOR* (XOR) construction is another prominent class of ciphers. The round function of such kind of ciphers involves only three operations—modular addition (the only non-linear operation), rotation (by a fixed amount), and XOR.

ARX ciphers typically offers few advantages such as

- Compact implementation,
- Fast performance in software,
- Due to its branchless nature, an ARX cipher does not leak *timing* side channel.

At the same time, it also suffers from the following limitations:

- It is arguably harder to analyse against classical attacks compared to a typical SPN cipher.
- It is not clear how to efficiently implement it in hardware.

Examples of ARX ciphers include CHASKEY [Mou15] (see Sect. 2.4.4 for its description), SIMON [BSS+13], SPECK [BSS+13].

2.4 Description of Exemplary Ciphers

Here we describe three ciphers, namely, AES (Sect. 2.4.1), PRESENT-80 (Sect. 2.4.2), GIFT-128 (Sect. 2.4.3), and CHASKEY (Sect. 2.4.4). These ciphers are analysed/utilized in multiple contexts in the subsequent parts, for example, GIFT-128 is used in Chaps. 5 and 8. Description for three more ciphers is given in Sect. 6.3.

2.4.1 ADVANCED ENCRYPTION STANDARD *(*AES*)*

AES is an SPN cipher with 128-bit state (thus takes 128-bit plaintext and generates 128-bit ciphertext during encryption) and can use keys of size 128, 192, and 256. The most common variant uses 128-bit key, which is described next. Refer to Fig. 2.12

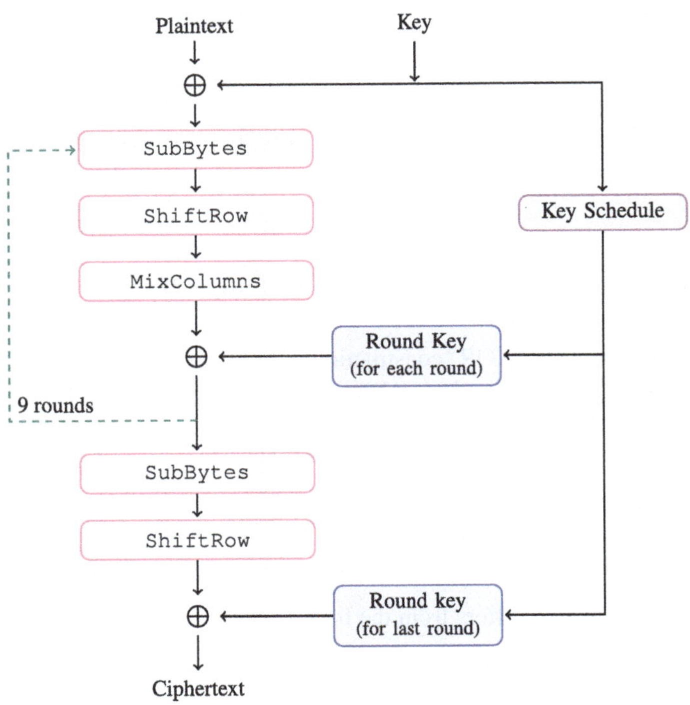

Fig. 2.12 Top level view of AES-128 encryption

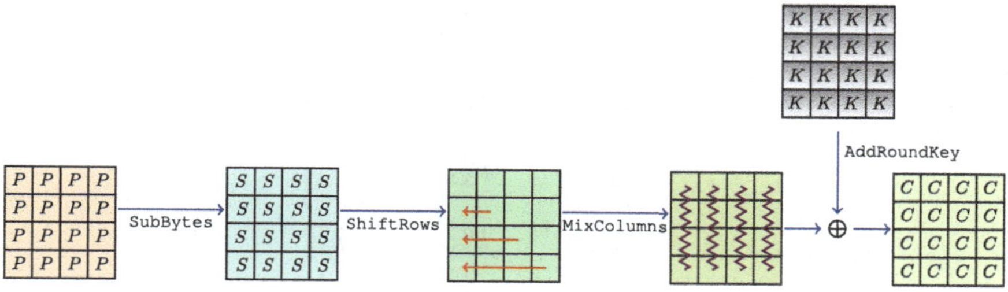

Fig. 2.13 One (non-last) round of `AES-128` encryption (schematic)

for a pictorial description of the work-flow of `AES-128` encryption, which is often referred to as `AES` for simplicity.

The state is arranged into a 4×4 matrix of bytes, which is arranged in the column-major fashion. The key is XORed with the state at the beginning. The key is also passed to the key scheduling subroutine (see Fig. 2.14 for a schematic of it), which generates 10 round keys each of which is 128-bit long. The state undergoes a total 10 rounds of modifications, not counting the initial key addition. The first nine rounds are identical, each consists of four steps in order: `SubBytes` (the only non-linear operation), `ShiftRows`, `MixColumns`, and `AddRoundKey` (see Fig. 2.13 for a depiction). The last (9th) round only has `SubBytes`, `ShiftRows`, and `AddRound Key` (`MixColumns` is missing). The `ShiftRows` and `MixColumns` together constitute the linear layer.

`SubBytes`

Each byte of the state is converted using the following 8×8 SBox to another byte :

```
63, 7C, 77, 7B, F2, 6B, 6F, C5, 30, 01, 67, 2B, FE, D7, AB, 76, CA, 82, C9, 7D, FA, 59, 47, F0, AD, D4, A2, AF, 9C, A4, 72, C0,
B7, FD, 93, 26, 36, 3F, F7, CC, 34, A5, E5, F1, 71, D8, 31, 15, 04, C7, 23, C3, 18, 96, 05, 9A, 07, 12, 80, E2, EB, 27, B2, 75,
09, 83, 2C, 1A, 1B, 6E, 5A, A0, 52, 3B, D6, B3, 29, E3, 2F, 84, 53, D1, 00, ED, 20, FC, B1, 5B, 6A, CB, BE, 39, 4A, 4C, 58, CF,
D0, EF, AA, FB, 43, 4D, 33, 85, 45, F9, 02, 7F, 50, 3C, 9F, A8, 51, A3, 40, 8F, 92, 9D, 38, F5, BC, B6, DA, 21, 10, FF, F3, D2,
CD, 0C, 13, EC, 5F, 97, 44, 17, C4, A7, 7E, 3D, 64, 5D, 19, 73, 60, 81, 4F, DC, 22, 2A, 90, 88, 46, EE, B8, 14, DE, 5E, 0B, DB,
E0, 32, 3A, 0A, 49, 06, 24, 5C, C2, D3, AC, 62, 91, 95, E4, 79, E7, C8, 37, 6D, 8D, D5, 4E, A9, 6C, 56, F4, EA, 65, 7A, AE, 08,
BA, 78, 25, 2E, 1C, A6, B4, C6, E8, DD, 74, 1F, 4B, BD, 8B, 8A, 70, 3E, B5, 66, 48, 03, F6, 0E, 61, 35, 57, B9, 86, C1, 1D, 9E,
E1, F8, 98, 11, 69, D9, 8E, 94, 9B, 1E, 87, E9, CE, 55, 28, DF, 8C, A1, 89, 0D, BF, E6, 42, 68, 41, 99, 2D, 0F, B0, 54, BB, 16.
```

For instance, if the input to the SBox is `00`, then it is converted to `63`. Likewise, the input `ff` is converted to `16`.

`ShiftRows`

The ith row of state is rotated to the left by ith locations, for $i = 0, 1, 2, 3$.

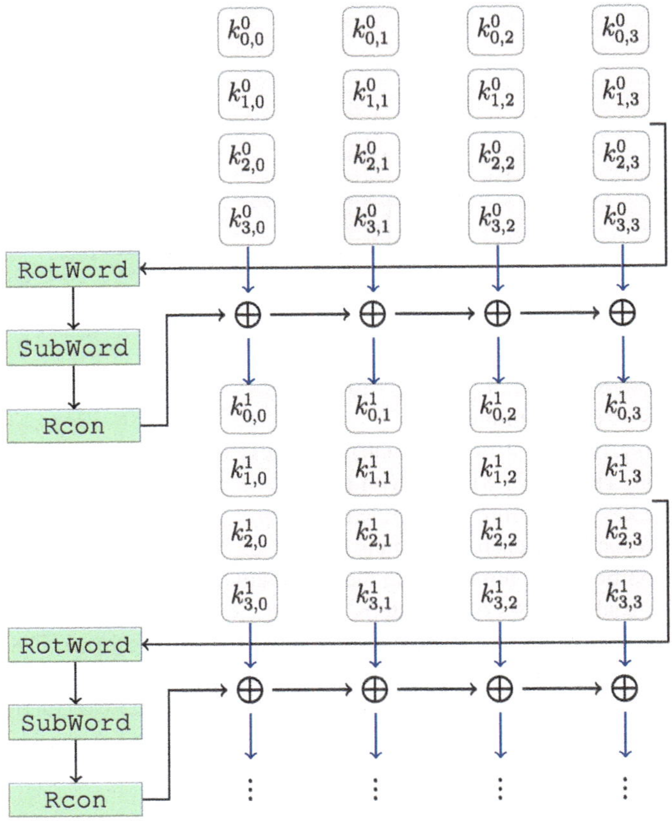

Fig. 2.14 AES-128 encryption key schedule (schematic)

MixColumns

This subroutine operates on the state column-wise. Each column of the state is considered a vector of length 4 defined over GF(2^8). The following matrix (defined over GF(2^8)),

$$M^{4\times4} = \begin{bmatrix} 02 & 03 & 01 & 01 \\ 01 & 02 & 03 & 01 \\ 01 & 01 & 02 & 03 \\ 03 & 01 & 01 & 02 \end{bmatrix}$$

is pre-multiplied to each column with the modulus $x^8 + x^4 + x^3 + x + 1$. The result thus obtained substitutes the old contents of the column.

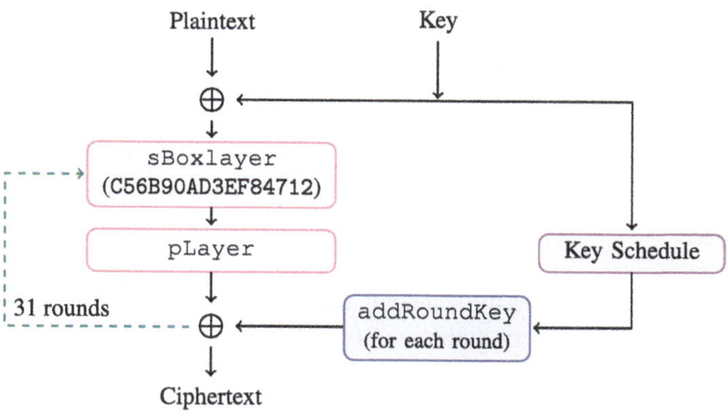

Fig. 2.15 Top level view of PRESENT-80 encryption and PRESENT-128 encryption

As stated in Remark 2.3, M can be written as the following 32×32 binary non-singular matrix with the encoding of [XZL+20]:

$$\begin{bmatrix}
0&0&0&0&0&0&0&1&1&0&0&0&0&0&0&1&1&0&0&0&0&0&0&0&1&0&0&0&0&0&0&0\\
1&0&0&0&0&0&0&1&1&1&0&0&0&0&0&1&0&1&0&0&0&0&0&0&0&1&0&0&0&0&0&0\\
0&1&0&0&0&0&0&0&0&1&1&0&0&0&0&0&0&0&1&0&0&0&0&0&0&0&1&0&0&0&0&0\\
0&0&1&0&0&0&0&1&0&0&1&1&0&0&0&1&0&0&0&1&0&0&0&0&0&0&0&1&0&0&0&0\\
0&0&0&1&0&0&0&1&0&0&0&1&1&0&0&1&0&0&0&0&1&0&0&0&0&0&0&0&1&0&0&0\\
0&0&0&0&1&0&0&0&0&0&0&0&1&1&0&0&0&0&0&0&0&1&0&0&0&0&0&0&0&1&0&0\\
0&0&0&0&0&1&0&0&0&0&0&0&0&1&1&0&0&0&0&0&0&0&1&0&0&0&0&0&0&0&1&0\\
0&0&0&0&0&0&1&0&0&0&0&0&0&0&1&1&0&0&0&0&0&0&0&1&0&0&0&0&0&0&0&1\\
1&0&0&0&0&0&0&0&0&0&0&0&0&0&0&1&1&0&0&0&0&0&0&1&1&0&0&0&0&0&0&0\\
0&1&0&0&0&0&0&0&1&0&0&0&0&0&0&1&1&1&0&0&0&0&0&1&0&1&0&0&0&0&0&0\\
0&0&1&0&0&0&0&0&0&1&0&0&0&0&0&0&0&1&1&0&0&0&0&0&0&0&1&0&0&0&0&0\\
0&0&0&1&0&0&0&0&0&0&1&0&0&0&0&1&0&0&1&1&0&0&0&1&0&0&0&1&0&0&0&0\\
0&0&0&0&1&0&0&0&0&0&0&1&0&0&0&1&0&0&0&1&1&0&0&1&0&0&0&0&1&0&0&0\\
0&0&0&0&0&1&0&0&0&0&0&0&1&0&0&0&0&0&0&0&1&1&0&0&0&0&0&0&0&1&0&0\\
0&0&0&0&0&0&1&0&0&0&0&0&0&1&0&0&0&0&0&0&0&1&1&0&0&0&0&0&0&0&1&0\\
0&0&0&0&0&0&0&1&0&0&0&0&0&0&1&0&0&0&0&0&0&0&1&1&0&0&0&0&0&0&0&1\\
1&0&0&0&0&0&0&0&1&0&0&0&0&0&0&0&0&0&0&0&0&0&0&1&1&0&0&0&0&0&0&1\\
0&1&0&0&0&0&0&0&0&1&0&0&0&0&0&0&1&0&0&0&0&0&0&1&1&1&0&0&0&0&0&1\\
0&0&1&0&0&0&0&0&0&0&1&0&0&0&0&0&0&1&0&0&0&0&0&0&0&1&1&0&0&0&0&0\\
0&0&0&1&0&0&0&0&0&0&0&1&0&0&0&0&0&0&1&0&0&0&0&1&0&0&1&1&0&0&0&1\\
0&0&0&0&1&0&0&0&0&0&0&0&1&0&0&0&0&0&0&1&0&0&0&1&0&0&0&1&1&0&0&1\\
0&0&0&0&0&1&0&0&0&0&0&0&0&1&0&0&0&0&0&0&1&0&0&0&0&0&0&0&1&1&0&0\\
0&0&0&0&0&0&1&0&0&0&0&0&0&0&1&0&0&0&0&0&0&1&0&0&0&0&0&0&0&1&1&0\\
0&0&0&0&0&0&0&1&0&0&0&0&0&0&0&1&0&0&0&0&0&0&1&0&0&0&0&0&0&0&1&1\\
1&0&0&0&0&0&0&1&1&0&0&0&0&0&0&0&1&0&0&0&0&0&0&0&0&0&0&0&0&0&0&1\\
1&1&0&0&0&0&0&1&0&1&0&0&0&0&0&0&0&1&0&0&0&0&0&0&1&0&0&0&0&0&0&1\\
0&1&1&0&0&0&0&0&0&0&1&0&0&0&0&0&0&0&1&0&0&0&0&0&0&1&0&0&0&0&0&0\\
0&0&1&1&0&0&0&1&0&0&0&1&0&0&0&0&0&0&0&1&0&0&0&0&0&0&1&0&0&0&0&1\\
0&0&0&1&1&0&0&1&0&0&0&0&1&0&0&0&0&0&0&0&1&0&0&0&0&0&0&1&0&0&0&1\\
0&0&0&0&1&1&0&0&0&0&0&0&0&1&0&0&0&0&0&0&0&1&0&0&0&0&0&0&1&0&0&0\\
0&0&0&0&0&1&1&0&0&0&0&0&0&0&1&0&0&0&0&0&0&0&1&0&0&0&0&0&0&1&0&0\\
0&0&0&0&0&0&1&1&0&0&0&0&0&0&0&1&0&0&0&0&0&0&0&1&0&0&0&0&0&0&1&0
\end{bmatrix}$$

2.4.2 PRESENT-80

PRESENT [BKL+07] family of block ciphers has two members, namely, PRESENT-128 (with 128-bit key) and the lightweight version PRESENT-80 (with 80-bit key). For a top level representation, one may refer to Fig. 2.15/Fig. 10.3b.

Both the variants use the same un-keyed permutation (Sect. 2.2.1) of 64 bits. The structure of PRESENT of rounds of this permutation can be seen from Fig. 8.6. It consists of 31 rounds, where each round contains three operations: AddRoundKey, sBoxLayer, and pLayer.

sBoxlayer

The sBoxLayer substitutes the state with the 4-bit SBox S=C56B90AD3EF84712. This SBox is applied to every nibble w_i of the state: $w_i \leftarrow S(w_i)$, $\forall i \in \{0, \ldots, f\}$. This SBox is affine equivalent to CB12354769A08DEF (which is the AE #14 representative in [DeC07, Chapter 5.4.2]).

pLayer

The pLayer is a bit permutation layer which maps state bits b_i of location i to bit location $P_{64}(i)$: $b_{P_{128}(i)} \leftarrow b_i$, $\forall i \in \{0, \ldots, 63\}$ and is given as: (0, 16, 32, 48, 1, 17, 33, 49, 2, 18, 34, 50, 3, 19, 35, 51, 4, 20, 36, 52, 5, 21, 37, 53, 6, 22, 38, 54, 7, 23, 39, 55, 8, 24, 40, 56, 9, 25, 41, 57, 10, 26, 42, 58, 11, 27, 43, 59, 12, 28, 44, 60, 13, 29, 45, 61, 14, 30, 46, 62, 15, 31, 47, 63).

Key Schedule and addRoundKey

The addRoundKey operation XORs the respective round keys to the state, and after 31 rounds it is applied one more time for post-whitening. In notation, given the state bits $b_0 \parallel b_1 \parallel \cdots \parallel b_{63}$ and the corresponding round key $k_0 \parallel k_1 \parallel \cdots \parallel k_{63}$, the state bits are updated by $b_i \leftarrow b_i \oplus k_i\ \forall i$.

For PRESENT-80, the key schedule subroutine is as given here. Let the master key is stored in the register $\mathcal{K}$ as $k_{79} \parallel k_{78} \parallel \cdots \parallel k_0$ and the 64-bit round key at the ith ($1 \le i \le 32$) round is denoted as $rk_{63} \parallel rk_{62} \parallel \cdots \parallel rk_0$. Initially, the round key is extracted the rightmost 64 bits of the master key: $rk_{63} \parallel rk_{62} \parallel \cdots \parallel rk_0 \leftarrow k_{79} \parallel k_{78} \parallel \cdots \parallel k_{16}$. After this, the register $\mathcal{K}$ is updated by the following rules (rc is the 5-bit round counter):

$$
\begin{aligned}
k_{79} \parallel k_{78} \parallel \cdots \parallel k_1 \parallel k_0 &\leftarrow k_{18} \parallel k_{17} \parallel \cdots \parallel k_{20} \parallel k_{19} \\
k_{79} \parallel k_{78} \parallel k_{77} \parallel k_{76} &\leftarrow S(k_{79} \parallel k_{78} \parallel k_{77} \parallel k_{76}) \\
k_{19} \parallel k_{18} \parallel k_{17} \parallel k_{16} \parallel k_{15} &\leftarrow k_{19} \parallel k_{18} \parallel k_{17} \parallel k_{16} \parallel k_{15} \oplus rc
\end{aligned}
$$

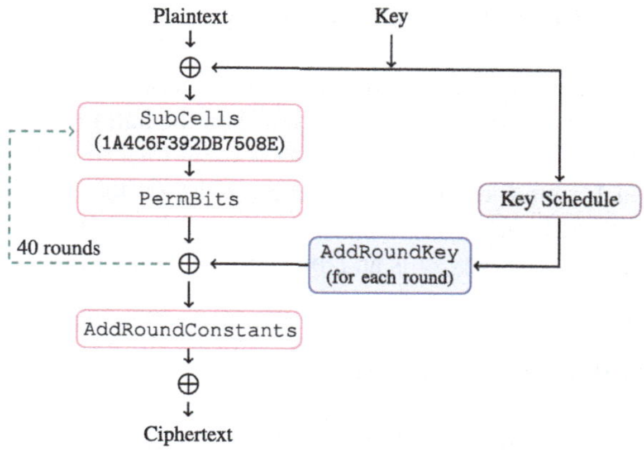

Fig. 2.16 Top level view of GIFT-128

2.4.3 GIFT-128

Motivated from PRESENT [BKL+07], GIFT [BPP+17] is a lightweight block cipher family optimized for hardware implementations due to its bit permutation layer. There are two ciphers in the family, namely, GIFT-128 and GIFT-64.

GIFT-128 is a 40-round SPN cipher that receives a 128-bit message as the state $X = b_{127}b_{126}\cdots b_0$, where b_0 is the least significant bit, and encrypts it with a 128-bit key. The state can be equivalently expressed as $X = w_{31} \parallel w_{30} \parallel \cdots \parallel w_0$, where w_i is a 4-bit (nibble) word. The round function of GIFT-128 consists of four steps (in order): SubCells—applying a 4-bit SBox to the state, PermBits—permute the bits of the state (acts as the linear layer), AddRoundKey—XORing the round key to the (half of) state, and AddRoundConstants—XORing a 6-bit constant as well as another bit to the state. A visual representation can be found in Fig. 2.16.

SubCells

It uses the 4-bit SBox $S =$ 1A4C6F392DB7508E. This SBox is applied to every nibble of the state: $w_i \leftarrow S(w_i)$, $\forall i \in \{0, \ldots, 31\}$.

PermBits

The bit permutation maps state bits from bit location i to bit location $P_{128}(i)$, see Sect. 8.3.3 for its description.

Key Schedule and `AddRoundKey`

The key schedule subroutine is as given here. The 128-bit key is divided into eight 16-bit words: $k_7 \parallel k_6 \parallel \cdots \parallel k_0$. Then the half of the cipher state is XORed with the round key: $b_{4i+2} \leftarrow b_{4i+2} \oplus k_5 \parallel k_4$ and $b_{4i+1} \leftarrow b_{4i+1} \oplus k_1 \parallel k_0$, for $i = 0(1)31$. Each round key is extracted first and then the next round key is produced by the update: $k_7 \parallel k_6 \parallel \cdots \parallel k_0 \leftarrow k_1 \ggg 2 \parallel k_0 \ggg 12 \parallel \cdots \parallel k_3 \parallel k_2$ followed by $k_7 \leftarrow k_7 \ggg 1$, where $\ggg$ is an i-bit right rotation within a 16-bit word.

`AddRoundConstants`

The state bits 3, 7, 11, 15, 19, 23 are XORed with the constants: 1, 3, 7, 15, 31, 62, 61, 59, 55, 47, 30, 60, 57, 51, 39, 14, 29, 58, 53, 43, 22, 44, 24, 48, 33, 2, 5, 11, 23, 46, 28, 56, 49, 35, 6, 13, 27, 54, 45, 26, and also the state bit 127 is flipped in each round.

The structure of `GIFT-128` for two rounds (out of 40 rounds) is given in Fig. 2.17. The SBoxes are numbered (from 0 to 31), so are the bit positions (from 0 to 127), for better clarity. The coloured lines indicate the permutation (linear) layer. By $\bigoplus$, we denote that the corresponding key bits are XORed. The round constant additions are shown by $\bigoplus$ (bits at 3, 7, 11, 15, 19, 23), and the bit at 127 is flipped in each round.

2.4.4 `CHASKEY`

`CHASKEY` [MMH+14, Mou15] is a MAC which is designed to have an edge on implementation on 32-bit microcontrollers. The permutation (π) is based on modular addition–rotation–XOR (ARX, see Sect. 2.3.3) design methodology similar to `SIPHASH` [AB12].

It takes a 128-bit key K and processes a message m in 128-bit blocks using the 128-bit permutation π. The permutation π which processes 128 is an ARX-based permutation and the round function is as given in Fig. 2.18; here, $v_0 \parallel v_1 \parallel v_2 \parallel v_3 \leftarrow \pi(v_0 \parallel v_1 \parallel v_2 \parallel v_3)$. This round function is iterated for $r = 12$ rounds [Mou15]. The key operations are addition over modulo 2^{32} ($\boxplus$), XOR ($\oplus$), and left rotation ($\lll$). Note that the $\boxplus$ is the only non-linear operation. The permutation π is repeatedly applied in a mode of operation to realize the MAC (see Fig. 2.19) and the whole process can be viewed as an *Even-Mansour* block cipher [EM91, EM97] with a $2n$-bit key and n-bit block size. For the key schedule, `CHASKEY` takes a 128-bit key K, from which two 128-bit sub-keys K_1 and K_2 are derived, each using a 128-bit shift and a 128-bit conditional XOR.

So far, the most significant cryptanalysis that has been carried out on `CHASKEY` can be found in [Leu16]. This attack was in fact on the older version [MMH+14],

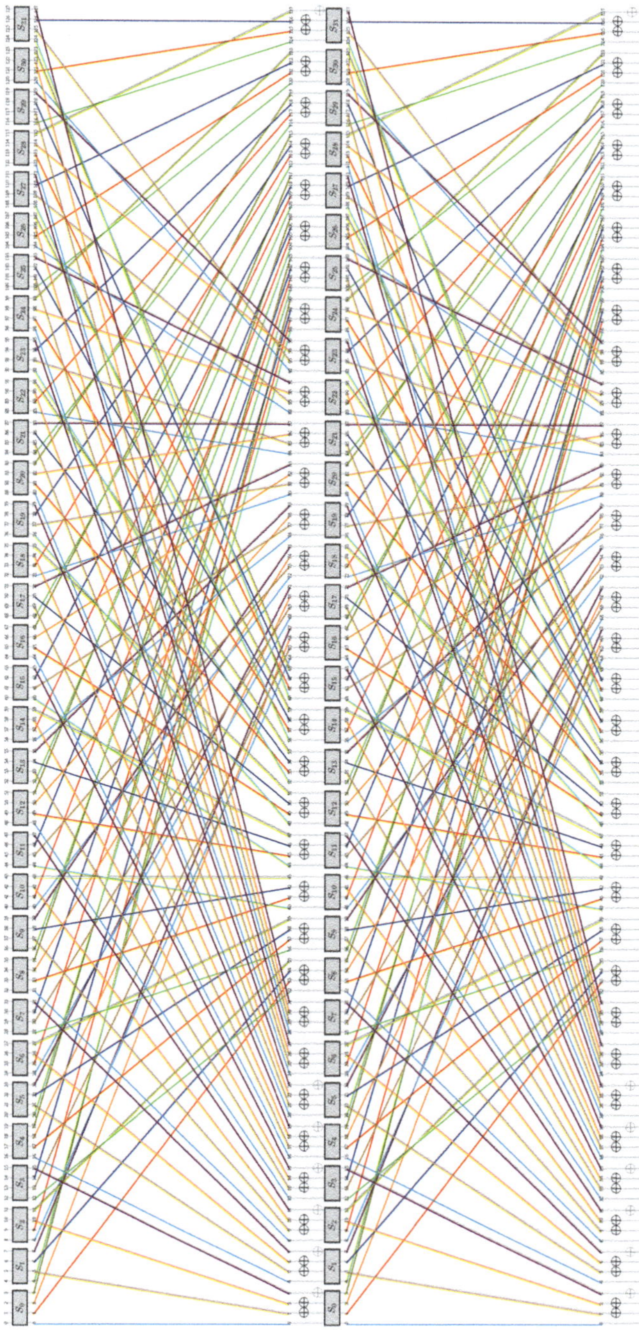

Fig. 2.17 Structure of `GIFT-128` (two rounds only)

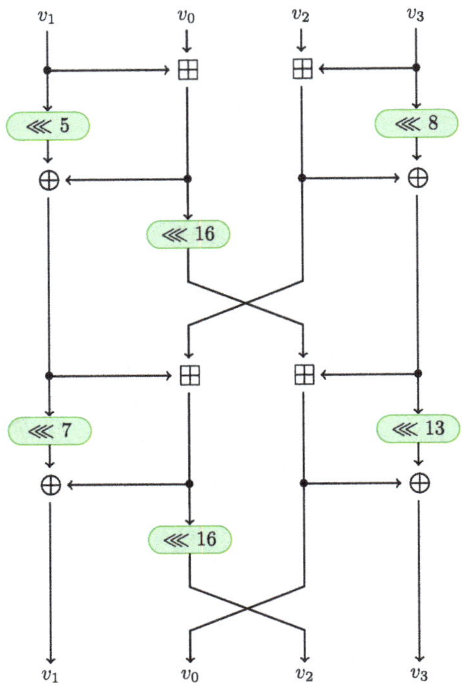

Fig. 2.18 One round of CHASKEY permutation π

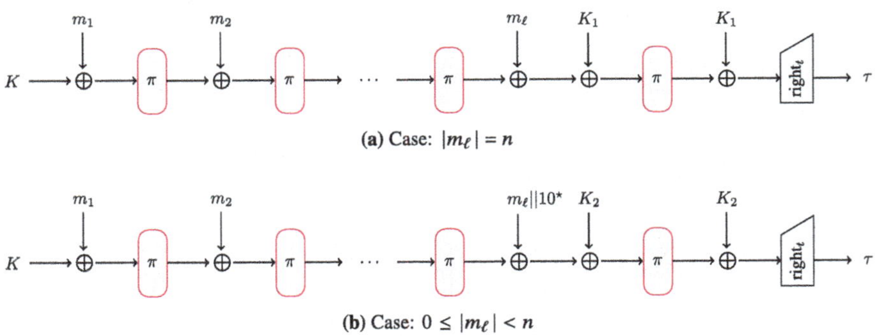

Fig. 2.19 CHASKEY mode of operation

where the number or rounds of the permutation π is set to $r = 8$. They used differential–linear cryptanalysis and succeed to attack for $r = 7$. After this attack, CHASKEY is revised with $r = 12$ [Mou15].

2.5 Formidability of the Attacker

The attacker tries to learn secret information regarding the cryptographic components. In general, the attacker tries to learn information on the secret key. Depending on the situation, the attacker may try to recover information on the (secret) plaintext, or find a collision in the hash function outputs, or forge the authentication tag.

2.5.1 Conventional Notions of Security

Coined by Shannon, *confusion* and *diffusion* are two fundamental principles for a secure cipher. Confusion refers to scrambling the message with the help of the secret key to remove any exploitable information. This refers to the complexity of finding the key/any information protected by the key. Diffusion, on the other hand, refers to absence of statistical information. Thus, for a cipher with good diffusion, there is no significant statistical relationship between its input and output. The update rules of the cipher (like SBox, linear layer, number of iterations) ensure diffusion.

Another important notion of security is due to Kerckhoffs. The *Kerckhoffs's principle* states that the security of a system should not be compromised even when the complete algorithm, save for the key , is public. Thus, the algorithmic description of the cipher can be released in public.

2.5.2 Power of the Attacker

The presumed power of the attacker varies from model to model. Depending on the context, it may be required that the cipher or the entire cryptographic system be capable of thwarting the attacker with the presumed power.

Plaintext/Ciphertext Setting

1. **Ciphertext Only.** The attacker passively observes the ciphertexts.
2. **Known Plaintext.** The attacker knows (a limited number of) plaintexts and the corresponding ciphertexts, but cannot choose the plaintexts.
3. **Chosen Plaintext.** The attacker chooses (a limited number of) plaintexts and asks for the corresponding ciphertexts. The choice of one plaintext is not affected by the ciphertexts for the previous plaintexts.
4. **Adaptive Chosen Plaintext.** The attacker chooses (a limited number of) plaintexts and asks for the corresponding ciphertexts. Based on the already obtained ciphertexts, the attacker decides the follow-up plaintexts.

5. **Chosen Ciphertext.** The attacker chooses (a limited number of) ciphertexts and asks for the corresponding plaintexts. The choice of one ciphertext is independent of the plaintexts for the already queried ciphertexts.
6. **Adaptive Chosen Ciphertext.** The attacker chooses (a limited number of) ciphertexts and asks for the corresponding plaintexts. While choosing a ciphertext, the attacker uses information on plaintexts for the already queried ciphertexts.

The adaptive chosen plaintext/cipher attacks assume maximum power to the attacker. Also, the attacker does not query in the ciphertext only and the known plaintext models.

Key/Nonce Setting

1. **Single Key.** Under this model, the attacker can only obtain information corresponding to a fixed secret key. This model is most commonly used in cipher design/cryptanalysis.
2. **Related Key.** This model is typically not intended to evaluate the security of a cipher but to break the cryptographic system which relies on that cipher. The attacker can obtain information on the relationship among the keys that are used in the cipher. Generally, the cipher designers tend not to claim any security under this model.
3. **Nonce Misuse.** The attacker can repeat the nonce (the is being fixed) for the encryption and the decryption queries.
4. **Nonce Respecting.** In this model, the attacker cannot repeat the same nonce (for a fixed key) for an encryption query. However, the nonce can be repeated for a decryption query [Rog04].

2.5.3 Objective of the Attacker

Distinguishing Attack

In this theoretical setting, it is assumed that a challenger (`ORACLE`) tosses a random coin. Based on the outcome choose either `RANDOM` (which is a true random source), or the (possibly round-reduced version of) `CIPHER`together with a uniformly chosen secret key. See Fig. 2.20 for the setup. Once chosen, the `ORACLE`will keep it fixed but secret. The attacker can query the `ORACLE`with a plaintext of her choice, and the `ORACLE`will either return an output from `RANDOM`, or the ciphertext obtained from the `CIPHER`by encrypting the plaintext with the previously chosen secret key. The goal of the attacker is to find out, with probability significantly $> \frac{1}{2}$ and with a polynomially bounded complexity, whether `ORACLE`has chosen `RANDOM` or `CIPHER`by analysing the outputs returned by the `ORACLE`and without repeating the same input.

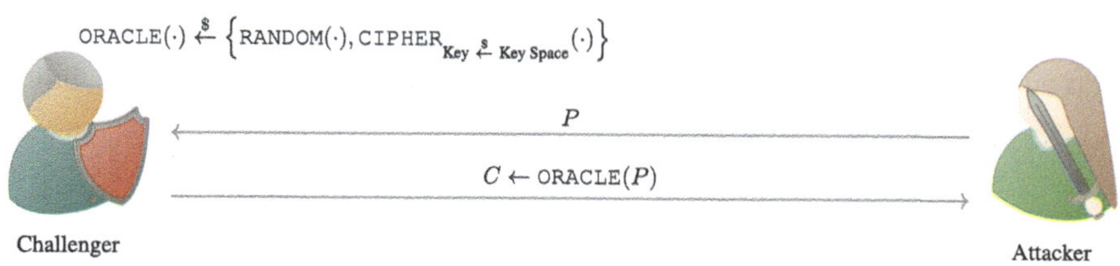

Fig. 2.20 Setup for challenger versus attacker model

No other access to the ORACLE, including the side channel information (e.g., how long does the ORACLEtake to return an output), is available to the attacker.

Notice that the description of CIPHERis public, hence the adversary is able to do some profiling before the actual challenge starts (the *off-line* phase). If any exploitable property is found during this time, it may help the adversary to deduce information at the time of the actual challenge (the *online* phase) when she does not know whether CIPHERor RANDOM is chosen by ORACLE.

Key/Plaintext Recovery Attack

In the key recovery attack, the attacker tries to find any information related to the secret key (other than the length of the key). This can be applied to a block cipher, stream cipher, or MAC.

In the plaintext recovery attack, similarly, the attacker tries to learn any information on the plaintext (except its length). This does not directly apply to stream ciphers, as plaintext is not directly processed in this case.

Collision Attack

In the context of a hash function, a *collision* can be loosely described as two distinct messages for which the hash values are same. Due to the *birthday problem* [Sti06, Chap. 4], such collisions can be found at the square root complexity of the search space of the hash values.

Forgery Attack

In a forgery attack, the attacker tries to forge the authentication tag:

1. **Universal Forgery.** The attacker creates a valid signature for a message chosen by her, with significant probability.
2. **Selective Forgery.** The message is chosen a priori by the attacker, for which she can create the signature with significant probability. This message may be chosen based on information regarding signatures of some previously chosen messages.

3. **Existential Forgery.** The attacker creates the signature for a given message which is not chosen by her, nor it has previously been signed, with significant probability.

The existential forgery allows maximum power to the attacker.

2.6 Major Classical Attacks

Some of the important classical attacks are briefly described here for completeness. Overall, it can be stated that these attacks share the common properties:

- A round-reduced version of the target cipher is considered. The results thus obtained are then generalized/extrapolated to the full round version.
- A divide-and-conquer approach is used to recover information about the key. Generally, it involves gathering some non-trivial information about the state just before the key insertion in the last round. Then, all the sub-keys are guessed for a manageable portion of state. Those sub-key candidates are thereafter filtered using the information previously gathered. In this way, the search space for each sub-key is reduced. This process can be repeated for other portions of the state. Eventually, this can reduce the search space for the entire key. If the exhaustive search on the reduced search space for the secret key is feasible on the full round cipher, then the cipher is considered broken.

To protect against the classical attacks, the cipher designers typically analyse the cipher with respect to the best known attacks with a small round version of the proposed cipher. Then, the actual number of rounds of the cipher is determined in a way that it blocks all those attacks with sufficient margin. During our literature survey, we empirically observe that the *Differential Attack* (Sect. 2.6.1) and the *Linear Attack* Sect. 2.6.2) are commonly the dominating attacks, meaning most classical attacks are blocked in less number of rounds than what would be required to block these two.

2.6.1 Differential Attack

One of the first cryptanalysis techniques, known as the Differential Attack (DA) or *Differential Cryptanalysis* (DC) [BS90], relies on the chosen plaintext model, and studies the relationship of the output difference (Δ) with respect to the input difference (δ). The term differential trail or *differential characteristic* indicates the path of the input differences traced over several rounds of the cipher, which holds for a significant probability. From (δ, Δ), the attacker can reduce the search space of the key candidates [Sti06, Chap. 3.4].

More precisely, the attacker chooses (random) inputs for the cipher, so that a pair of inputs follow certain difference δ ($\neq 0$). Say, the attacker chooses P as the input randomly, then she constructs another input $P \oplus \delta$. By querying these two inputs to the ORACLE, she receives the corresponding outputs, say C and $C \oplus \Delta$. Given a certain input difference δ, it is possible to formulate its probabilistic propagation through multiple rounds of cipher. For example, in case of the SBox layer, the values in the DDT (see Sect. 2.1.2) directly indicate the probability at which a certain output difference appears given a certain input difference. As the round keys are typically XORed, the effect is nullified in the difference relations. If the challenger chooses RANDOM, such formulation would not be possible. In that case, the probability distribution of Δ given δ is indistinguishable from the uniform distribution. On the other hand, if it happens that the distribution is distinguishable from uniform then the attacker can distinguish between the CIPHERand RANDOM. This profiling can help the attacker to win the distinguishing game (in Sect. 2.5.3). During the offline phase (i.e., prior to the challenge), the attacker computes the probability distribution and sees if it is distinguishable from uniform (she is able to do this as the description of the CIPHERis public). She is able to use this information during the online phase (i.e., the actual challenge) to distinguish between RANDOM and the CIPHER.

Additionally, the non-uniformity of the probability distribution can be used to recover the secret key. Because of this non-uniformity, the search space for the state at the last round but before the last round key addition may be reduced for a typical block cipher. Using this information, the search space for the last round key can be reduced to the point of finally recovering the last round key with a significantly less complexity than exhaustive search. In case of a typical stream cipher, this non-uniformity may reduce the search space for the state, from which the key can be recovered (as typically the state update is reversible and the state is initialized with the key).

Given the input difference–output difference pair (δ, Δ), the relative frequency of number of such input pairs with the difference δ to the total number of possible inputs is denoted as the expected differential probability. The *Maximum Expected Differential Probability* (MEDP) measures the maximum of all expected differential probabilities. The MEDP for a random permutation is the inverse of total number of permutations.

Remark 2.10 The higher entries in the DDT of an SBox correspond to higher probability distribution (i.e., away from the uniform distribution), more discussion can be found in Chap. 5. If DU $= 2^n$ for an n-bit SBox (i.e., it has a non-zero linear structure), it performs poorly against differential attack compared to all other SBoxes that do not have a (non-zero) linear structure.

Remark 2.11 The concept of DA is closely related to that of the derivative of a Boolean function.

2.6.2 Linear Attack

One important cryptanalysis method, known as the Linear Attack (LA) or the *Linear Cryptanalysis* (LC), is based on the known plaintext model. It is first reported in [Mat93]. Here, the attacker tries to find affine approximation for the (possibly round-reduced) cipher. If such a good linear approximation is found, the attacker tries to find linear equations which holds with a high probability. The *Piling-up Lemma* [Sti06, Chap. 3.3.1] is an important theorem used in this context. Once such equations are found, these can be used to reduce the search space for the key.

Assume the relation $\Pr[p \cdot \gamma \oplus c \cdot \Gamma = 0] = \frac{1}{2} \pm \epsilon$ holds, where the input to the cipher is p, the corresponding output is c, γ and Γ are two suitably chosen bit vectors, $\cdot$ is the dot product, and $\epsilon \neq 0$ is known as the *bias* (the probability is taken over all possible count of the permutation). Such a relation can be computed by the attacker during the offline phase, and can be used to recover information on the output during the online phase. The propagation through the linear layer is relatively straightforward and that of the SBox is done through the LAT. Finally, the piling-up lemma computes the propagation through the rounds by assuming independence between the consecutive rounds (see Chap. 5 for more details).

The aforementioned probability is also known as the expected linear probability. The maximum of all it is known as the *Maximum Expected Linear Probability* (MELP).

Remark 2.12 Similar to the differential case, the higher (absolute) values in the LAT of an SBox indicate that the probability for an affine relation would be higher than some other SBox which has lower (absolute) value in its LAT. Therefore, an SBox with lower (absolute) values in the LAT is preferred when designing a cipher.

2.6.3 Algebraic Attack

The *algebraic attack/algebraic cryptanalysis* expresses the cipher in terms of algebraic relations and thereafter uses some automated solver to gain information about the secret components. Typically, the key and the plaintext can be described by symbolic variables over Boolean polynomial ring, so that the ciphertext can be described as an algebraic expression. For this purpose, the coordinate functions of the SBox (Sect. 2.1.2) can be used. The algebraic properties for the cipher output, if feasible to compute, would be complicated (e.g., high algebraic degree) and grow fast as the round increments. Ideally, for the full-round version of the cipher, it should be infeasible to compute the cipher output algebraically.

Example 2.6 We take the stream cipher-based AEAD CLXOTE [WH19]. It is based on a 160-bit NLFSR, and its update function has algebraic degree of 1 (which is in the form of a NAND gate of 106th and 93th state bits), as shown in Fig. 2.21. We show three algebraic properties of it in Fig. 2.22. In order to do this, all the state

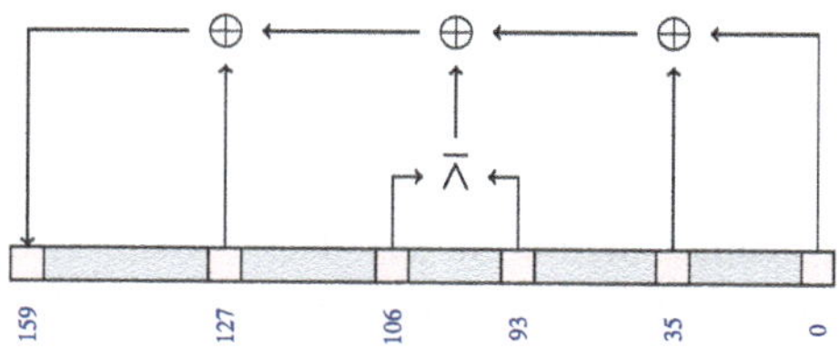

Fig. 2.21 Structure of `CLX-128`

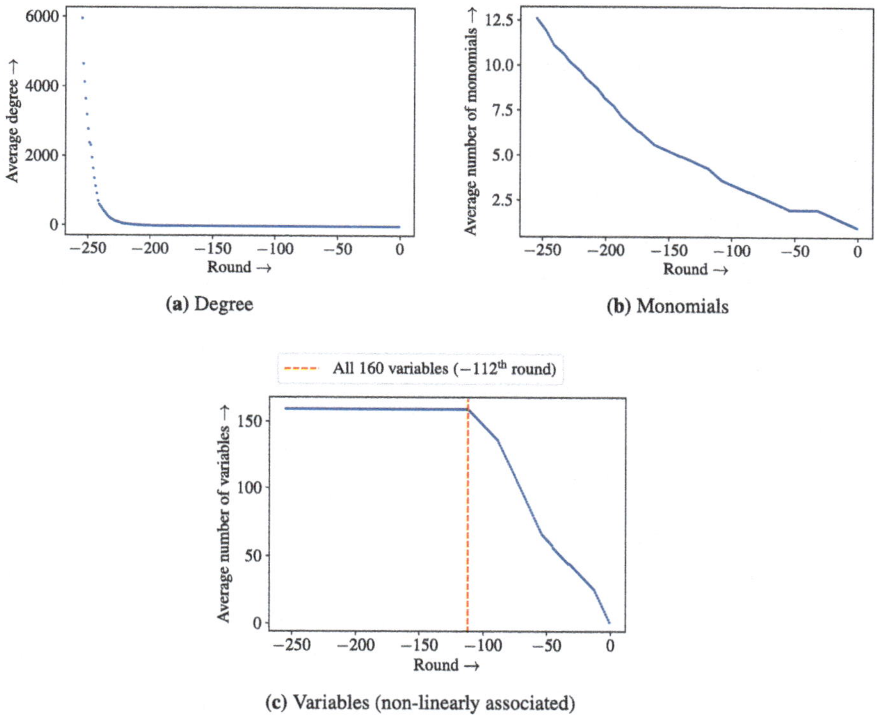

Fig. 2.22 Algebraic properties of last 32 bits of tag of `CLXOTE`over varying rounds

bits are substituted by a symbolic variable over the Boolean polynomial ring with Sage. Starting from the 255th round before the last 32-bits of the tag are released, we show over the follow-up rounds the average algebraic degree, average number of monomials, and average number of variables which are non-linearly associated in the expression (the average is taken over all the 32 bits). As shown in Fig. 2.22c, all the 160 state variables become non-linearly associated in the expression of the last 32 bits of tag at the −112th round. Beyond the −255th round, the computation becomes expensive so that it is no longer possible to do in a reasonable time. Therefore, it can be argued that the algebraic attack becomes practically ineffective beyond this.

2.6.4 Integral/Cube Attack

The *cube attack/integral attack*[3] is proposed in [KW02, Vie07, DS08] which acts as a powerful attack against stream ciphers. Here, a subset of n the public variables is chosen. For all possible 2^n combinations of the public variables, the outputs returned by cipher are XORed. This eliminates the effect of the subset of public variables and reveals a relatively simple (typically an affine). For example, consider the Boolean function (from [Roh19, Chap. 2.4.4.2]) $f(k_0, k_1, k_2, v_0, v_1) = v_0k_0 \oplus v_0v_1k_2 \oplus v_0v_1 \oplus k_0k_1 \oplus v_1k_2 \oplus k_2 \oplus 1$ where k_0, k_1, k_2 are the secret variables and v_0, v_1 are the public variables. With the chosen subset v_0, v_1 of the public variables and XORing all of those, i.e., computing $f(k_0, k_1, k_2, 0, 0) \oplus f(k_0, k_1, k_2, 0, 1) \oplus f(k_0, k_1, k_2, 1, 0) \oplus f(k_0, k_1, k_2, 1, 1)$ will give $k_0 \oplus k_2 \oplus 1$. Thus, the attacker recovers information about the secret variables, from which the search space of the secret key can be reduced.

More formally, consider the function $f(\vec{k}, \vec{v})$ with the private variable $\vec{k} = (k_0, k_1, \ldots, k_{n-1})$ and the public variable $\vec{v} = (v_0, v_1, \ldots, v_{m-1})$ produces 1-bit output. Suppose it can be written as $f(\vec{k}, \vec{v}) = t_I \cdot p(\vec{k}, \vec{v}) \oplus q(\vec{k}, \vec{v})$ where $I = \{i_0, i_1, \ldots, i_{\mu-1}\} \subseteq \{0, 1, \ldots, \mu - 1\}$ is the set of chosen indices of the public variable $\vec{v}$, $J = \{v_{i_0}, v_{i_1}, \ldots, v_{\mu-1}\}$, $t_I = \prod_{j \in J} j$, $p(\vec{k}, \vec{v})$ is a polynomial that does not contain any j belonging from J and known as the *superpoly*, and $q(\vec{k}, \vec{v})$ misses at least one j from J. All possible 2^μ combinations of the μ elements from J are given by C_I, and the remaining $n + m - \mu$ elements are set of some (known) constants. Now taking XOR over C_I, we see $\bigoplus_{C_I} f(\vec{k}, \vec{v}) = \left(\bigoplus_{C_I} t_I \cdot p(\vec{k}, \vec{v})\right) \oplus \left(\bigoplus_{C_I} q(\vec{k}, \vec{v})\right)$. As $q(\vec{k}, \vec{v})$ misses at least one j from J, it will be XORed an even number of times thus will vanish. Also, $t_I = 1$ for exactly one combination (where each element of J is 1) and 0 for the rest. Therefore, $\bigoplus_{C_I} = p(\vec{k}, \vec{v})$. Thus, if the superpoly $p(\cdot, \cdot)$ is of low algebraic degree, it can be recovered.

The distinguishing variant of the cube attack is mentioned as the *cube tester* [ADM+09]. With the previously described setup, if the superpoly is of high algebraic degree, it will likely show non-randomness property which can be used as a distinguisher . For example, the probability distribution of the affine function $x_0 \oplus x_1$ is not distinguishable from RANDOM, but the probability distribution of x_0x_1 can be distinguished. The latter is explored in cube tester.

The *Division property* is introduced by Todo in [Tod15]. Here the integral attack is modelled as an MILP problem.

Remark 2.13 The integral attack can be described as a special case of the differential attack and (similar to Remark 2.11) derivative of Boolean function.

Remark 2.14 The integral attack, in some sense, is related to the algebraic attack (as it is related to the algebraic property of the superpoly).

---

[3] Also referred to as, *square attack*, *saturation attack*, or *multiset attack*.

2.6.5 *Impossible Differential Attack*

The *Impossible Differential Attack/Impossible Differential Cryptanalysis* (IDA/IDC) is related to DA [Knu98, BBS99]. Instead of trying to find an output difference, this attack makes use of an output difference that does not appear. Therefore, this acts as a distinguisher since the output difference appears with probability 0, whereas it appears with non-zero probability for RANDOM.

To the best of our knowledge, the only generic modelling for finding impossible differential is based on fixing input and output differences and check if the model is feasible [CJF+16, ST17'2].

2.7 Device Implementation

For practical application, a cipher needs to be implemented on a device. The cryptographic community generally resorts to the following approaches:

1. Reduce the cost for an already analysed cipher for various platforms.
2. Propose a variant of an already analysed cipher that reduces the cost.
3. Propose new ciphers that are designed specifically to have a low device footprint.
4. Design/improve a tool/method which can find better implementation for a given cipher component.

The device footprint of a cipher can be measured by the following metrics:

1. **Area.** For ASIC, the area can be measured in terms of NAND gate equivalent. Also, the actual area footprint (typically measured in square micro-metre) can also be reported. For FPGA, it can be measured by number of flip flops (F/Fs) and number of look-up tables (LUTs). In case of a software implementation, it would refer to the code size.
2. **Clock Cycles/Operating Frequency/Time**
3. **Throughput.** Throughput measures the amount of output that is generated per unit time (typically measured in bits per second).
4. **Power/Energy.** The energy consumption (typically measured in micro-Joule or micro-Joule per bit) is given by the power consumption (typically measured in micro-Watt) of a device over a certain time period.
5. **Efficiency.** The efficiency is given by the ratio of throughput to area.

Lightweight Cryptography

The term *Lightweight Cryptography* refers to a relatively new genre in symmetric key cryptography and highly active at this moment. The objective of lightweight cryptography is mainly to custom-design ciphers that have typically low device foot-

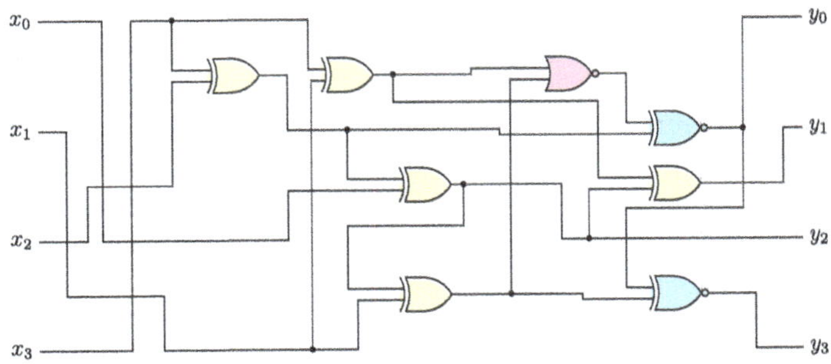

Implementation 1 Hardware (ASIC, TSMC 65 nm) implementation for SBox `126CDE39F58BA047`: 18.5 GE

print. The *Lightweight Cryptography Competition* (LWC)[4] is a currently ongoing competition organized by NIST and aims at finding secure AEAD. This competition is at its second round with 32 candidates.

Benchmarking

Different tools/methods can be used to find an optimized for a complete cipher or a component of the cipher. Such optimized implementation is used for benchmarking for comparison with other ciphers. For example, the method from [XZL+20] finds an implementation of the `AES MixColumns` matrix in 92 XOR gates. Similarly, the LIGHTER tool [JPS+17] can be used to find implementation of a 4×4 SBox for CPU or ASIC given the costs for various logic gates in the library.

Example 2.7 In Implementation 1, the optimized ASIC (TSMC 65 nm) implementation for the SBox `126CDE39F58BA047`, as obtained from LIGHTER [JPS+17], is shown. The hardware consists of five XOR gates, two XNOR gates, and one NOR gate. The optimized software implementation using the same tool with eight instructions is shown in Implementation 2.

Pitfalls

In general, ensuring a fair benchmarking is difficult, as multiple extraneous factors come into play. We list a few problems here.

Software. An optimized implementation of `AES` which uses large look-up tables [DR20, Chap. 4.2.1] may not be suitable for some low-end microcontrollers like ATmega328P (e.g., the implementation of `AEGIS` [WP15] given by the designers[5] uses large look-up tables and not directly usable for a low-end microcontroller).

[4] https://csrc.nist.gov/projects/lightweight-cryptography.

[5] https://www.ntu.edu.sg/home/wuhj/research/caesar/finalist_code/aegis_code.zip.

Implementation 2 Software (CPU) implementation for SBox `126CDE39F58BA047`: eight instructions

```
F[0] = X[2];
F[1] = X[0];
F[2] = X[1];
F[3] = X[3];

F[0] = XOR2(F[0], F[3]);
F[3] = XNOR2(F[3], F[2]);
F[1] = XOR2(F[1], F[0]);
F[2] = XOR2(F[2], F[1]);
F[0] = XOR2(F[0], OR2(F[2], F[3]));
F[3] = XNOR2(F[3], F[1]);
F[2] = XNOR2(F[2], F[0]);

Y[0] = F[2];
Y[1] = F[3];
Y[2] = F[1];
Y[3] = F[0];
```

Further, as described in [BSS20, Sect. 3.4]/Sect. 9.3.4, the FELICS tool [DCK+15] intrinsically takes the test vectors in bytes. This causes overhead if the cipher is designed at the bit level, since bit-to-byte conversion at the beginning and byte-to-bit conversion at the end would be required.

Hardware. It is hard to fix a generalized API for hardware. Increasing the number of input/output pins would allow faster read/write, but at the same time increases the fabrication cost. Also, the number of clocks allowed to read the key/plaintext can be changed that results in various costs.

2.8 Additional Topics

2.8.1 Black Box–Grey Box–White Box Models

The *black box* model only allows the output for the attacker, but no further access to the ORACLEis allowed, and the *grey box* model allows further access/control like the time taken, power consumption, etc. The side channel and fault attacks form the grey box model. Under this model, limited access to the device performing the cryptographic operation or the communication medium.

A relatively new concept, the so-called *white box* [CEJ+02], allows further access/control for the attacker. In particular, the entire software implementation (binary file) is made available to the attacker. However, the attacker cannot retrieve the key as it is not directly hardcoded but hidden inside the code. The binary file allows API access only for the encryption function. To the best of our knowledge, whether or not a secure white box implementation exists is not known till date.

2.8.2 Mixed Integer Linear Programming (MILP)

The Mixed Integer Linear Programming (MILP) refers to a class of optimization problems, which is frequently used in cryptanalysis/cipher design (Chap. 5). In an MILP problem, a linear objective function is optimized (maximized or minimized) subject to a family of linear constraints.

More formally, an MILP problem with n variables $x_0, x_1, \ldots, x_{n-1}$ and with m constraints can be described by

- Linear constraints of the form: $\sum_{j=0}^{n-1} a_{i,j}x_j \leq b_i$, for $i = 0(1)m - 1$, where $a_{i,j}, b \in \mathbb{R}\ \forall i, j$ are constants.
- Objective function of the form: Maximize $\sum_{j=0}^{n-1} c_j x_j$, where $c_j \in \mathbb{R}$ is a constant.
- The variables can be of various types; such as Boolean (binary, as per terminology), integer, or real.

It is to be noted that an MILP instance can be equivalently stated with the minimization objective, and it can handle $\geq$ and $=$ type constraints as well. However, non-strict-type constraints (i.e., $>$ or $<$) are not permitted, to the best of our knowledge.

Example 2.8 Consider the following MILP instance ($m = n = 2$):

$$\text{Maximize } 8x_0 + 5x_1$$

$$\text{Subject to } \begin{cases} x_0 + x_1 \leq 6 \\ 9x_0 + 5x_1 \leq 45 \end{cases}$$

x_0, x_1 are non-negative integers.

It can shown that the optimum occurs at (5, 0) with the objective solution of 40. If the variables are of type real, then the optimum occurs at (3.75, 2.25) (which is the intersection of the straight lines obtained by taking equality for the constraints) with solution of 41.25. A plot of the constraints with the corresponding feasible regions (with overlap) is shown in Fig. 2.23.

Indicator Constraint

The *indicator constraint* (also known as the *big M* method) is a well-known concept in the context of MILP. Consider the linear constraint, $\vec{a}^\top \vec{x} \geq b$, where $\vec{x}$ is the vector of n variables, and with the constants $\vec{a} \in \mathbb{R}^n, b \in \mathbb{R}$. The idea here is to

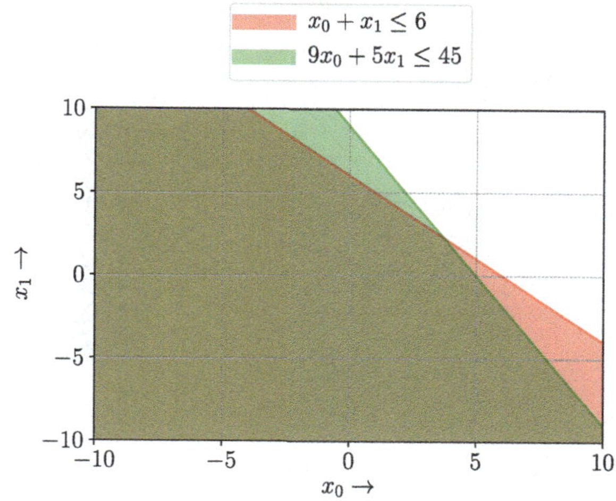

Fig. 2.23 Example of linear constraints

introduce a Boolean variable y, multiplied by a large constant (M). Depending on other constraints or a penalty in the objective function, y indicates whether the constraint is active (the feasible solution must satisfy it) or not (the constraint is redundant in the sense that the feasible solution of may not satisfy it): $\vec{a}^\top \vec{x} \geq b + My$. Theoretically, an arbitrarily large M can be chosen, but doing so could slow down the solution process.

2.8.3 *Machine Learning (ML)*

Machine Learning (ML) refers to classes of optimization or classification algorithms which are based on/inspired from natural processes, such as the evolution of the species, neuron-based computations in human brains, and so on. ML-based algorithms are almost ubiquitously used in a plethora of domains like image processing, biometrics, speech processing, etc.

Types

1. **Supervised Learning.** For this category, the already labelled data are used for training the ML model. Based on the training data, the model tries to learn intrinsic parameters. After the learning is done, the testing phase takes place, where the model predicts corresponding classes for data which are not labelled a priori.
2. **Unsupervised Learning.** The algorithms in this category work without the labelled data. The algorithms can classify data based on the association among the data.

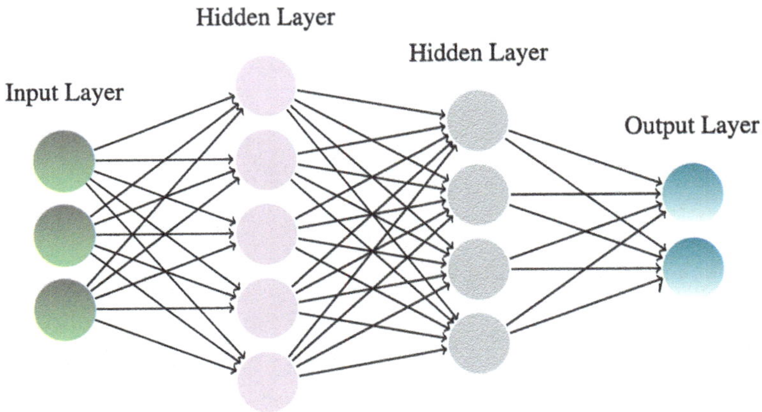

Fig. 2.24 Multi-layer perceptron (schematic)

Artificial Neural Network (ANN)

The *Artificial Neural Network* (ANN) [Hay08] is a common type of ML technique. Neurons are organized in multiple interconnected layers, with each neuron with an individual firing rule. From a top level view, the overall network takes the labelled data during training and set multiple parameters based on some update rules. Those parameters converge to certain values as the training progresses. When the training is done, a testing phase with not previously labelled data takes place. During this testing phase, the already trained ANN model predicts the classes for the data. Figure 2.24 shows the schematic structure of the *Multi-Layer Perceptron* (MLP), one of the earliest ANN models, where each neuron in a layer is connected to every neuron in the subsequent layer.

ANNs are capable of finding inherent characteristics of a user-provided dataset (often called the *training* data) by iterating through it repeatedly and gradually adjusting its parameters, until these parameters are finally stabilized (the process is called *convergence*).

The basic processing unit of an ANN is termed as a *neuron*. An ANN can have from dozens to thousands of neurons. The neurons are arranged in a series of *layers*. One layer interacts with the next layer is a certain predefined mathematical model (called *activation*). More depth of layers generally makes the ANN capable of handling more complex data. An input data thus go through all the layers of an ANN, finally producing a human understandable output.

Neuron. A neuron is the basic computing unit of an ANN. For a neuron in ith layer, it generally has a connection from several neurons in layer $(i-1)$. Each connection has a weight parameter associated to it. The output from each connection and its corresponding weight are multiplied and summed over all the connections to give a value.

Layer. A Layer is a general term used to denote a collection of neurons operating together at a specific depth within a neural network. The *input layer* takes the raw data. The *hidden layer(s)* are the layers which internally learn the hidden pattern in the data. Each hidden layer tries to "learn" different aspects of the data by minimizing cost function (termed as the *loss function*). Generally, the number of hidden layers needs to increase as the patterns in the data become more subtle. The *output layer* is the simplest, often consisting of a small number of neurons. Depending on the firing rate of the neurons at the output layer, the prediction is done. A *dense layer* is a common type of layer where each neuron receives input from all the neurons in the previous layer, thus densely connected. By *dropout*, a fraction of randomly selected neurons is set to 0 during training (thus are ignored). This is done usually to prevent over-fitting.

Activation Function. An activation function is used to determine the output of neural network. Depending on the function, it maps the resulting values in between 0 and 1 or -1 and 1, etc. A neuron takes multiple inputs (and probably another hidden input for the *bias* term) and outputs a value based on its activation function. Activation functions are also used in between layers to introduce non-linearity.

- *ReLU.* Also known as the rectifier function, it is a popular choice for the initial layers. It is perhaps the most commonly used activation function, and is given by

$$f(x) = \max(0, x).$$

- *Softmax.* This function takes in input of n numbers that might be negative, or greater than 1, and might not add up to 1, but after application of the function, each value lies in $[0, 1)$ and sums to 1. It is typically used in the output layer in multi-classification problems. The mathematical form of the Softmax function is given below, here $z = (z_1, z_2, \ldots, z_n)$ is a vector of real numbers (which is the input to the output layer):

$$f(z_i) = \frac{e^{z_i}}{\sum_{j=1}^{n} e^{z_j}} \text{ for } i = 1, 2, \ldots, n.$$

Epoch. An epoch is a one pass over the entire data. Typically, training requires several epochs for good testing accuracy, or equivalently low loss function (close to 0).

Training Accuracy, Testing Accuracy. When training a neural network, the model tries to fit the data. After every epoch, the training accuracy is simply how correctly the model fits the dataset. This gives an estimate on how the ML model is learning the hidden parameters of the input data. Once training phase is complete, the testing phase will test the model on new and previously unseen datasets. If the new datasets are already classified, then the testing accuracy is reported based on how accurately the ML model works on this new data.

Cross-entropy Loss Function. A loss function measures the deviation of the model from the dataset. In this work, we employ *sparse categorical cross-entropy* which is defined as the following:

$$f(y) = \sum_{c=1}^{M} y_{(o,c)} \log(p_{(o,c)}).$$

The notations used above have the following significance:

M : the number of classes;

y : the binary indicator (0 or 1) if class label c is the correct classification for the observation o;

p : the probability predicted that observation o belongs from class c.

2.8.4 Competitions

In order to boost cipher design, implementation, and cryptanalysis, competitions are held from time to time. Here we note the competitions which are relevant with respect to the book:

- **Advanced Encryption Standard (AES).** This competition was organized by the United States National Institute of Standards and Technology (NIST)[6] around 1997–2000. The winner was a block cipher named `RIJNDAEL`, which is adopted as the standard and is now referred to as `AES`[7] (its description can be found in Sect. 2.4.1).
- **ECRYPT Stream Cipher Project (eSTREAM).** Focusing on stream ciphers, this competition was held around 2004–2008. In total, seven (four in the software category and three in the hardware category) ciphers were chosen.
- **Differential Power Analysis (DPA) Contest.** More details can be found in Sect. 4.2.2.
- **Secure Hash Algorithm (SHA-3).** Organized by NIST during 2007–2012, this competition aimed at finding a secure hash function. The cipher, `KECCAK` was chosen as the winner and now it is referred to as `SHA-3`.[8]
- **Competition for Authenticated Encryption: Security, Applicability, and Robustness (CAESAR).**[9] This competition was held around 2013–2017 to promote research in the field of AEAD. It chose a total of six winners.
- **Lightweight Cryptography Competition (LWC).** See Sect. 2.7 for more information.

[6] https://www.nist.gov/.

[7] https://nvlpubs.nist.gov/nistpubs/FIPS/NIST.FIPS.197.pdf.

[8] https://www.nist.gov/news-events/news/2015/08/nist-releases-sha-3-cryptographic-hash-standard.

[9] https://competitions.cr.yp.to/.

References

[BKL+07] A. Bogdanov, L.R. Knudsen, G. Leander, C. Paar, A. Poschmann, M.J. Robshaw, Y. Seurin, C. Vikkelsoe, PRESENT: An ultra-lightweight block cipher, in *CHES*, vol. 4727 (Springer, 2007), pp. 450–466

[BCG+12] J. Borghoff, A. Canteaut, T. Güneysu, E. Bilge Kavun, M. Knezevic, L.R. Knudsen, G. Leander, V. Nikov, C. Paar, C. Rechberger, P. Rombouts, S.S. Thomsen, T. Yalçin, PRINCE - A Low-latency Block Cipher for Pervasive Computing Applications (Full version), *IACR Cryptology ePrint Archive*, 2012:529 (2012), http://eprint.iacr.org/2012/529

[GPP+12] J. Guo, T. Peyrin, A. Poschmann, *IACR Cryptology ePrint Archive*, 2012:600 (2012), http://eprint.iacr.org/2012/600

[BJK+16] C. Beierle, J. Jean, S. Kölbl, G. Leander, A. Moradi, T. Peyrin, Y. Sasaki, P. Sasdrich, S.M. Sim, The SKINNY family of block ciphers and its lowlatency variant MANTIS. *IACR Cryptology ePrint Archive*, 2016:660 (2016), http://eprint.iacr.org/2016/660

[BDM+10] K. Browning, J. Dillon, M. McQuistan, A. Wolfe, An apn permutation in dimension six. Finite Fields: Theory and Appl. **518**, 33–42 (2010)

[CDP16] A. Canteaut, S. Duval, L. Perrin, A generalisation of dillon's apn permutation with the best known differential and nonlinear properties for all fields of size $2^4k + 2$. Cryptology ePrint Archive, Report 2016/887 (2016), https://eprint.iacr.org/2016/887

[SS17] S. Sarkar, H. Syed, Bounds on differential and linear branch number of permutations. Cryptology ePrint Archive, Report 2017/990 (2017), https://eprint.iacr.org/2017/990

[SMS19] S. Sarkar, K. Mandal, D. Saha, On the relationship between resilient Boolean functions and linear branch number of s-boxes. Cryptology ePrint Archive, Report 2019/1427 (2019), https://eprint.iacr.org/2019/1427

[BBB+18] A. Baksi, S. Bhasin, J. Breier, M. Khairallah, T. Peyrin, Protecting block ciphers against differential fault attacks without re-keying (extended version). Cryptology ePrint Archive, Report 2018/085 (2018), https://eprint.iacr.org/2018/085

[HJM06] M. Hell, T. Johansson, W. Meier, Grain - a stream cipher for constrained. eSTREAM (2006). https://www.ecrypt.eu.org/stream/p2ciphers/grain/Grain_p2.pdf

[KBC97] H. Krawczyk, M. Bellare, R. Canetti, HMAC: Keyed-hashing for message authentication. Technical report (1997)

[PR00] E. Petrank, C. Rackoff, CBC MAC for real-time data sources. J. Cryptol. **13**(3), 315–338 (2000)

[Wu16] H. Wu, ACORN: A lightweight authenticated cipher (v3). Candidate for the CAESAR Competition (2016), https://competitions.cr.yp.to/round3/acornv3.pdf

[WP15] H. Wu, B. Preneel, AEGIS: A Fast Authenticated Encryption Algorithm (v1.1) (2015), https://competitions.cr.yp.to/round3/aegisv11.pdf. Accessed: 2018-01-22

[HR10] V.T. Hoang, P. Rogaway, On generalized feistel networks. Cryptology ePrint Archive, Report 2010/301 (2010), https://eprint.iacr.org/2010/301

[LR88] M. Luby, C. Rackoff, How to construct pseudorandom permutations from pseudorandom functions. SIAM J. Comput. **17**(2), 1988–04

[Mou15] N. Mouha, Chaskey: a MAC algorithm for microcontrollers - status update and proposal of chaskey-12 -. IACR Cryptology ePrint Archive 1182/201 (2015)

[XZL+20] Z. Xiang, X. Zeng, D. Lin, Z. Bao, S. Zhang, Optimizing implementations of linear layers. Cryptology ePrint Archive, Report 2020/903 (2020), https://eprint.iacr.org/2020/903

[MMH+14] N. Mouha, B. Mennink, A.V. Herrewege, D. Watanabe, B. Preneel, I. Verbauwhede, Chaskey: an efficient MAC algorithm for 32-bit microcontrollers. In Selected Areas in Cryptography - SAC 2014 - 21st International Conference, Montreal, QC, Canada, August 14-15, 2014, Revised Selected Papers, pp. 306–323 (2014), https://doi.org/10.1007/978-3-319-13051-4%5C_19

[AB12] J. Aumasson, D.J. Bernstein, Siphash: A fast short-input PRF. In Progress in Cryptology - INDOCRYPT 2012, in *13th International Conference on Cryptology in India, Kolkata, India, Proceedings*, pp. 489–508 (2012), https://doi.org/10.1007/978-3-642-34931-7%5C_28. Accessed 9-12 Dec 2012

[EM91] S. Even, Y. Mansour, A construction of a cipher from a single pseudorandom permutation, in *Advances in Cryptology - ASIACRYPT '91, International Conference on the Theory and Applications of Cryptology*, Fujiyoshida, Japan, Proceedings, pp. 210-224 (1991), https://doi.org/10.1007/3-540-57332-1%5C_17. Accessed 11–14 Nov 1991

[Leu16] G. Leurent, Improved differential-linear cryptanalysis of 7-round chaskey with partitioning, in *Advances in Cryptology - EUROCRYPT 2016 - 35th Annual International Conference on the Theory and Applications of Cryptographic Techniques*, Vienna, Austria, Proceedings, Part I (2016), pp. 344–371, https://doi.org/10.1007/978-3-662-49890-3%5C_14. Accessed 8–12 May 2016

[WH19] H. Wu, T. Huang, Clx: a family of lightweight authenticated encryption algorithms (2019), https://csrc.nist.gov/CSRC/media/Projects/Lightweight-Cryptography/documents/round-1/spec-doc/CLX-spec.pdf

[Roh19] R. Rohit, Design and Cryptanalysis of Lightweight Symmetric Key Primitives. Ph.D. thesis, University of Waterloo (2019), https://uwspace.uwaterloo.ca/bitstream/handle/10012/15556/rohit_raghvendra.pdf?sequence=5&isAllowed=y

[Tod15] Y. Todo, Structural evaluation by generalized integral property, in *Advances in Cryptology - EUROCRYPT 2015 - 34th Annual International Conference on the Theory and Applications of Cryptographic Techniques*, Sofia, Bulgaria, Proceedings, Part I, pp. 287–314 (2015), https://doi.org/10.1007/978-3-662-46800-5_12. Accessed 26–30 April 2015

[Knu98] L. Knudsen, Deal - a 128-bit block cipher, in *NIST AES Proposal* (1998)

[ST17'2] Y. Sasaki, Y. Todo, New impossible differential search tool from design and cryptanalysis aspects - revealing structural properties of several ciphers, in *Advances in Cryptology - EUROCRYPT 2017 - 36th Annual International Conference on the Theory and Applications of Cryptographic Techniques*, Paris, France, Proceedings, Part III, pp. 185–215 (2017), https://doi.org/10.1007/978-3-319-56617-7%5C_7. Accessed April 30–May 4 2017

[CJF+16] T. Cui, K. Jia, K. Fu, S. Chen, M. Wang, New automatic search tool for impossible differentials and zero-correlation linear approximations. IACR Cryptology ePrint Archive, 2016:689 (2016), http://eprint.iacr.org/2016/689

[JPS+17] J. Jean, T. Peyrin, S.M. Sim, J. Tourteaux, Optimizing implementations of lightweight building blocks. IACR Trans. Symmetric Cryptol. *2017*(4), 130- 168 (2017), https://doi.org/10.13154/tosc.v2017.i4.130-168

[BSS20] A. Baksi, D. Saha, S. Sarkar, To infect or not to infect: a critical analysis of infective countermeasures in fault attacks. J. Cryptogr. Eng. (2020), https://link.springer.com/article/10.1007%2Fs13389-020-00224-9

[CEJ+02] S. Chow, P. A. Eisen, H. Johnson, and P. C. van Oorschot. White-box cryptography and an AES implementation. In K. Nyberg and H. M. Heys, editors, Selected Areas in Cryptography, 9th Annual International Workshop, SAC 2002, St. John's, Newfoundland, Canada, August 15-16, 2002. Revised Papers, volume 2595 of Lecture Notes in Computer Science, pages 250-270. Springer, 2002. https://doi.org/10.1007/3-540-36492-7%5C_17

[Hay08] S. Haykin, *Neural Networks and Learning Machines*, 3rd edn. (Pearson, 2008)

[BPP+17] S. Banik, S.K. Pandey, T. Peyrin, Y. Sasaki, S.M. Sim, Y. Todo, Gift: a small present. Cryptology ePrint Archive, Report 2017/622 (2017), https://eprint.iacr.org/2017/622

[BSS+13] R. Beaulieu, D. Shors, J. Smith, S. Treatman-Clark, B. Weeks, L. Wingers. The simon and speck families of lightweight block ciphers. Cryptology ePrint Archive, Report 2013/404 (2013), https://eprint.iacr.org/2013/404

[DeC07] C. De Cannière, Analysis and Design of Symmetric Encryption Algorithms. Katholieke Universiteit Leuven, Belgium (2007), https://www.esat.kuleuven.be/cosic/publications/thesis-139.pdf

[EM97] S. Even, Y. Mansour, A construction of a cipher from a single pseudorandom permutation. *J. Cryptol.* **10**(3), 151–162 (1997), https://doi.org/10.1007/s001459900025

[Sti06] D.R. Stinson, *Cryptography - Theory and Practice*. Discrete Mathematics Applications Series (CRC Press, 2006). ISBN: 978-1-58488-508-5

[Can16] A. Canteaut, Lecture notes on cryptographic boolean functions. version: March 10, 2016 (2016)

[BS90] E. Biham, A. Shamir, Differential cryptanalysis of des-like cryptosystems, in *Advances in Cryptology - CRYPTO'90, 10th Annual International Cryptology Conference, Santa Barbara, California, USA, August 11-15, 1990, Proceedings*, pp. 2–21 (1990), https://doi.org/10.1007/3-540-38424-3_1

[Rog04] P. Rogaway, Nonce-based symmetric encryption, in *Fast Software Encryption*, ed. by Roy, B., Meier, W., pp. 348–358 (Springer, Berlin, Heidelberg, 2004). 978-3-540-25937-4

[KW02] L.R. Knudsen, D. Wagner, Integral cryptanalysis, in *FSE 2002, Lecture Notes in Computer Science*, vol. 2365 (Springer, 2002), pp. 112–127

[Vie07] M. Vielhaber, Breaking one.fivium by aida an algebraic iv differential attack. Cryptology ePrint Archive, Report 2007/413 (2007), https://eprint.iacr.org/2007/413

[DS08] I. Dinur, A. Shamir, Cube attacks on tweakable black box polynomials. Cryptology ePrint Archive, Report 2008/385 (2008), https://eprint.iacr.org/2008/385

[ADM+09] J.-P. Aumasson, I. Dinur, W. Meier, A. Shamir, Cube testers and key recovery attacks on reduced-round md6 and trivium, in *Fast Software Encryption*, ed. by O. Dunkelman (Heidelberg. Springer, Berlin, Heidelberg, 2009), pp. 1–22. 978-3-642-03317-9

[BBS99] E. Biham, A. Biryukov, and A. Shamir. Cryptanalysis of skipjack reduced to 31 rounds using impossible differentials, in *Advances in Cryptology - EUROCRYPT '99, International Conference on the Theory and Application of Cryptographic Techniques, Prague, Czech Republic, May 2-6, 1999, Proceeding*, pp. 12–23 (1999), https://doi.org/10.1007/3-540-48910-X_2

[DR20] J. Daemen, V. Rijmen. *The Design of Rijndael: The Advanced Encryption Standard (AES)*, 2nd edn. (Springer, Berlin, Heidelberg, 2020)

[DCK+15] D. Dinu, Y.L. Corre, D. Khovratovich, L. Perrin, J. Großschädl, A. Biryukov, Triathlon of lightweight block ciphers for the internet of things. Cryptology ePrint Archive, Report 2015/209 (2015), https://eprint.iacr.org/2015/209

[PUB16] L. Perrin, A. Udovenko, A. Biryukov. Cryptanalysis of a theorem: decomposing the only known solution to the big APN problem (full version). *IACR Cryptology ePrint Archive*, 2016:539 (2016), http://eprint.iacr.org/2016/539

[Nyb93] K. Nyberg, Differentially uniform mappings for cryptography. In T. Helleseth, editor, *Advances in Cryptology - EUROCRYPT '93, Workshop on the Theory and Application of of Cryptographic Techniques, Lofthus, Norway, May 23-27, 1993, Proceedings, volume 765 of Lecture Notes in Computer Science* (Springer, 1993), pp. 55–64, https://doi.org/10.1007/3-540-48285-7_6

[BN00] M. Bellare, C. Namprempre, Authenticated encryption: relations among notions and analysis of the generic composition paradigm. *IACR Cryptology ePrint Archive*, 2000:25 (2000), http://eprint.iacr.org/2000/025

[Mat93] M. Matsui, Linear cryptanalysis method for DES cipher. In T. Helleseth, editor, *Advances in Cryptology - EUROCRYPT '93, Workshop on the Theory and Application of of Cryptographic Techniques, Lofthus, Norway, May 23-27, 1993, Proceedings*, volume 765 of Lecture Notes in Computer Science (Springer, 1993), pp. 386–397, https://doi.org/10.1007/3-540-48285-7%5C_33

Chapter 3
Fault Attack

3.1 Introduction

With the emergence of the Internet of things in modern era, small-scale devices are becoming more ubiquitous. These devices often perform cryptographic operations, which are vulnerable to device-specific attacks. One such attack, referred to as *Fault Attack* (FA), is shown to be powerful against common ciphers; including the high-profile ones, which are considered secure against classical cryptanalysis methods. This concept is first presented to the cryptographic community in the context of public key cryptography and adopted to symmetric key soon afterwards. In this type of attack, the attacker/adversary, Eve, is assumed to have physical access to the device under operation. She deliberately injects disturbances into the device while it is running some cryptographic computation. The disturbances could be injected by means of, for example, voltage glitches, Light Amplification by Stimulated Emission of Radiation (LASER) shots, sudden temperature variations. By utilizing the information from the correct (*non-faulty*) and the incorrect (*faulty*) outputs, she is normally able to deduce some information on the secret key. Under certain models, attacker does not need the non-faulty output, as the faulty output only is enough for her to recover the secret information.

On the device, an injected fault can modify/skip the normal flow of operation, alter/erase data, access non-permitted location of memory, perform a non-permitted operation, bypass certain countermeasures, and so on. Basically, fault attacks allow the adversary to bypass certain security claims in contrary to classical cryptanalysis methods where the cipher is considered a black box and therefore, it cannot be tampered with. As the recent research trend shows, FA is a very popular cryptanalysis technique in the community; thanks to its wide usability, practicality (often possible by a non-expert with cheap equipment), easy computations, and the attacker's ability to simulate.

A. Baksi, *Classical and Physical Security of Symmetric Key Cryptographic Algorithms*, Computer Architecture and Design Methodologies,
https://doi.org/10.1007/978-981-16-6522-6_3

3.2 Fault Models

By a fault model we mean the nature, efficiency, repeatability, and feasibility of the faults injected. Nature of a fault can be described by the impact on the target device (such as bits or a nibble/byte/word) and precision of location (targets one particular bit, etc.). Efficiency indicates attacker's precision (how precisely she can choose) the fault value. Repeatability means whether the attacker can repeat the identical fault (the fault value may be unknown) in either the temporal or the spatial domain. Feasibility refers to the practical aspects (cost, equipment, time, expertise required) of the fault attack. Generally, it is accepted that a more restrictive/precise model is more difficult/impractical in the real-life scenario.

3.2.1 *Precise Bit Flip*

The proponents of fault attacks try to show the effectiveness of their attacks by assuming as little as possible (e.g., the fault location is assumed to be unknown, which is later recovered through rigorous analysis). In terms of *Differential Fault Attack* (DFA, Sect. 3.5.1), the strongest model would be to flip one particular bit (the round of the cipher as well as the location of the bit is chosen by the attacker). Such model, to the best of our knowledge, is never considered.

Consider the effect of changing one input bit of an AND operation: $y = x_0 \wedge x_1$. If, say, Eveis able to precisely flip x_0, then by judging whether the output changes or not, she is able to retrieve x_1 (no change in output means $x_1 = 0$, otherwise $x_1 = 1$). This way, any AND gate can be targeted. It is also interesting to notice, this model (which we call as, *precise bit flip*), if/when exists, we break certain countermeasures in their current forms; such as the *detection* (Sect. 3.7.1) or *infection* (Sect. 3.7.2). Even though the cipher is "protected" by detection or infection (which effectively stops her to get the faulty output she desires); the attacker is able to check whether flipping one bit of an AND operation changes the output (whether the output is suppressed in case of detection; or, a random value, instead of the non-faulty output, is returned in case of infection) or not. One way to protect against this model in such countermeasures would be to ensure that there is no non-linear operation that takes two bits as inputs; this can be done, e.g., implementing all the SBoxes as look-up tables. Further discussion can be found in Sect. 8.2.1/Sect. 8.9.

3.2.2 *Single/Multiple Fault Adversary*

Majority of published works in fault attacks assumes *single fault adversary* model. Under this model, the attacker, Evecan inject faults at most once during one execution of the cipher (that may result in affecting multiple locations). So, if she injects two sets of faults, say, at the first as well as the last round of encryption, that will be considered a violation of the model. Relaxed model, where attacker can inject multiple sets of faults during one execution, has been proposed on rare occasions, such as *Internal Differential Fault Analysis* (InDFA, see Sect. 3.5.2 for more details).

3.2.3 Random/Deterministic Fault Model

The *random fault model* is the most commonly used model in the literature, e.g., [PQ03, SH13]. Here, the attacker can control in which round the faults can be injected; but cannot control the value changed by the fault injection. Basically, here the injected fault results in flipping one/more targeted bit(s) of the operand value(s). Generally, attacker can control the intensity, location, and duration of the external disturbance, but cannot control the fault precision (the precise effect of fault is assumed to be unknown to her). Under certain cases, the exact fault location could also be assumed to be unknown (the attack complexity is multiplied by the number of fault locations), e.g., [JKP12'1]. Depending on the particular attack, the target for fault injection can be a word (byte/nibble) or a series of bits. Other than flipping bits, different models can be employed where particular bits are set to 1 or reset to 0 [BS02]. For byte-oriented ciphers like `AES` , the random byte fault is most commonly assumed.

Slight variation of the above models can also be found in the literature. For example, the *neighbourhood fault model* `TRIVIUM` [DA14] assumes that any number of bits in a neighbourhood of few bits of a stream cipher state can be flipped—this type of assumption is meaningful when the implementation is bit-sliced (which is normally the case), and a LASER induced *instruction skip* attack (such as the practical attack described in [BJC15]) can do it.

3.2.4 Information Theoretic View

Few works attempt to interpret fault attacks (so far, these works focus only on differential fault attack which is described in Sect. 3.5.1) from an information theoretic point of view. The objective is to find out whether the information leaked by the faults are optimal or there is any scope to utilize the information further. Examples include [SLI+12, KSK14].

3.2.5 Other Aspects

As noted in [LCF+17, BBB+18, Sect. 7.1]; symmetric modes have properties that can be useful to resist certain type of fault attacks, DFA, in particular (information on the block cipher modes can be found in Sect. 2.2.2). Except these two works, rest inherently assume ECB mode; to the best of our knowledge. Besides, fault attacks on block ciphers can target the main cipher routine or the key schedule if it contains non-linear operations (such as the attack on AES key scheduling [AM11, Kim12]).

Nonce-based primitives generally can be considered relatively more difficult as the non-repeatability of nonce makes certain fault attacks (like DFA, see Sect. 3.5.1) not possible. The InDFA model (see Sect. 3.5.2) is proposed for such a case.

Another aspect to consider is for authenticated encryption ciphers that block the output if the verification fails (assuming the attacker cannot attack the decryption/verification circuit). This also makes DFA impossible but the ineffective faults (see Sect. 3.5.2) are still possible. The attacker can bypass this limitation by targeting the decryption/verification circuit.

3.3 Data Alteration Methods

3.3.1 Volatility

Based on volatility of faults, it can be classified either as *hard* (also known as *permanent*) or *transient* faults. Hard faults assume that one/a few particular bit(s) in the device are permanently set some value . Even though this type is not so popular, it has been applied against, for example, the stream cipher-based AEAD ACORN [DRA16]. Transient faults, on the other hand, allow the device to get back to its normal operation flow when the source of fault is revoked. Generally, all the hard fault models assume too much power for the attacker; hard to achieve in practice; and also easier to detect. One possible way to attain this scenario could be a manufacturing defect or causing an irreparable damage to the chip. These attacks are apparently impossible to protect from a purely cipher design point of view, since the cipher description is essentially changed. A new attack model, termed as the *persistent* fault model is proposed (refer to Sect. 3.5.4), which lies somewhat in between the transient and the hard fault models.

3.3.2 Modification of Operation

Under this model, the attacker is allowed to alter (skip, misinterpret or execute multiple times) a number of operations in a microcontroller. Skipping an instruction is an effective way to retrieve secret information from the system, Fig. 3.1 shows how the key can be recovered where the last round key addition is skipped. This model is shown practically in [BJC15], where the authors show how a well-aimed instruction skip can reveal a full secret key of AES. Similarly, the attack on CHACHA in [KPB+17]

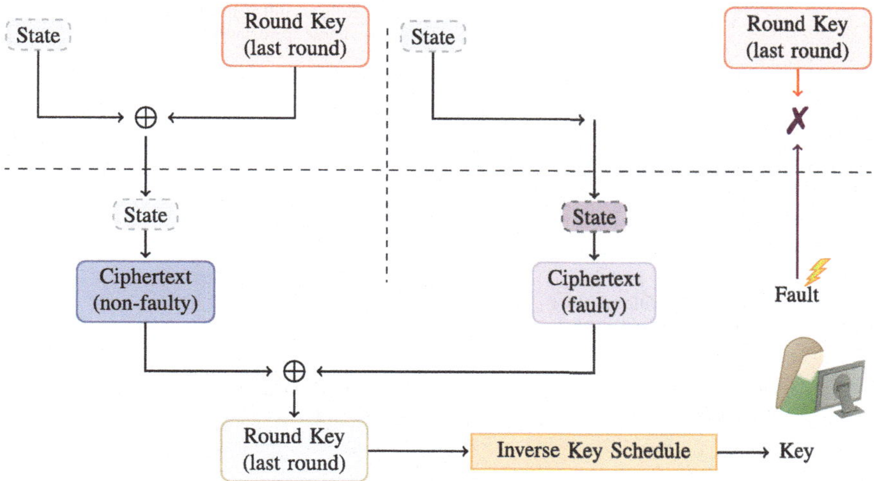

Fig. 3.1 An exemplary use of instruction skip at last round

substitutes one instruction with another—addition operation is effectively changed into subtraction, allowing fault analysis.

These models are too powerful to protect from a cipher design perspective, as the attacker can effectively alter the cipher description and recover the key easily. In the most extreme case, it may reduce to a situation where the operation actually performed hardly qualifies as a cryptographic one, e.g., skipping the call to the entire `encrypt()` subroutine. Nonetheless, this type of research paradigm is not uncommon in the community, e.g., [BdSG+14, EHH+14]. The *Internal Differential Fault Attack* (InDFA, Sect. 3.5.2) works by altering round counter, hence belongs to this category.

3.3.3 Modification of Operand

In this fault model, Eveis able to manipulate the operands during any phase of the cipher execution. Such type of attacks is most commonly used in practice.

The fault attacks that belong to the modification of operand category are Differential Fault Attack (DFA, Sect. 3.5.1), *Algebraic Fault Attack* (AFA, Sect. 3.5.1), *Impossible Differential Fault Attack* (IDFA, Sect. 3.5.1), *Collision Fault Attack* (CFA, Sect. 3.5.2); *Linear Fault Attack*, and *Integral Fault Attack* (described in Sect. 3.5.1). These analysis methods work based on the analysis of the propagation of the difference injected at the internal state (generally a few rounds before the final round) and are somewhat similar.

In essence, all these methods result in a weakened (easier for Eve) version of the corresponding classical attack, as shown in Fig. 3.2. More precisely, a classical attack

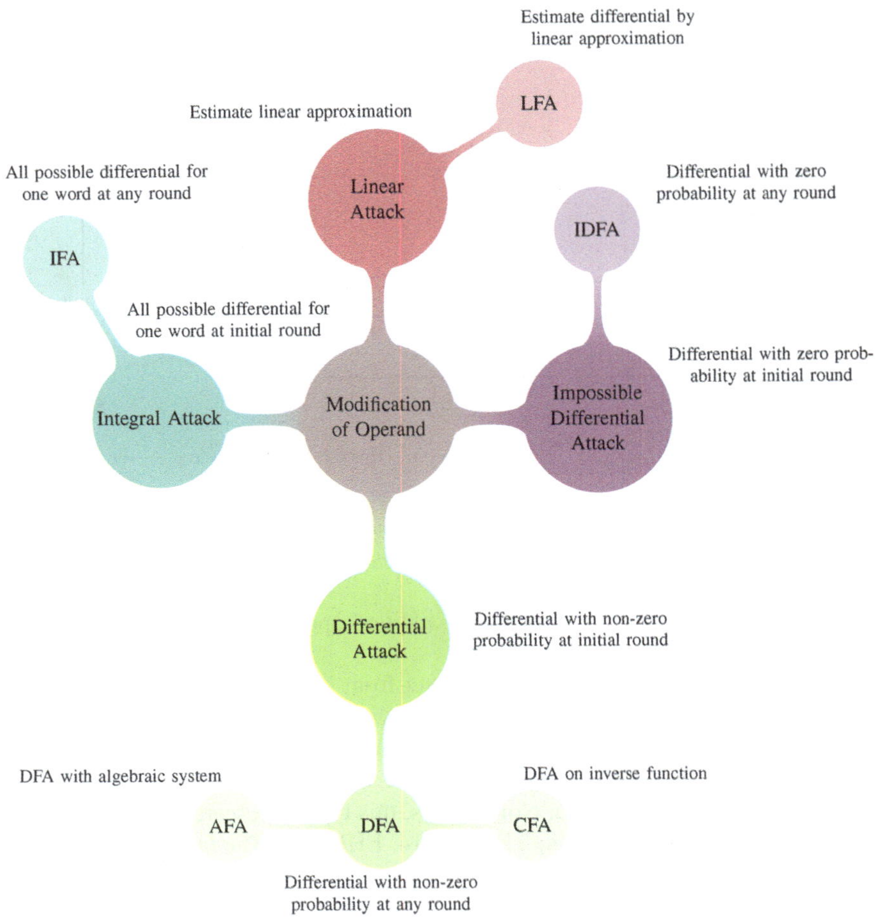

Fig. 3.2 Attacks (classical and fault) based on modification of operand models

can inject difference at one/more words at the very beginning of the cipher execution (such as difference in the plaintext). This difference passes through all the rounds of the cipher. In contrast, in the corresponding fault attacks, the attacker can inject the difference at any round during the cipher execution, thus relaxing the model for her. As modern ciphers that are in use are designed to withstand classical attacks, putting the difference at the beginning would not yield any exploitable information to Eve. When she puts the differences near the end of the cipher execution (e.g., within the last three rounds for SPN block ciphers, see Sect. 3.5.1), she can lower the entropy in the ciphertext, giving her some information on the secret key. As a result, by using a few faults (sometimes with a single fault), she can recover the entire secret key. One may also refer to [SSL15, Chaps. 5.5, 6] for the analysis of such fault attacks from a classical attack point of view. Since the attacks in modification of operand category closely follow classical attacks; it may be possible to design a cipher which, by design, is capable of thwarting such fault attacks; at least to some extent. As far as we know, no such cipher is reported in the literature. It may be mentioned that a recent cipher CRAFT [BLM+19], attempts to protect against DFA (to some extent) inherently; although it falls under the category of detective countermeasures (see Sect. 3.7.1 for more details).

Generally, the analysis procedure in these types of fault attacks consists of two orthogonal terms; namely, fault complexity and search complexity. It is commonly possible to extract the secret key by using many faults and a low-cost searches; the same result can be obtained by using less faults and high cost searches. As the cost of computation up-to a certain complexity is negligible, whereas a fault might not be easy to perform at some cases; the attacker usually tries to keep the fault complexity minimal (say, within few 10's) while keeping the search complexity within a doable limit (say, less than 2^{40}). However, certain attacks require a few hundred or even a few thousand faults (refer to Sect. 3.5.1 for examples).

3.4 Sources of Fault Injection

In this part, we briefly discuss on the equipment used to inject faults. Figure 3.3 shows the schematic of a fault injection setup. A comparison of common tools is given in Table 3.1 (which is adopted with slight modification from [BBK+12]), where the most frequently used tools are highlighted.

The fault injection tools can be compared based on their properties. The first property is the effect of the fault injection tool. In this case, the effect could be global

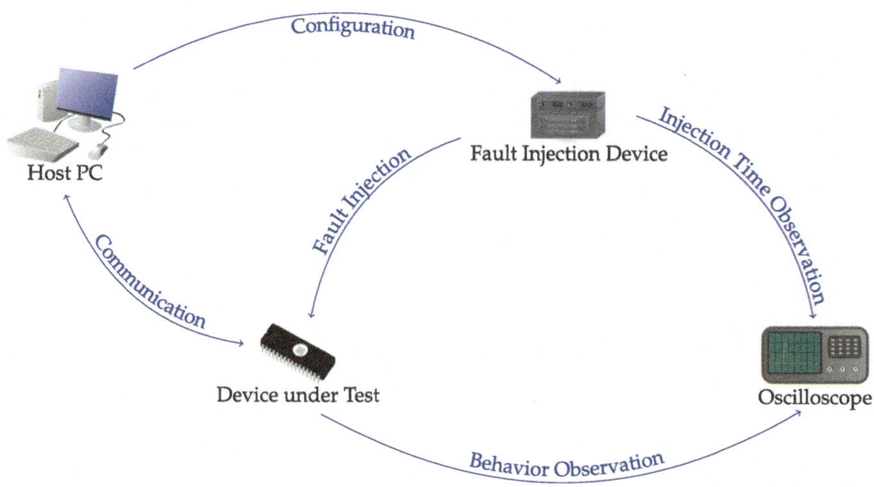

Fig. 3.3 Schematic view of a fault injection setup

Table 3.1 Comparison of various fault injection equipment

Method	Effect	Device			Accuracy			
		De-packaging	Damage	Design Details	Time	Position	Cost of Equip-ment	Technical Skill
Clock Glitch	Global	No	No	Required	High	Low	Low	Moderate
Voltage Glitch	Global	No	No	Partially required	Moderate	Low	Low	Moderate
Underfeeding	Global	No	No	Not required	None	High	Low	Low
Heating	Global	No	Possible	Required	None	Low	Low	Low
EM Pulse	Local	No	Possible	Partially required	Moderate	Low	Low	Moderate
Localized EM Pulse	Local	May be	Possible	Required	Moderate	Moderate	Moderate	Moderate
Light Pulse	Local	Yes	Possible	Required	Moderate	Moderate	Moderate	Moderate
LASER Beam	Local	Yes	Possible	Required	High	High	High	High
Light Radia-tion	Local	Yes	Yes	Not required	Low	Low	Low	Moderate
Focused Ion Beam	Local	Yes	Yes	Required	High	Very high	Very high	Very high

or local. Localized fault indicates that the adversary could precisely limit the effect of fault on the specific location or spot in the device, as compared to global fault where the attacker cannot really control where the fault is injected.

The second comparison is a requirement on the target device. De-packaging means that the outer layer of the device will have to be opened either through chemical (e.g., acid) or mechanical means, in order to access the chip inside. Damage refers to the state of the device after fault injection. For some fault injection tools, the target device might be destroyed. Hence, the choice of tools can be adjusted depending on the availability and importance of the target device. The last property under this category is the detail of the device design. Some of the tools required the knowledge of the device in order to properly inject the fault, while in other cases, partial or no knowledge is required.

The third comparison is about the accuracy of the fault injection tool. The time parameter refers to the timing precision when the fault is injected, and the position refers to the precision of fault location in the target device. The next comparison is regarding the cost of the fault injection equipment. The cost is highly related to the accuracy, in this case, in order to achieve higher accuracy, the fault injection equipment will be more expensive. Lastly, the comparison is regarding the technical or expertise skill required to operate the equipment. In this case, for some specialized tools (such as LASER or focused ion beam) might be more dangerous to work with and can cause potential harm if mishandled; and hence, a higher level of knowledge/expertise is required to safely operate the equipment.

(a) Voltage glitch on a smart card

(b) Electromagnetic pulse injection on a microcontroller

(c) LASER fault injection on a microcontroller

Fig. 3.4 Fault injection equipment on different targets

Figure 3.4 shows various practical setups used to inject faults. A short discussion on most common setups is given next. For further information on these fault injection setups, one may refer to [Kor16, Chap. 3.1].

Clock/Power Glitch

If the device uses an external clock, then a sudden variation in the clock can result in a transient fault in the device. For example, the technique is demonstrated in [DEK+18]. A glitch in the power can also result in a similar fault [OFl16]. Glitch is an efficient, cost-effective, and commonly employed method for fault injection in practice, which also does not require too much technical expertise. It can be applied on both the pins as well as the die. However, it may be hard for a glitch-based setup to control the fault, for example, to inject all possible 255 faults in a byte fault setting; and also the faults injected may not be uniform.

Optics (LASER/UV-ray/X-ray)

Induction of faults by optical sources is another commonly used method. Often, this type of attack is done with a focused LASER beam at a particular location of the chip, e.g., [ABC+17, SBH+15]. In this technique, the chip has to be de-packaged. Also, the correct location for injecting the fault has to be found by inspection, which requires specific expertise, time, and expensive equipment. The advantage of this approach is that the fault can be precisely injected into the device, and thus, wider variety of fault model can be realized. A comprehensive categorization of various optical fault analysis setup can be found at [PDL18].

Electromagnetic Emanation

We can further divide this technique into two categories. *Pulse injection* targets the digital blocks of integrated circuits, such as logic and memories. The cells in the chip can be effectively forced to produce erroneous output by a high energy EM pulse for a short period of time. One may refer to [MDH+13] for more information. It has a few advantages over other techniques. For example; effect of glitch is generally global to the chip, whereas EM emanation can be made local to a particular location by suitably placing the probe; it does not require de-packaging of the chip as in the case of optical fault, while maintaining almost similar level of fault precision. *Harmonic injection* targets the analog blocks, such as internal clock generator or random number generators. In this case the attacker generates a stable sinusoidal signal at a given frequency, injecting a harmonic wave to create a parasitic signal biasing the behaviour of the block. Such attack is more global and the duration of the signal is relatively long compared to pulse injection. More details on such setup can be found for example in [HHM+12].

Others

Besides the commonly used fault injection methods described before, there are also other techniques for fault injection include, for example, *row hammer* [KDK+14]. Row hammer is a known side effect in Dynamic Random Access Memory (DRAM). By repeatedly accessing a row in a DRAM, the attacker can potentially flip bits in adjacent rows. It is inducible by software and can be repeated. Another source can be a hardware Trojan, as shown in [BH15]. Other examples include *Temperature variation*, where forcing a chip to work outside its operating temperature can cause faulty operations [HS13], as well as *under powering*, where the faults are induced by reducing the supply voltage [SGD08].

3.5 Analysis Methods

Broadly, the fault analysis techniques can be classified in three major categories—differential (Sect. 3.5.1), collision based (Sect. 3.5.2) and statistical (Sect. 3.5.3). Figure 3.5 shows a general overview of the classification of various fault analysis methods. The *Statistical Ineffective Fault Attack* (SIFA, see Sect. 3.5.4) falls somewhat in between collision and difference based. Two other new methods, namely, *Persistent Fault Attack* and *Fault Intensity Map Analysis* (FIMA) are hard to fit in an exact category. These three methods are kept separate (Sect. 3.5.4).

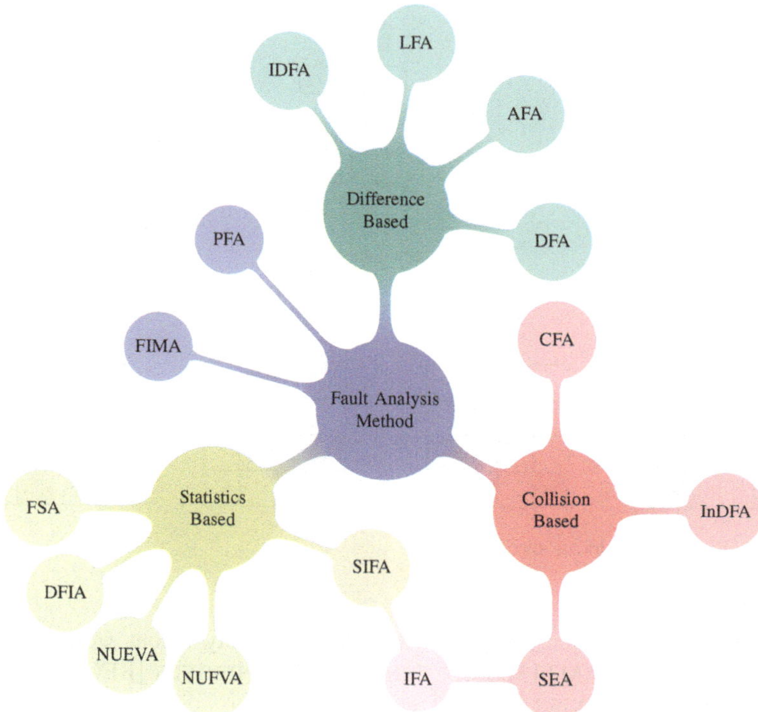

Fig. 3.5 Classification of fault analysis methods

3.5.1 Difference-Based Fault Analysis

All the methods in the difference-based analysis category are somewhat similar and can be thought of as an extension of DFA (which itself can be thought of as an extension of the classical differential attack). These methods exploit the confusion and diffusion characteristics of a cipher to inject a well-aimed fault that allows to obtain the desired difference distribution at the output of the cipher. In general, DFA and variants are more practical (such as require less number of faults) than ineffective or statistical models.

Differential Fault Attack (DFA)

The concept of DFA is introduced in [BS97]. As already mentioned (Sect. 3.2), the attacker injects a difference at a round of her choosing. A schematic version of DFA is given in Fig. 3.6, here the cipher runs for η rounds and η' is a round close to the final round η. Hence, DFA requires effective faults (where the faults change the normal execution of the cipher). DFA is possibly the most commonly

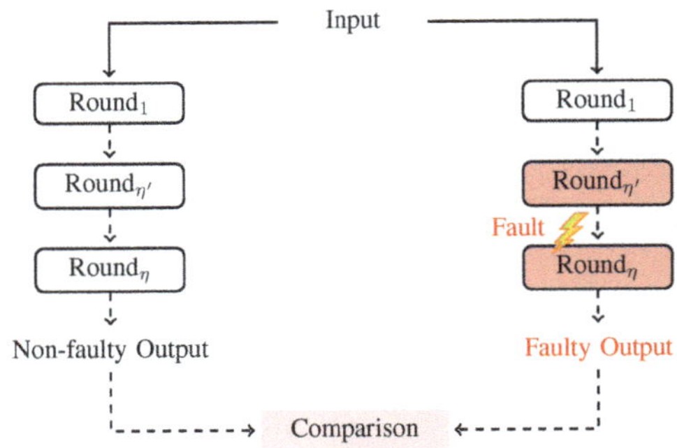

Fig. 3.6 Schematic for differential fault attack

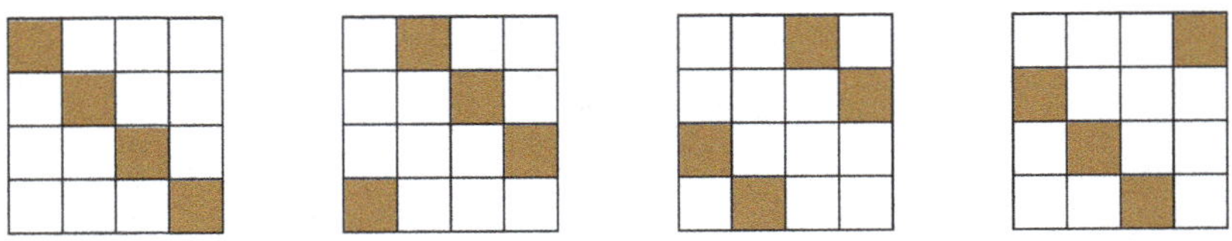

Fig. 3.7 Diagonal fault locations of AES

employed fault technique. Here we list a few relatively recent works on high-profile ciphers: AES [AMT13], TWOFISH [AM13'1], PRESENT [BEG13, JLS+13], PRINCE [SH13], MIDORI [CZS16], etc.

To the best of our knowledge, the earliest round DFA on AES works at the 4th last round [DFL11]. The *diagonal fault model* [SMC09] proposes the best-known DFA on AES (without using information from the key schedule), where the computational complexity is reduced to a nominal bound by inserting four related differences in one diagonal of the final state; as shown in Fig. 3.7 (see [SMC09, Fig. 8] for the inter-relations of the diagonals being faulty).

In general, the underlying philosophy of attacking SPN ciphers by various fault methods are somewhat similar (particularly those use a diffusion matrix, instead of bit permutation, as the linear layer), and can be (to some extent) generalized from a theoretical perspective. We omit a thorough analysis here, as interested readers may find it in [SSL15, Sects. 5.5, 6].

In order to give an overall idea, we present the simplest case of DFA on SPN. Here, the attacker injects faults at the last round SBoxes one by one. For each SBox S, she injects a fault δ and obtains a relation of the form: $S(x) \oplus S(x \oplus \delta) = \Delta(x, \delta)$, where Δ is known to her, x is the unknown input she wants to retrieve and δ is either known (deterministic model) or unknown (random model). Nonetheless, using the input difference—output difference relation of S, she is able to narrow down the solutions for x. Finally, with a few such relations, she is able to solve x uniquely. Now that she knows the state before the last SBox layer, she can compute the last round key by XORing the ciphertext with the output of the last SBox layer. In this approach, the fault complexity is maximized, and the search complexity is minimized.

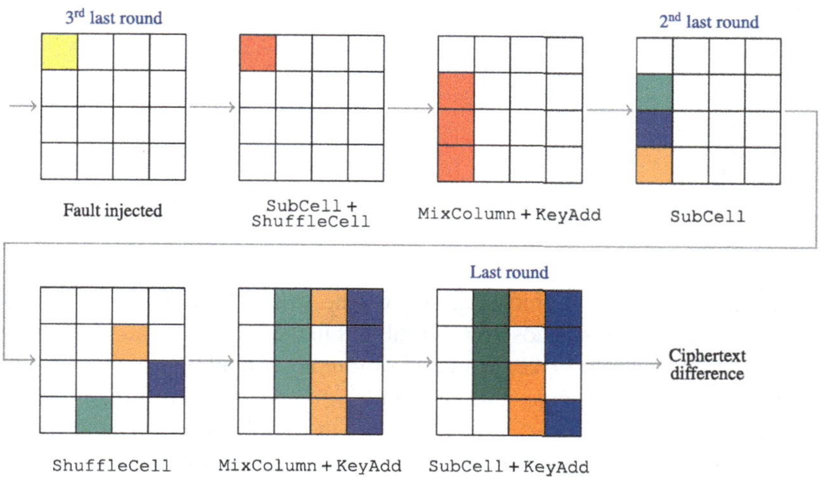

Fig. 3.8 DFA on MIDORI encryption at third last round

The fault complexity of the approach mentioned above can be reduced if the fault is injected a few rounds earlier. Let us assume the fault is injected at the beginning of the 2nd last round SBox . Then, the difference will spread through one round of the linear layer, where it enters another layer of SBoxes. Thus, one fault in a round effectively induces multiple faults on the next round, the number depends on the nature of the linear layer (e.g., in case of a standard 4×4 SubBytes-ShiftRows-MixColumns construction; one column, i.e., 4 faults are induced). However, in this situation, the differential relation becomes more complicated, which results in more computational complexity. In other words, injecting fault(s) at the penultimate round reduces the number of required faults at the expense of more computation. Now, consider the effect of injecting fault at the beginning of 3rd last round SBox. In a typical SPN block cipher, the fault passes through 2 linear layers before reaching the last round SBox; which generally results in activating all SBoxes at the last round. Thus, generally one fault in 3rd last round is equivalent to the naïve approach on each individual SBox with the overhead of computational cost (which is generally within the doable range). Figure 3.8 shows one such case with the MIDORI block cipher, based on [CZS16, Fig. 2]. Here, the fault is injected at the (0, 0)th word of the MIDORI state at the beginning of the 3rd last round. This fault propagates through the following SubCell, ShuffleCell, MixColumns and AddKey to other 3 words in that column. Thus, at the beginning of the 2nd last round, the attacker enjoys the effect of injecting 3 faults. These 3 faults induce a total of 9 differences at the ciphertext. So, the attacker effectively reduces the required number of faults to one-ninth from the last round attack.

As noted earlier, the SPN ciphers with the bit permutation as the linear layer act a little differently. For example, in DFA on PRESENT (refer to Sect. 2.4.2 for a quick description) the fault propagation pattern spreads bit-wise through the state because

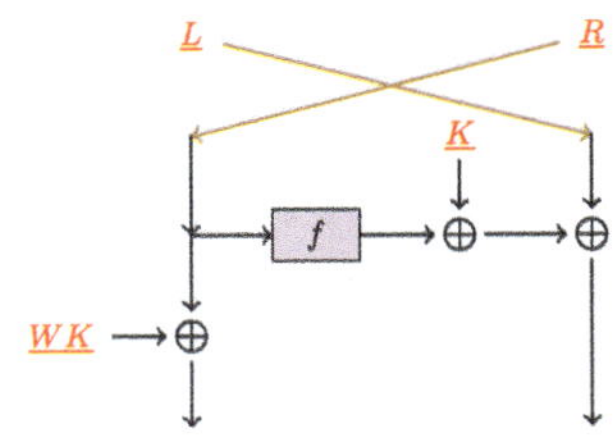

Fig. 3.9 Last round of many Feistel network-based ciphers

of the cipher's `pLayer` (which depends on the value) [BH15]. If a nibble gets affected in `PRESENT`, it spreads into 2–4 nibbles in the following round. Therefore, if the attacker manages to make the output of four chosen nibbles in round i faulty, it might spread into the whole state in round $i + 1$. Together with the differential characteristic of the `sBoxLayer`, it will reduce the number of input candidates per nibble to 2–4. Then, if the fault is injected into different four nibbles, and therefore, different input bits of the SBox in round $i + 1$ get affected, the input candidates of each nibble might be known. After this step, the attacker simply recovers the last round key and computes the inverse key schedule algorithm to get the secret key.

The FAs on the Feistel block ciphers are of more variety (see Sect. 2.3.2 for more information on Feistel network), because the analysis of these attacks depends on the following (non-exhaustive) list of factors:

- Choice of the f function,
- How the round keys are inserted (through a linear function or a non-linear function),
- Number of branches.

Hence, it is difficult to generalize all the attacks here. One may refer to [BTL+14] for DFA on a few Feistel block ciphers; namely, `DES`, `MIBS`, `TWINE`, and `CLEFIA`.

However, the last round of several high-profile Feistel network-based ciphers (such as `CLEFIA` [AM13'2], `TWOFISH` [AM13'1], and `PICCOLO` [Jeo12]) can be modelled as in Fig. 3.9. As for the notations, L and R denote the left and right branches, K as the round key and WK as the whitening key, f as the Feistel function, and red indicates it is unknown to attacker. One can view $K \oplus L$ as the final round key addition in an SPN cipher, and thus DFA can be launched as in the case of a SPN. Once the input of the Feistel function f is identified, WK can be uniquely determined. Now, the attacker can perform a similar strategy on the previous round to determine the value of L and thus recover K from the previous round as well, etc.

DFA or its variants can be successfully applied to any unprotected symmetric key cipher. Interestingly though, we notice that there are several stream ciphers on which no DFA or its variants (see AFA) are reported so far. In [KY10], the authors present a DFA on the stream cipher `HC-128` where the attacker is allowed to inject fault at a random word of the state of the cipher but cannot control its exact location or the resulted value of the word. Normally, DFA/AFA on stream ciphers is considered difficult to carry out in practice, compared to block ciphers, as the analysis complexity often goes beyond the computational requirements of automated tools such as satisfiability (SAT) solvers. One may check eSTREAM finalist SALSA or its variant CHACHA in this regard [BGV17].

As an interesting side note, it may be observed that, the complexity of a DFA (or its variant) may be increased if the following design choices are incorporated in a stream cipher design:

- Suppress several key-stream words (say, output every two hundredth key-stream bit in TRIVIUM).
- Use constructions with bigger state which are (possibly) difficult to solve by SAT solvers.
- Value—position swap (as in RC4) such that fault equations are hard to formulate.
- Use block ciphers with CTR mode (it inherently works like a protocol level countermeasure, see Sect. 3.2.5).

It is not long since the faults attacks on hash functions or other primitives that use a hash function appear in the literature. As hash functions are commonly underrepresented in fault attacks, here we present a compact overview of major fault attacks.

The first such attack is reported by Li et al. in [LLG09] against SHACAL1, a block cipher employing SHA-1 without the final addition operation. In [HH11], the authors argue in favour of analysing public primitives like hash functions or the internal compression function in the light of DFA, citing reasons that the impact of the result will get amplified in applications like HMAC or key derivation functions. They extend the attack of [LLG09] to include the final addition and demonstrated a DFA on SHA-1 compression function using 1000 faulty outputs. The authors conclude that applications using the SHA-1 compression function having fixed input and output which are known to the attacker are particularly vulnerable to their attack.

In [FR12], the authors demonstrate a DFA on the hash function GRØSTL which is an AES -based finalist of the SHA-3 competition (see Sect. 2.8.4). They indirectly make use of the classical DFA on AES along with some pre-computation to invert the output transformation using 16 faults and required 280 faults on average to invert each compression step attacking both GRØSTL-256 and GRØSTL-224. They are able retrieve the key of HMAC-GRØSTL using 300 faults and also present a possible countermeasure using a modification of the truncation function. In [SC15], a diagonal fault analysis is reported on GRØSTL when used in dedicated MAC mode that makes use of the notion of fault repeatability introduced in [RLK11]. The number of faults to invert the output transformation is reduced from the previous record of 16–8.

The first fault attack on the SHA-3 standard is presented in [BGS15]. The main result in this paper is a DFA on SHA-3 variants SHA-3-512 and SHA-3-384. The authors show that using a random bit fault model, 1592 bits among the total of 1600 bits of the internal state of SHA-3 can be extracted using 80 faulty messages given the first 320 least significant bits. This gives the indication that it is indeed possible to recover (most of) the state of a compression function even if the full state is not available to the attacker (only a truncated state is available). The idea of the attack is based on the DFA models used in GRØSTL [FR12] and STREEBOG [AY15]. The ciphers, SHA-3-224 and SHA-3-256, are attacked in [LFZ+16], where they adopt more realistic single byte fault model to mount DFA. In [LAF+17], the first algebraic fault analysis of SHA-3 family is reported, where they argue in favour of AFA being a more effective technique as it recovers more bits per fault injection and also more efficient since it required lesser fault injections compared to the DFA results. Moreover, they report that AFA is more suited for SHA-3 as it can exploit the inherent algebraic properties of SHA-3. The main result is to recover the internal state of the penultimate round using a random byte model. The work is extended in [LAF+18] where the authors use more relaxed fault models using 16 and 32-bit faults and found the attack to be more effective for shorter message digests.

STREEBOG, a Russian hash function standard, is analysed in the light of differential fault analysis attacks [AY15]. The authors use the random bit-flip model and also discussed the scenario where the fault location in the state can be controlled. They demonstrate a two-stage attack which recovers the compression function inputs with number of fault injection ranging between 330 and 1640 based on certain assumptions.

DFA is also applied against the AEAD MINALPHER in [CDN17], which was a second round CAESAR (Sect. 2.8.4) candidate. Using a single fault at a fixed position, almost universal forgeries on the third round CAESAR candidates, CLOC and SILC are shown in [RCC+16]. In [BDM+19], DFA on AEGIS-128L (which is in the final portfolio) and TIAOXIN-346, two candidates of CAESAR are reported. The authors comment that the ciphers following a stream cipher design paradigm and a particular form of update function are vulnerable to similar attack.

Algebraic Fault Attack (AFA)

The notion of AFA [CJW10] is similar to that of DFA and related to the algebraic attack (see Sect. 2.6.3)). The difference between AFA and DFA lies in the process analysing the faulty and non-faulty outputs. DFA relies on astute observation done manually on the difference propagation through the cipher, thereby solving the unknown state of the cipher. AFA relies on forming algebraic equations and commonly feeding the resulting system of equations typically to a SAT solver thereafter. AFA has been successfully applied against block ciphers (e.g., LED [JKP12'2, ZGZ+13], PICCOLO [ZZG+13]) and hash functions such as SHA-3 [LAF+17].

It may be noted that AFA is particularly useful against stream ciphers and stream cipher-based designs; as it is relatively straightforward to get the algebraic equations using computers rather than manually exploiting structure of the cipher. Observing differential trails manually is tedious in such cases. For example, AFA against ciphers in GRAIN family [SBM15, DCA+15], TRIVIUM [DA14], SPROUT [MSB+15], etc. are reported. ACORN is also analysed in [SSM+17] (where few bits are exhaustively searched, and the rest are solved by the SAT solver) together with GRAIN-v1 and LIZARD (in the latter case, the state recovery at a particular round is shown to be possible). The first step in these attacks is to deduce optimal places for fault injection. Generally, precise location and round of fault injection is considered unknown to Eve(to be determined the difference pattern in the key-stream, which is referred to as the *fault signature*) although Evecontrols which (part of) LFSR–NLFSR is affected. However, in these stream cipher related attacks, the authors describe their works as DFAs (not AFAs), possibly because the typical DFA approach that works (without involving automated solvers) on block ciphers does not work in stream ciphers.

As for the AEADs, the finalist in the first CAESAR portfolio (lightweight applications); `ACORN` is analysed by AFA (by assuming few bits are known to the attacker) in [SSM+17], and also in [ZFL18] where the feasibility of a DFA is discussed and verified with a toy version of `ACORN`. Two other CAESAR candidates; `AEGIS` which is a finalist in the second portfolio of CAESAR (high performance applications), and `TIAOXIN` which is a third round candidate are analysed by DFA (technically, AFA) in [DRS+16].

Impossible Differential Fault Attack (IDFA)

The underlying concept of IDFA [BGN05] is similar to classical impossible differential attack (IDA, refer to Sect. 2.6.5), where the attacker looks for probability zero differentials instead of high probability ones. Once such a differential is chosen, the attacker checks which sub-key guesses produce this differential. Those sub-key guesses are then discarded as they cannot lead to a correct key guess. IDA utilizes input difference to get a differential, whereas IDFA makes use of fault to achieve the same. IDFA is successfully used against `AES`, e.g., in [PY06, DFL11].

Linear Fault Attack/ Integral Fault Attack

Two other types of fault models were proposed which resemble two classical cryptanalysis techniques. The linear fault attack (LFA) [LGL+12] is a fault-based generalization of the linear cryptanalysis (seeSect. 2.6.2). Similarly, the integral fault attack, described in [PY06] and later in [SSL15, Chap. 6.3], is the fault-based generalization of the classical integral cryptanalysis (see Sect. 2.6.4). In the LFA scenario, Evetries to recover the information regarding the fault through the output from a (non-linear) component appended to it by means of linear approximation. The number of pairs of output required for the attacker to successfully apply LFA is of the order of $c \times \epsilon^{-4}$, where ϵ is the linear bias ($c \times \epsilon^{-2}$ pairs of output would be needed to apply linear cryptanalysis) and c is a small constant (typically, 4). For the integral fault case, she repeats all the possible faults for a word, effectively simulating the integral attack. Note that, this model can be hard to achieve in practice and requires sophisticated equipment; also it is worth mentioning that such model has never been reported with any practical attack setups up to date.

3.5.2 Collision-Based Fault Analysis

Collision Fault Attack (CFA)

The collision fault attack (CFA) is proposed in [BK06] and independently in [Hem04].[1] The concept is used against `IDEA` [CGV08]. In CFA, the attacker, Eve, gains information about the secret key if the outputs, obtained from a normal (non-faulty) encryption and from a faulty encryption, are equal. She arbitrarily chooses the first plaintext P_0 and obtains the corresponding faulty ciphertext: $C' = E'_K(P_0)$ (here E_K is the encryption operation and where E'_K denotes the faulty invocation of the encryption operation with secret key K). Then, she asks for encryption of another input, P, which might give an output identical to C' (i.e., $E(P) = C = C'$). Such collision can be then used to get an information about the intermediate state of the cipher, and consequently, about the secret key. Unlike DFA, the fault is normally introduced in the early rounds of the cipher, exploiting the property that the propagation of the fault in the intermediate value will result in a collision with the normal output.

One natural problem that arises is, how to find two such colliding inputs; as the cipher is expected to have a very good diffusion. Evesolves this problem by utilizing differential properties at early rounds (to avoid much diffusion). Therefore, unlike DFA, the fault is normally introduced in the early rounds of the cipher, exploiting the property that the propagation of the fault in the intermediate value will result in the collision with the normal output. Hence, one can consider a CFA on E_K to be similar to the DFA on E_K^{-1}.

Internal Differential Fault Attack (InDFA)

The Internal Differential Fault Attack (InDFA) [SC16, SC17] mainly leverages upon the internal symmetry of a cipher. The ciphers which are employing a parallel mode of operation are the most vulnerable to this kind of analysis. The idea of InDFA is to overcome the nonce barrier that acts as an implicit guard against DFA, as the nonce is not allowed to repeat. This guard can be removed when faults are injected at some counters of the parallel branches (which differ minimally at the inputs). This can result in both branches having same inputs, as the faulty counter gets the same content as the non-faulty counter. The rest of the analysis can be done similarly to DFA. In other words, the attacker injects a second fault in the internal state of any of the (currently identical) branches. This second fault propagates through the rest of the state.The effect of this fault can be utilized by DFA to recover secret information. Refer to Fig. 3.10 for a visual representation of the scenario. The InDFA model is applied on the CAESAR candidate `PAEQ` [SC16].

[1] The work against `3-DES` in [Hem04] is described as a DFA; though with the current terminology, it probably makes better sense to classify this work as CFA.

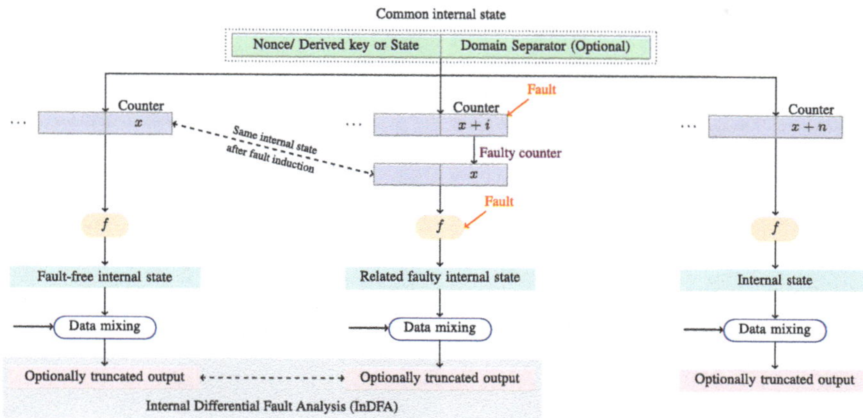

Fig. 3.10 Generic model of internal differentials in fault analysis of parallelizable ciphers in counter mode

Safe Error Analysis (SEA)/ Ineffective Fault Analysis (IFA)

Similar to the CFA model, the *Safe Error Analysis* (SEA) [YJ00, YKL+01, JQY+02] model makes use of the situation when the fault introduced in the cipher does not alter the output. To achieve this, intermediate bits are changed and the information whether the output changed or not is used to determine the non-faulty content at the time of the fault injection. Generally, this attack works by recovering one bit at a time; so, as many faults as the state size may be required. It may be noted that the SEA requires a more stringent adversarial model. The ineffective fault analysis (IFA) [Cla07] is a special type of SEA where the injected fault has no effect.

SEA attacks are able to break many countermeasures that do not withhold the output in case a fault is safe (that is, the fault does not change the output). In that case, SEA is able to utilize the information whether the output is withheld or not.

IFA is used to recover the secret SBox of the block cipher `KUZNYECHIK` (a Russian standard) [ADY15]. In addition, the authors also presented a DFA with the random byte fault model in the same research paper.

3.5.3 *Statistics-Based Fault Analysis*

In this third category of fault analysis, the statistical distribution of a variable is considered, which is biased because of the nature of fault injection. Generally, the underlying models assume more powerful adversary compared to difference-based

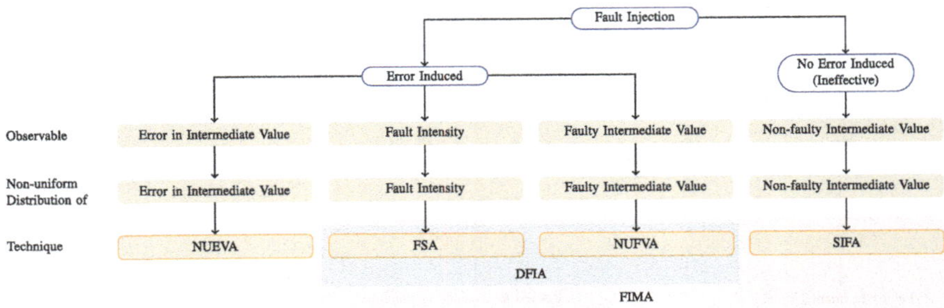

Fig. 3.11 Overview of statistics-based fault analysis methods

method, and are of *stuck-at*[2] type. The details on practical implementations on such fault models can be found in [Nah15]. Further, the authors also showed the connection among these models and gave a comparison on efficiency. More information on each of the analysis methods are given later.

An abridged taxonomy of statistics-based fault attacks (adopted with slight modifications from [RAD19'2, Fig. 1]) is given in Fig. 3.11. The terms, *Non-Uniform Error Value Analysis* (NUEVA) and *Non-Uniform Fault Value Analysis* (NUFVA), are adopted following [Nah15, RAD19'2, RAD19'1].

The following description of the terminologies is adopted (with slight modification) from [GYS15, Sect. 2.1]:

- **Fault Intensity.** Fault intensity is the strength by which a device is forced to work under non-favourable conditions (that may effectively induce a fault). For example, in case of clock glitch as the source of fault injection, the fault intensity refers to the shortened clock cycle.
- **Fault Sensitivity.** The fault sensitivity is the minimal fault intensity at which a device starts to show faulty behaviour. For example, in case of clock glitches, the fault sensitivity generally corresponds to the critical path.
- **Biased Fault.** A biased fault refers to the gradual change in the faulty behaviour of the device, which is a result of gradual change in fault intensity.

The *Differential Fault Intensity Analysis* (DFIA) uses the biased fault model described earlier that is based on minimal bias (e.g., flip of one or two bits); and it relies on side channel leakage. Here, side channel information generated by a countermeasure to a fault attack (such as starting time of an alarm or time of premature stop of a cipher execution) is used. Hence, the knowledge on the fault and the fault-free computations is not necessarily required. Notice that *Fault Sensitivity Analysis* (FSA) is considered one variant of DFIA under the current taxonomy. On another note, FSA and NUFVA do not require the knowledge of the faulty output.

[2] The hard fault model (Sect. 3.3.1) is different from the stuck-at fault, as the hard fault model assumes more control for the attacker. In the hard fault model, the effect of fault does not disappear even after the device is reset. This does not happen for a stuck-at fault.

Non-Uniform Error Value Attack (NUEVA)

In *Non-Uniform Error Value Attack* (NUEVA) [LRD+12], the fault value (which acts as a mask) is considered as the variable of interest and the non-uniform distribution of it is observed. This non-uniformity in distribution can be caused by, e.g., the Hamming weight of the fault value.

Non-Uniform Faulty Value Attack (NUFVA)

Non-Uniform Faulty Value Attack (NUFVA) is proposed in [FJL+13], where the biased distribution of the state is considered. This bias is caused due to the result of the fault injection technique. For example, it is shown in [Nah15] that certain fault values will more likely appear than the rest.

Fault Sensitivity Analysis (FSA)

Fault Sensitivity Analysis (FSA) [LSG+10] exploits the behaviour of underlying hardware when exposed to faults. This behaviour changes with the change of the intensity of the fault injection. This behaviour is a potential source of side channel information to the attacker. She does not need the output to analyse, instead she checks the data dependency for each intensity. The attack procedure is therefore related to side channel analysis. It tests the circuit by varying the intensity of the injection and checks the data dependency for each case. In the actual attack procedure, Evecollects the input to the cipher, non-faulty output, and the side channel information. After this, the correlation coefficient is used to find the most likely sub-keys.

Differential Fault Intensity Analysis (DFIA)

Differential Fault Intensity Analysis (DFIA) [GYT+14] combines fault injection with side channel attack setup. Similar to FSA, it tests the system responses under different fault intensity and takes advantage of a non-uniform distribution of the faults (biased fault model). Afterwards, it treats the biased fault model as the source of leakage and applies side channel methods. However, unlike FSA, it does not require a fault sensitivity profile of the attacked device. DFIA has been successfully applied against `AES` [GYT+14], `PRESENT` and `LED` [GYS15].

3.5.4 Others

Statistical Ineffective Fault Attack (SIFA)

Statistical Ineffective Fault Attack (SIFA) [DEK+18]/Chap. 10 lies somewhere in between the IFA (Sect. 3.5.2) and SFA (Sect. 3.5.3). Similar to IFA, SIFA utilizes the probability that a fault injection may not result in a successful alteration in the device (ineffective fault). The main observation that works in SIFA is that the probability of error as a result of a successful ineffective fault injection is non-uniform. The bias can be observed in case of a correct key guess, but not for the cases where the key guesses are wrong. A SIFA on the ASCON AEAD, which is a finalist in the CAESAR portfolio for the use case 1 (lightweight applications) is reported in [RAD19'1]. Another SIFA is reported on GIMLI in [GPT19]. SIFA can be used to break existing fault countermeasures (see Sect. 3.7) that are used to protect against DFA and other common fault methods.

Persistent Fault Attack (PFA)

Recently, another category of biased fault injection is introduced, based on the fault duration. The fault model, known as the *persistent* fault, is proposed as a hybrid fault model between commonly used transient and permanent faults. Unlike transient fault, it affects several calls of the target function, however, persistent fault is not permanent, and disappears with a device reset/reboot. Based on this fault model, a statistical analysis technique is developed, called *Persistent Fault Analysis* (PFA) [ZLZ+18], to attack block ciphers. The attack targets the (common) SBox table stored in memory for a serialized implementation, with one fault injection altering any one element in SBox. The adversary collects several ciphertexts resulting from this faulty SBox, to statistically recover the key. PFA is shown to break redundancy-based fault countermeasures. Application of PFA can be found in [CB19] and [ZZY+19].

Fault Intensity Map Analysis (FIMA)

Fault Intensity Map Analysis (FIMA) [RAD19'2] generalizes the statistics-based fault models, taking into account all the relevant variables and the side channel information. So, it can be considered as the common method for FSA, DFIA, NUEVA, and SIFA. This method utilizes the bias that is introduced by faults in the distribution of any of the relevant variables (faulty, non-faulty, or the fault value), and the information on the fault intensity. As such, it can work with less amount of data than the predecessor methods.

3.6 Generalized Fault Attack Automation Frameworks

As fault attacks are becoming a more serious threat, efforts have been made to construct a generalized and automated framework that can evaluate the effectiveness of the attack and provide recommendations for cipher implementers, on which they can spot vulnerabilities and judge if they want to make their algorithm more robust. Conceptually, it can be described as follows: Given a cipher (and possibly a platform-specific implementation), such tools will automatically look for vulnerability; thereby cutting off the manual labour and expertise required to perform such attacks which is often error-prone [BHB19].

The overall trend in this area, along with the possible challenges, can be deduced from the recent publications. Basically, the methods that focus on fully automated analysis always utilize some sort of automated solver, such as a SAT solver. While this solver can provide a formally proven output, the quality of this output always depends on the input that has to be constructed by the analysis framework. It is an open problem to have a universal transformation framework which would take a cipher representation in any form and construct set of equations for a solver. Another challenge is to provide automated analysis on protected implementations—so far, the above-mentioned works only focus on unprotected implementations since those only consider a single fault adversary. To be able to bypass most of the countermeasures, it is necessary to have a multiple fault adversary model; e.g., in case the attacker wants to break a simple redundancy countermeasure only with fault injection attack, the attacker needs either to inject the same fault into both branches of the execution, or inject a fault into one branch and then into the comparison module. Next, we will explain the working principle of the main works in this area, divided into cipher level techniques and implementation level techniques.

3.6.1 Cipher Level Approaches

The first idea of automating cipher-specific fault analysis is aimed at AFA [ZGZ+16]. It requires manual interpretation of the cipher representation in an algebraic form and then a SAT solver is used to find the vulnerable spots. The first automated framework focused on DFA, proposed in [KRH17], requires a manual identification of non-linear operations in a cipher. It then proceeds with finding a fault propagation pattern through the cipher rounds which is represented by colours—if a value served as an input to a non-linear operation, the output of this operation is assigned a unique colour so that the number of these operations can be determined from the fault until the output of the cipher.

Later, several fully automated cipher level frameworks emerge in the literature. Some of the proposed works have adopted the machine learning methods to identify vulnerable spots in the ciphers, based on existing fault attacks observed or learned on different ciphers. The work in [SKM+17] adopts the association rule mining to explore fault vulnerabilities on DFA, whereas the work in [SJP+19] uses random forest to identify solvable algebraic representation for AFA. Later, [SMD18] propose a framework that estimates the attack complexity required to recover the secret. To find a distinguisher, a data-mining approach is utilized in this case. While works generally focus on block ciphers, the authors in [SDA+17] attempted to fill the gap for stream ciphers. Recently, this work has been improved in [BSS+20] by substituting correlation coefficient (which is used in [SDA+17]) by machine learning.

3.6.2 Implementation Level Approaches

In case of *software level* techniques, there are two works focusing on analysing assembly implementations. The work in [BHL18] aims at unrolled microcontroller implementations and shows that some vulnerabilities are not visible from the cipher level perspective. For example, when permutation layer is combined with the substitution layer in block ciphers into a form of look-up tables, it introduces new vulnerabilities that cannot be estimated from the algorithmic representation. However, this work leaves the final step of the attack, namely, the analysis of DFA equations, on user. This task is later completed in [HBZ+19] which utilizes Satisfiability Modulo Theory (SMT) solver to automate this last step.

The *hardware level* approach is presented in [BGE+17, GPU+19]. This framework takes the hardware description of the cipher and creates a fault list based on exploited fault model. This is represented in CNF which is analysed by a SAT solver. For implementation-based approaches; the authors argue that the requirement for cryptanalyst expertise could be minimized, since the algebraic representation solely depends on the implementations themselves, rather than the cipher description.

3.7 Countermeasures

The fault attacks are usually not prevented at the cipher design level. Instead, the cipher designer has to rely on the circuit implementer or the secure software developer or the network administrator to ensure that an adequate countermeasure is deployed. Hence, countermeasures are designed and analysed separately from the ciphers, as we discuss here. For the sake of simplicity, we do not include the combined fault and side channel countermeasures. A compact view of the countermeasures is presented in Fig. 3.12. Overall, the countermeasures can be divided into two main categories. In the first category, the countermeasures try to prevent the fault happening in the first place (achieved through a specialized device). In the second, the countermeasures try to eliminate the effect of the fault either by means of redundancy or protocol.

All the countermeasures that mitigate fault effect utilize some form of redundancy to achieve protection against FA. The following classification shows the three classes based on the part where the redundancy is introduced:

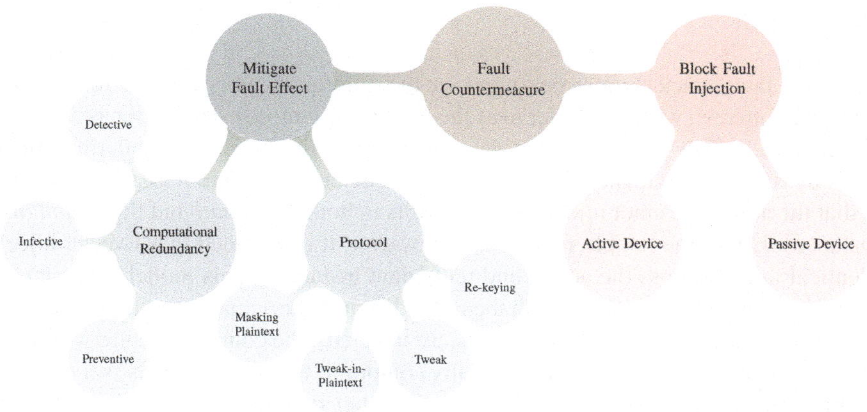

Fig. 3.12 Generic overview of fault countermeasures

1. **Using a separate, dedicated device.** These devices are not related to the cipher. These devices can be *passive*, where a shield is used to block external interference [BCN+04]. It may be possible to peel off the passive layer using chemicals or precise milling tools, such as Ultra Tec ASAP-1. Alternatively, *active* devices can be used (such as, a sensor that detects any potential stress [ZDT+14, HBB16, HBB+16'2]).
2. **Using redundancy in computation.** The common procedure for these types of countermeasures is to duplicate (fully/partially) the circuit. This duplication can be in the spatial domain or the temporal domain, or a precomputed signature can be used to check integrity [Kor16, Chap. 3.4]. Then, some predefined procedure to handle the situation of a fault detection is used, such as not producing any output or producing a random output.
3. **Using protocol level technique.** The communication protocol between Alice and Bob ensures that the probability for a successful fault injection is negligibly small.

The countermeasures in the 2nd category that use redundant computations are quite well studied. It can be classified into *Detection* (Sect. 3.7.1), *Infection* (Sect. 3.7.2), *Prevention* (Sect. 3.7.3), which are described next. The protocol level countermeasures are generally one among *Re-keying*, *Tweak*, and *Tweak-in-Plaintext*, *Masking plaintext*, which are combined subsequently (Sect. 3.7.4).

Some specialized countermeasures are proposed in the literature, which are briefly described in Sect. 3.7.6/Sect. 10.3.2. Those are proposed to overcome the inability of thwarting the inability of the common duplication-based countermeasures (see Sect. 10.3.1 for more details).

Double Fault

Given the fault attack countermeasures (see the subsequent parts), a particular model may be of interest, which we refer to as the *double fault* model. The notion is adopted from [BSS20, Sect. 2.6]. Note that the detective and the infective countermeasures work by running two identical instances of the cipher. The inherent assumption here is that the attacker cannot inject identical faults in both the *actual* and the *redundant* instances. Under the double fault model, however, it is assumed that Evecan inject identical faults in both the actual and redundant instances. This model is shown to be practical in [SHS16]. Since both the redundant and the actual computation results in the same output (since identical faults are injected), the countermeasure will treat this case as no fault, and return the (faulty) output to the attacker. This is a generic problem against detection and infection—if higher than double redundancy is not used. If such models are considered within the scope, *protocol level* or *preventive* countermeasures can be used.

Another interesting aspect of the double fault model is whether it is to be considered within the single fault or the multiple fault adversary model (see Sect. 3.2.2). Basically, the answer depends on the precise definition of the term "one execution of the cipher". To be more specific; consider by "one execution", we refer to the complete system (both the actual and the redundant instances)—in this case, we consider the double fault model outside the scope of single fault adversary model. On the contrary, consider the scenario where we refer to a single execution of the cryptographic algorithm designed by a cryptographer as "one execution of the cipher"; here the attacker injects faults once per one execution; so the single fault adversary model is maintained. For the sake of consistency, we stick to the second notion, so we do not consider the double fault model as a violation of the frequently used single fault adversary model. The recently proposed `CRAFT` [BLM+19] cipher has an in-built mechanism to protect against the double fault (see Sect. 3.7.1 for more information on this cipher). This model of double fault can be generalized for more faults, such as the triplication-based countermeasure proposed in [BBK+10]. Note that, this notion of "double fault" is different from [RAD19'1, RAD19'2], where the same term refers to attacking a pair of SBoxes at the same round.

3.7.1 Detection

Detection-based countermeasures[3] aim at detecting an anomaly. After it is detected, a predefined recovery procedure can be launched to stop the attacker from getting any information regarding the faulty output (the output from the cipher is suppressed or a random output is returned). These methods try to detect any data modification due to fault injection, deploying methods from coding and information theory, such as

[3] Sometimes referred to as, Concurrent Error Detection (CED).

linear parity, non-linear (n, k) codes, representing data in *wave dynamic differential logic* [BJB16], etc.

In [LRT12], it is commented that detection-based countermeasures that rely on error correcting codes are difficult to adopt in cases where side channel security (such as masking) is needed. Therefore, researchers try to design a countermeasure which is capable of protecting against both [SMG16, BH17]. Ways to find efficient codes and to evaluate the encoding-protected cipher implementations are studied in [BHL19].

The authors of [BLM+19] propose a tweakable block cipher named CRAFT. This cipher is claimed to be the first cipher level protection against DFA. The underlying approach here is a detective countermeasure which is able to detect the double fault, thanks to an error detection code used in conjunction with duplication. Also, because of the overlapping nature of error detecting code and the (non-duplicated) cipher, the cost is generally less than what would be for other major ciphers. The idea of using coding theory is previously used in [KGK+18], where an architecture capable of detecting and correcting certain fault induced errors is proposed.

3.7.2 *Infection*

Infective countermeasures are proposed as an enhancement to detection. A detection mechanism generally relies on the logical state of only one particular bit (such as the zero flag of a microcontroller). The idea of infection first emerges in the context of public key cryptography, which is later extended to symmetric key. Examples include [SBG+09, GSS+15, GST12, TBM14]. It is noted in [LRT12] that such countermeasures would require randomness.[4] All the infective proposals following [LRT12] use randomness.

Infection-based countermeasures introduce a diffusion caused by the faulted intermediate value that spreads further into entire cipher state to avoid explicit detection. Overall, it can be classified into two types. In one type, the two computations (actual and redundant) are allowed to run for full rounds. Then, the XOR between the two computations (Δ) is computed; which is passed into another function, τ, such that $\tau(0) = 0$ and $\tau(\Delta)$ is a random value if $\Delta \neq 0$ (τ uses a hidden random parameter). The output from τ is then used with the actual computation of the cipher (such as XORed). Therefore, if both outputs are equal, the circuit continues by giving the result. Examples of this type include [LRT12, GSS+15]. In the other type, introduced in [GST12], XOR of the actual and the redundant computations is done as these are being computed at each round. If the XOR result is non-zero at some round, then the diffusion of the underlying cipher is used to propagate the difference through the subsequent rounds, finally making the output unusable for the attacker. A more comprehensive analysis, together with state-of-the art results can be found in [BSS20]/Chap. 9.

[4] However, [LRT12] is more of a case study on AES rather than a formal proof for a general case.

As already mentioned, infective countermeasures do not work against the double faults . Besides, infective countermeasures cannot provide safeguard against collision (Sect. 3.5.2), ineffective (Sect. 3.5.2) attacks; as those utilize situations where injection of faults does not alter the output.

Interestingly, quite a few infective countermeasures were broken soon afterwards by standard DFA technique. Nonetheless, to the best of our knowledge, the following three have not been broken by DFA: GF(2^{128}) multiplication proposed in [LRT12] (the lightweight variant of this protection is broken in [BG13]); modification of [GSS+15] proposed in [GSS+17, Sect. 5] (the original proposal in [GSS+15] is broken in [BB15]); and the one in [TBM14, PCM15][5] (see Sect. 3.7.5 for non-DFA attacks on this proposal).

It has been shown in [BSS20]/Chap. 9 that the countermeasure proposed in [TBM14] does not follow the infection paradigm, as it suffers from the one-bit judgment condition similar to detection. A similar claim is made in [FCL+20, FCL$^+$20 Sect. VIII.A].

3.7.3 Prevention

This type of countermeasure works at the software implementation by inserting dummy instructions, randomization of execution, etc. For example, usage of idempotent instruction sequences against an instruction skip attack on a microcontroller [MHE+14] or *intra-instruction redundancy* [PYG+16] fall into this category. These countermeasures can be adopted on top of other fault (such as infection) or side channel (such as masking) countermeasures.

3.7.4 Re-keying, Tweak and Tweak-in-Plaintext, Masking Plaintext

Re-keying [MSG+10] is a protocol level approach; here it is assumed that a relatively smaller portion of the device can be protected, whereas the rest of it is still vulnerable to fault attack. The concept is extended to multi-party setup in [MPR+11]. The basis of re-keying is the *re-keying function*, g, which generates a different session key $K_i^\star$ for each session i with the help of the (fixed) master key K and a random nonce (different for every session); and has a low hardware footprint. Later, this session key is used in an underlying block cipher E with plaintext P_i as $C_i = E_{K_i^\star}(P_i)$. Even if g is not secure cryptographically, as per the authors claim, this scheme is capable of thwarting DFA and side channel attacks. The basic principle is, g guarantees that the effective key is changing at every session, while these attacks require the key to remain fixed for successive sessions.

[5] The two proposals presented in these two papers are conceptually the same.

However, this scheme suffers from a birthday attack and can only provide security up to the birthday bound on the session key size, as pointed out in [DEM+14]. As an improvement, it is later proposed to replace the re-keying function with a tweakable block cipher, with a block counter acting as a tweakable input [DKM+15]. This work extended the security of fresh re-keying scheme beyond birthday-bound security against side channel attacks. In a parallel work, it is shown in [PRM16] that a tweakable block cipher with a random (unknown) tweak can prevent fault attacks.

An approach entitled tweak-in-plaintext is proposed to solve some of the issues with the aforementioned works [BBB+18]. By reducing the bandwidth linearly, a sub-exponential safeguard against DFA is achieved, without the need of the synchronization between the communicating parties. One similar concept of masking the plaintext is proposed in [GSD+10, Algorithm 1], where the bandwidth is halved. However, in both the cases, decryption circuit cannot be protected against fault attacks (the authors in [BBB+18] indeed argue that this is not a serious limitation).

3.7.5 Attacks on Countermeasures

Several fault attacks specifically target some fault countermeasure. Normally, redundancy-based countermeasures can be bypassed by injecting identical faults on the redundant operations. For example, in [PCN+15], the authors introduced bias in the fault distribution, which could be exploited to inject same faults in the branches of a time redundancy countermeasure. In [SHS16], identical faults are injected to bypass the hardware redundancy countermeasures. In [HBB+16'1, BJB17], the authors show the practical experiments on how to bypass information redundancy countermeasure (encoding approach). Also, recently, in [SJB+18], the authors show how to exploit the redundancy-based countermeasure itself to leak some information.

Some of the infective countermeasures are shown to be broken by DFA approach (see Sect. 3.7.2). Some non-DFA techniques, such as modifying the round counter or stuck-at fault, are also proposed in [BG16] against [TBM14]. Also, in [DEK+18], it is shown that even with the knowledge of only the correct outputs (the technique is called SIFA, see Sect. 3.5.4), it is still possible to recover the secret information when the cipher is protected by the aforementioned countermeasure. The FIMA model [RAD19'2] (Sect. 3.5.4) can be used to break both the detective and the infective countermeasures.

3.7.6 Specialized Countermeasures Against Statistical Ineffective Fault Attack

The countermeasures discussed so far are incapable of thwarting SIFA, in general. Specialized countermeasures are needed for protecting against SIFA, as listed here (see also Sect. 10.3.2).

Repetition Code

The authors in [BKH+19] propose an error correction which is based on binary repetition code. Taking majority voting, this countermeasure corrects the fault injection (which can alter at most one bit) and stops the attacker from getting information on whether the fault has occurred or not.

Masking and Repetition Code

A two-phase approach is proposed in [SJR+20]. At the first phase, masking is used that protects against faults at the state. At the second phase, attacks on individual sub-operations like SBox are prevented by error correction.

Error Detection Through Toffoli Gate and Masking

Based on reversible computing paradigm, the authors in [DDE+19] propose an error detection mechanism that targets at most one bit. The non-linear components are designed using Toffoli gates in such a way that a single bit flip would result in a garbage output.

Error Correction

An error correcting code is used by [SRM19] which is able to protect against SIFA. The authors propose a more general code than the repetition code used earlier.

Removing Bias by Duplication

While the other countermeasures rely on non-standard gate like Toffoli or at least triplicate the cost, a recent countermeasure relies on duplication [BKK+20'2]. This way, the authors are able to provide the least expensive SIFA countermeasure.

References

[PQ03] G. Piret, J. Quisquater, A differential fault attack technique against SPN structures, with application to the AES and KHAZAD, in *Cryptographic Hardware and Embedded Systems - CHES 2003, 5th International Workshop, Cologne, Germany, September 8-10, 2003, Proceedings*, pp. 77–88 (2003), https://doi.org/10.1007/978-3-540-45238-6_7

[SH13] L. Song, . Hu, Differential fault attack on the PRINCE block cipher, in*Lightweight Cryptography for Security and Privacy - Second International Workshop, LightSec 2013, Gebze, Turkey, May 6-7, 2013, Revised Selected Papers*, pp. 43–54 (2013), https://doi.org/10.1007/978-3-642-40392-7_4

[JKP12'1] P. Jovanovic, M. Kreuzer, I. Polian, A fault attack on the LED block cipher, in *Constructive Side-Channel Analysis and Secure Design - Third International Workshop, COSADE 2012, Darmstadt, Germany, May 3-4, 2012. Proceedings*, pp. 120–134 (2012), http://dx.doi.org/10.1007/978-3-642-29912-4_10

[BS02] J. Blömer, J. Seifert, Fault based cryptanalysis of the advanced encryption standard. Cryptology ePrint Archive, Report 2002/075, 2002. http://eprint.iacr.org/2002/075

[DA14] P. Dey, A. Adhikari, Improved multi-bit differential fault analysis of Trivium, in *INDOCRYPT 2014, New Delhi, India, Proceedings*, pp. 37-52, 2014, https://doi.org/10.1007/978-3-319-13039-2_3

[BJC15] J. Breier, D. Jap, C.-N. Chen, Laser profiling for the back-side fault attacks (with a practical laser clock glitch attack on aes), in *First Cyber-Physical System Security Workshop (CPSS 2015)*, pages 99-103, Singapore. ACM, 2015-04 (pages: 59 and 60)

[LCF+17] B. Lac, A. Canteaut, J. Fournier, R. Sirdey, DFA on ls-designs with a practical implementation on SCREAM (extended version). IACR Cryptology ePrint Archive, 2017:76 (2017), http://eprint.iacr.org/2017/076

[BBB+18] A. Baksi, S. Bhasin, J. Breier, M. Khairallah, T. Peyrin, Protecting block ciphers against differential fault attacks without re-keying (extended version). Cryptology ePrint Archive, Report 2018/085 (2018), https://eprint.iacr.org/2018/085

[AM11] S. Ali D. Mukhopadhyay, Adifferential fault analysis on AES key schedule using single fault, in *2011 Workshop on Fault Diagnosis and Tolerance in Cryptography, FDTC 2011, Tokyo, Japan, September 29, 2011*, pp. 35–42 (2011), https://doi.org/10.1109/FDTC.2011.10

[Kim12] C.H. Kim, Improved differential fault analysis on AES key schedule. IEEE Trans. Inf. Forens. Secur. **7**(1), 41–50 (2012), https://doi.org/10.1109/TIFS.2011.2161289

[DRA16] P. Dey, R.S. Rohit, A. Adhikari, Full key recovery of ACORN with a single fault. J. Inf. Sec. Appl. **29**, 57–64 (2016) https://doi.org/10.1016/j.jisa.2016.03.003

[KPB+17] S.V.D. Kumar, S. Patranabis, J. Breier, D. Mukhopadhyay, S. Bhasin, A. Chattopadhyay, A. Baksi, A practical fault attack on arx-like ciphers with a case study on chacha20:33-40 (2017), https://doi.org/10.1109/FDTC.2017.14

[BdSG+14] J. Blömer, R.G. da Silva, P. Günther, J. Krämer, J. Seifert, A practical secondorder fault attack against a real-world pairing implementation, in *2014 Workshop on Fault Diagnosis and Tolerance in Cryptography, FDTC 2014, Busan, South Korea, September 23, 2014*, pp. 123–136 (2014), https://doi.org/10.1109/FDTC.2014.22

[EHH+14] S. Endo, N. Homma, Y.-i. Hayashi, J. Takahashi, H. Fuji, T. Aoki, A multiple-fault injection attack by adaptive timing control under black-box conditions and a countermeasure, in *Constructive Side-Channel Analysis and Secure Design: 5th International Workshop, COSADE 2014, Paris, France, April 13-15, 2014*. Revised Selected Papers (Springer International Publishing, Cham, 2014), pp. 214–228. ISBN: 978-3-319-10175-0. https://doi.org/10.1007/978-3-319-10175-0_15

[SSL15] K. Sakiyama, Y. Sasaki, Y. Li, *Security of Block Ciphers - From Algorithm Design to Hardware Implementation* (Wiley, 2015). isbn: 978-1-118-66001-0, http://eu.wiley.com/WileyCDA/WileyTitle/productCd-1118660013.html

[BLM+19] C. Beierle, G. Leander, A. Moradi, S. Rasoolzadeh, CRAFT: lightweight tweakable block cipher with efficient protection against DFA attacks. IACR Trans. Symmet. Cryptol. **2019**(1), 5–45 (2019), https://doi.org/10.13154/tosc.v2019.i1.5-45

[BBK+12] A. Barenghi, L. Breveglieri, I. Koren, D. Naccache, Fault injection attacks on cryptographic devices: theory, practice, and countermeasures. Proc. IEEE **100**(11), 3056–3076 (2012), https://doi.org/10.1109/JPROC.2012.2188769

[Kor16] R. Korkikian, Side-channel and fault analysis in the presence of countermeasures : tools, theory, and practice. Theses, PSL Research University, 2016-10, https://tel.archives-ouvertes.fr/tel-01762404

[DEK+18] C. Dobraunig, M. Eichlseder, T. Korak, S. Mangard, F. Mendel, R. Primas, Sifa: exploiting ineffective fault inductions on symmetric cryptography. Cryptology ePrint Archive, Report 2018/071 (2018), https://eprint.iacr.org/2018/071

[OFl16] C. O'Flynn, Fault injection using crowbars on embedded systems. Cryptology ePrint Archive, Report 2016/810 (2016), https://eprint.iacr.org/2016/810

[ABC+17] S. Anceau, P. Bleuet, J. Clédière, L. Maingault, J.-l. Rainard, R. Tucoulou, Nanofocused x-ray beam to reprogram secure circuits, in *Cryptographic Hardware and Embedded Systems - CHES 2017* ed. by W. Fischer, N. Homma (Springer International Publishing, Cham, 2017), pp. 175–188. ISBN: 978-3-319-66787-4

[SBH+15] B. Selmke, S. Brummer, J. Heyszl, G. Sigl, Precise laser fault injections into 90 nm and 45 nm sram-cells, in *Smart Card Research and Advanced Applications - 14th International Conference, CARDIS 2015, Bochum, Germany, November 4-6, 2015*. Revised Selected Papers, pp. 193–205 (2015), https://doi.org/10.1007/978-3-319-31271-2%5C_12

[PDL18] D. Petryk, Z. Dyka, P. Langendoerfer, Optical fault injections: a setup comparison (2018), https://www.researchgate.net/publication/330702668_Optical_Fault_Injections_a_Setup_Comparison

[MDH+13] N. Moro, A. Dehbaoui, K. Heydemann, B. Robisson, E. Encrenaz, Electromagnetic fault injection: towards a fault model on a 32-bit microcontroller, in *2013 Workshop on Fault Diagnosis and Tolerance in Cryptography, Los Alamitos, CA, USA, August 20, 2013*, pp. 77–88 (2013), https://doi.org/10.1109/FDTC.2013.9

[HHM+12] Y.-I. Hayashi, N. Homma, T. Mizuki, T. Aoki, H. Sone, Transient iemi threats for cryptographic devices. IEEE Trans. Electromag. Compatibil. **55**(1), 140–148 (2012)

[KDK+14] Y. Kim, R. Daly, J. Kim, C. Fallin, J. Lee, D. Lee, C. Wilkerson, K. Lai, O. Mutlu. Flipping bits in memory without accessing them: an experimental study of DRAM disturbance errors, in *ACM/IEEE 41st International Symposium on Computer Architecture, ISCA 2014, Minneapolis, MN, USA, June 14-18, 2014*, pp. 361–372 (2014), https://doi.org/10.1109/ISCA.2014.6853210

[BH15] J. Breier, W. He, Multiple fault attack on present with a hardware Trojan implementation in fpga, in *2015 International Workshop on Secure Internet of Things (SIoT)*, pp. 58–64, 2015-09

[HS13] M. Hutter, J. Schmidt, The temperature side channel and heating fault attacks, in *Smart Card Research and Advanced Applications - 12th International Conference, CARDIS 2013, Berlin, Germany, November 27-29, 2013*. Revised Selected Papers, pp. 219–235 (2013), https://doi.org/10.1007/978-3-319-08302-5%5C_15

[SGD08] N. Selmane, S. Guilley, J. Danger, Practical setup time violation attacks on AES, in *Seventh European Dependable Computing Conference, EDCC-7 2008, Kaunas, Lithuania, 7-9 May 2008*, pp. 91–96 (2008), https://doi.org/10.1109/EDCC-7.2008.11

[BS97] E. Biham and A. Shamir. Differential fault analysis of secret key cryptosystems, in *Advances in Cryptology - CRYPTO '97* ed. by B.S. Kaliski (Springer, Berlin, Heidelberg, 1997), pp. 513–525. isbn: 978-3-540-69528-8

[AMT13] S. Ali, D. Mukhopadhyay, M. Tunstall. Differential fault analysis of AES: towards reaching its limits. J. Cryptogr. Eng. **3**(2), 73–97 (2013), https://doi.org/10.1007/s13389-012-0046-y

[AM13'1] S.S. Ali, D. Mukhopadhyay, Differential fault analysis of twofish, in *Information Security and Cryptology: 8th International Conference, Inscrypt 2012, Beijing, China, November 28-30, 2012* ed. by M. Kutyíowski and M. Yung, Revised Selected Papers (Springer, Berlin, Heidelberg, 2013), pp. 10–28. isbn: 978-3-642-38519-3, http://dx.doi.org/10.1007/978-3-642-38519-3_2

[BEG13] N. Bagheri, R. Ebrahimpour, N. Ghaedi, New differential fault analysis on PRESENT. EURASIP Journal on Advances in Signal Processing, 2013(1):145, 2013. ISSN: 1687-6180, http://dx.doi.org/10.1186/1687-6180-2013-145

[JLS+13] K. Jeong, Y. Lee, J. Sung, S. Hong, Improved differential fault analysis on PRESENT-80/128. Int. J. Comput. Math. **90**(12), 2553–2563 (2013), https://doi.org/10.1080/00207160.2012.760732

[CZS16] W. Cheng, Y. Zhou, L. Sauvage, Differential fault analysis on midori, in *Information and Communications Security - 18th International Conference, ICICS 2016, Singapore, November 29 - December 2, 2016, Proceedings*, pp. 307–317 (2016), https://doi.org/10.1007/978-3-319-50011-9_24

[DFL11] P. Derbez, P.-A. Fouque, D. Leresteux, Meet-in-the-middle and impossible differential fault analysis on aes, in *Cryptographic Hardware and Embedded Systems - CHES 2011* ed. by B. Preneel, T. Takagi (Springer, Berlin, Heidelberg, 2011), pp. 274–291. ISBN: 978-3-642-23951-9

[SMC09] D. Saha, D. Mukhopadhyay, D.R. Chowdhury, A diagonal fault attack on the advanced encryption standard. IACR Cryptology ePrint Archive, 2009:581 (2009), http://eprint.iacr.org/2009/581

[BTL+14] H. L. Bouder, G. Thomas, Y. Linge, A. Tria, On fault injections in generalized feistel networks, in *2014 Workshop on Fault Diagnosis and Tolerance in Cryptography, FDTC 2014, Busan, South Korea, September 23, 2014*, pp. 83–93 (2014), https://doi.org/10.1109/FDTC.2014.18

[AM13'2] S. Ali, D. Mukhopadhyay, Improved differential fault analysis of CLEFIA, in *2013 Workshop on Fault Diagnosis and Tolerance in Cryptography, Los Alamitos, CA, USA, August 20, 2013*, pp. 60–70 (2013), https://doi.org/10.1109/FDTC.2013.11

[Jeo12] K. Jeong, Differential fault analysis on block cipher piccolo. IACR Cryptology ePrint Archive, 2012:399 (2012) http://eprint.iacr.org/2012/399

[KY10] A. Kircanski, A.M. Youssef, Differential fault analysis of hc-128, in *Progress in Cryptology - AFRICACRYPT 2010* ed. by D.J. Bernstein, T. Lange (Springer, Berlin, Heidelberg, 2010), pp. 261–278. ISBN: 978-3- 642-12678-9

[BGV17] A. Beckers, B. Gierlichs, I. Verbauwhede, Fault analysis of the chacha and salsa families of stream ciphers, in *Smart Card Research and Advanced Applications - 16th International Conference, CARDIS 2017, Lugano, Switzerland, November 13-15, 2017*, Revised Selected Papers, pp. 196–212 (2017), https://doi.org/10.1007/978-3-319-75208-2%5C_12

[LLG09] R. Li, C. Li, C. Gong, Differential fault analysis on SHACAL-1, in *Sixth International Workshop on Fault Diagnosis and Tolerance in Cryptography, FDTC 2009, Lausanne, Switzerland, 6 September 2009*, pp. 120–126 (2009), https://doi.org/10.1109/FDTC.2009.41

[HH11] L. Hemme, L. Hoffmann, Differential fault analysis on the SHA1 compression function, in *2011 Workshop on Fault Diagnosis and Tolerance in Cryptography, FDTC 2011, Tokyo, Japan, September 29, 2011*, pp. 54–62 (2011), https://doi.org/10.1109/FDTC.2011.16

[FR12] W. Fischer, C.A. Reuter, Differential fault analysis on grostl, in *2012 Workshop on Fault Diagnosis and Tolerance in Cryptography, Leuven, Belgium, September 9, 2012*, pp. 44–54 (2012), https://doi.org/10.1109/FDTC.2012.14

[SC15] D. Saha, D.R. Chowdhury, Diagonal fault analysis of grostl in dedicated MAC mode, in *IEEE International Symposium on Hardware Oriented Security and Trust, HOST 2015, Washington, DC, USA, 5-7 May, 2015*, pp. 100–105 (2015), https://doi.org/10.1109/HST.2015.7140246

[RLK11] T. Roche, V. Lomné, K. Khalfallah, Combined fault and side-channel attack on protected implementations of aes, in *Smart Card Research and Advanced Applications* ed. by E. Prouff (Springer, Berlin, Heidelberg, 2011), pp. 65–83. ISBN: 978-3-642-27257-8

[BGS15] N. Bagheri, N. Ghaedi, S.K. Sanadhya, Differential fault analysis of SHA-3, in *Progress in Cryptology - INDOCRYPT 2015 - 16th International Conference on Cryptology in India, Bangalore, India, December 6–9, 2015, Proceedings*, pp. 253–269 (2015)

[SLI+12] K. Sakiyama, Y. Li, M. Iwamoto, K. Ohta, Information-theoretic approach to optimal differential fault analysis. IEEE Trans. Inf. Forens. Sec. **7**(1), 109–120 (2012), https://doi.org/10.1109/TIFS.2011.2174984

[KSK14] J. Krmer, A. Stber, Á. Kiss, On the optimality of differential fault analyses on CLEFIA. IACR Cryptology ePrint Archive, 2014:572 (2014) http://eprint.iacr.org/2014/572

[AY15] R. AlTawy, A.M. Youssef, Differential fault analysis of streebog, in *Information Security Practice and Experience - 11th International Conference, ISPEC 2015, Beijing, China, May 5-8, 2015. Proceedings*, pp. 35–49 (2015), https://doi.org/10.1007/978-3-319-17533-1

[LFZ+16] P. Luo, Y. Fei, L. Zhang, A.A. Ding, Differential fault analysis of SHA3-224 and SHA3-256, in *2016 Workshop on Fault Diagnosis and Tolerance in Cryptography, FDTC 2016, Santa Barbara, CA, USA, August 16, 2016*, pp. 4–15 (2016), https://doi.org/10.1109/FDTC.2016.17

[LAF+18] P. Luo, K. Athanasiou, Y. Fei, T. Wahl, Algebraic fault analysis of sha-3 under relaxed fault models. IEEE Trans. Inf. Forens. Secur. **13**(7), 1752–1761 (2018). ISSN: 1556-6013

[LAF+17] P. Luo, K. Athanasiou, Y. Fei, T. Wahl, Algebraic fault analysis of SHA-3, in *Design, Automation & Test in Europe Conference & Exhibition, DATE 2017, Lausanne, Switzerland, March 27-31, 2017*, pp. 151–156 (2017), https://doi.org/10.23919/DATE.2017.7926974

[CDN17] A. Chakraborti, N. Datta, M. Nandi, Practical fault attacks on minalpher: how to recover key with minimum faults? in *Security, Privacy, and Applied Cryptography Engineering - 7th International Conference, SPACE 2017, Goa, India, December 13-17, 2017, Proceedings*, pp. 111–132 (2017), https://doi.org/10.1007/978-3-319-71501-8_7

[RCC+16] D.B. Roy, A. Chakraborti, D. Chang, S.V.D. Kumar, D. Mukhopadhyay, M. Nandi, Fault based almost universal forgeries on CLOC and SILC, in *Security, Privacy, and Applied Cryptography Engineering* ed. by C. Carlet, M.A. Hasan, V. Saraswat (Springer International Publishing, Cham, 2016), pp. 66–86. isbn: 978-3-319-49445-6

[BDM+19] H. Bartlett, E. Dawson, H.Q.A. Mahri, M.I. Salam, L. Simpson, K.K. Wong, Random fault attacks on a class of stream ciphers. Secur. Commun. Netw. **2019**, 1680263:1–1680263:12 (2019), https://doi.org/10.1155/2019/1680263

[CJW10] N.T. Courtois, K. Jackson, D. Ware, Fault-algebraic attacks on inner rounds of des, in *e-Smart'10 Proceedings: The Future of Digital Security Technologies* (2010)

[JKP12'2] P. Jovanovic, M. Kreuzer, I. Polian, An algebraic fault attack on the led block cipher. Cryptology ePrint Archive, Report 2012/400, (2012), http://eprint.iacr.org/2012/400

[ZGZ+13] X. Zhao, S. Guo, F. Zhang, Z. Shi, C. Ma, T. Wang, Improving and evaluating differential fault analysis on LED with algebraic techniques, in *2013 Workshop on Fault Diagnosis and Tolerance in Cryptography*, pp. 41–51 (2013)

[ZZG+13] F. Zhang, X. Zhao, S. Guo, T. Wang, Z. Shi, Improved algebraic fault analysis: A case study on piccolo and applications to other lightweight block ciphers, in *Constructive Side-Channel Analysis and Secure Design - 4th International Workshop, COSADE 2013, Paris, France, March 6-8, 2013*, Revised Selected Papers, pp. 62–79 (2013), https://doi.org/10.1007/978-3-642-40026-1_5

[SBM15] S. Sarkar, S. Banik, S. Maitra, Differential fault attack against grain family with very few faults and minimal assumptions. IEEE Trans. Comput. **64**(6), 1647–1657 (2015), https://doi.org/10.1109/TC.2014.2339854

[DCA+15] P. Dey, A. Chakraborty, A. Adhikari, D. Mukhopadhyay, Improved practical differential fault analysis of grain-128, in *Proceedings of the 2015 Design, Automation & Test in Europe Conference & Exhibition, DATE 2015, Grenoble, France, March 9-13, 2015*, pp. 459–464 (2015), http://dl.acm.org/citation.cfm?id=2755858

[MSB+15] S. Maitra, S. Sarkar, A. Baksi, P. Dey, Key recovery from state information of sprout: application to cryptanalysis and fault attack. IACR Cryptology ePrint Archive, 2015:236 (2015), http://eprint.iacr.org/2015/236 (page: 70)

[SSM+17] A. Siddhanti, S. Sarkar, S. Maitra, A. Chattopadhyay, Differential fault attack on grain v1, ACORN v3 and lizard, in *Security, Privacy, and Applied Cryptography Engineering - 7th International Conference, SPACE 2017, Goa, India, December 13-17, 2017, Proceedings*, pp. 247–263 (2017), https://doi.org/10.1007/978-3-319-71501-8

[ZFL18] X. Zhang, X. Feng, D. Lin, Fault attack on ACORN v3. Comput. J. **61**(8), 1166–1177 (2018) https://doi.org/10.1093/comjnl/bxy044

[DRS+16] P. Dey, R.S. Rohit, S. Sarkar, A. Adhikari, Differential fault analysis on tiaoxin and AEGIS family of ciphers, in *Security in Computing and Communications - 4th International Symposium, SSCC 2016, Jaipur, India, September 21-24, 2016, Proceedings*, pp. 74–86 (2016), https://doi.org/10.1007/978-981-10-2738-3%5C_7

[BGN05] E. Biham, L. Granboulan, P.Q. Nguyen, Impossible fault analysis of RC4 and differential fault analysis of RC4, in *Fast Software Encryption: 12th International Workshop, FSE 2005, Paris, France, February 21–23, 2005*, Revised Selected Papers, pp. 359–367 (2005), https://doi.org/10.1007/11502760_24

[PY06] R.C. Phan, S. Yen, Amplifying side-channel attacks with techniques from block cipher cryptanalysis, in *Smart Card Research and Advanced Applications, 7th IFIP WG 8.8/11.2 International Conference, CARDIS 2006, Tarragona, Spain, April 19–21, 2006, Proceedings*, pp. 135–150 (2006), https://doi.org/10.1007/11733447_10

[LGL+12] Z. Liu, D. Gu, Y. Liu, W. Li, Linear fault analysis of block ciphers, in *Applied Cryptography and Network Security - 10th International Conference, ACNS 2012, Singapore, June 26-29, 2012. Proceedings*, pp. 241–256 (2012), https://doi.org/10.1007/978-3-642-31284-7_15

[BK06] J. Blomer, V. Krummel. Fault based collision attacks on AES, in *Fault Diagnosis and Tolerance in Cryptography, Third International Workshop, FDTC 2006, Yokohama, Japan, October 10, 2006, Proceedings*, pp. 106–120 (2006), https://doi.org/10.1007/11889700%5C_11

[Hem04] L. Hemme. A differential fault attack against early rounds of (triple-)des, in *Cryptographic Hardware and Embedded Systems - CHES 2004: 6th International Workshop Cambridge, MA, USA, August 11-13, 2004. Proceedings*, pp. 254–267 (2004), https://doi.org/10.1007/978-3-540-28632-5_19

[CGV08] C. Clavier, B. Gierlichs, I. Verbauwhede, Fault analysis study of IDEA, in *Topics in Cryptology - CT-RSA 2008, The Cryptographers' Track at the RSA Conference 2008, San Francisco, CA, USA, April 8-11, 2008. Proceedings*, pp. 274–287 (2008), https://doi.org/10.1007/978-3-540-79263-5_17

[SC16] D. Saha, D.R. Chowdhury, Encounter: on breaking the nonce barrier in differential fault analysis with a case-study on PAEQ, in *Cryptographic Hardware and Embedded Systems - CHES 2016 - 18th International Conference, Santa Barbara, CA, USA, August 17-19, 2016, Proceedings*, pp. 581–601 (2016), https://doi.org/10.1007/978-3-662-53140-2

[SC17] D. Saha, D.R. Chowdhury, Internal differential fault analysis of parallelizable ciphers in the counter-mode. J. Cryptogr. Eng. (2017) https://doi.org/10.1007/s13389-012-0031-5

[YJ00] S. Yen, M. Joye, Checking before output may not be enough against faultbased cryptanalysis. IEEE Trans. Comput. **49**(9), 967–970 (2000) https://doi.org/10.1109/12.869328

[YKL+01] S. Yen, S. Kim, S. Lim, S. Moon, A countermeasure against one physical cryptanalysis may benefit another attack, in *Information Security and Cryptology - ICISC 2001, 4th International Conference Seoul, Korea, December 6-7, 2001, Proceedings*, pp. 414–427 (2001), https://doi.org/10.1007/3-540-45861-1%5C_31

[JQY+02] M. Joye, J. Quisquater, S. Yen, M. Yung, Observability analysis - detecting when improved cryptosystems fail, in *Topics in Cryptology - CT-RSA 2002, The Cryptographer's Track at the RSA Conference, 2002, San Jose, CA, USA, February 18-22, 2002, Proceedings*, pp. 17–29 (2002), https://doi.org/10.1007/3-540-45760-7%5C_2

[Cla07] C. Clavier, Secret external encodings do not prevent transient fault analysis, in *Cryptographic Hardware and Embedded Systems - CHES 2007, 9th International Workshop, Vienna, Austria, September 10-13, 2007, Proceedings*, pp. 181–194 (2007), https://doi.org/10.1007/978-3-540-74735-2_13

[ADY15] R. AlTawy, O. Duman, A.M. Youssef, Fault analysis of kuznyechik. IACR Cryptology ePrint Archive, 2015:347, (2015), http://eprint.iacr.org/2015/347

[Nah15] P. Schaumont, N.F. Ghalaty, B. Yuce. Analyzing the efficiency of biased fault based attacks. Cryptology ePrint Archive, Report 2015/663 (2015), https://eprint.iacr.org/2015/663

[RAD19'2] K. Ramezanpour, P. Ampadu, W. Diehl, Fima: fault intensity map analysis, in *International Workshop on Constructive Side-Channel Analysis and Secure Design* (Springer, 2019), pp. 63–79

[RAD19'1] K. Ramezanpour, P. Ampadu, W. Diehl, A statistical fault analysis methodology for the ascon authenticated cipher, in *IEEE International Symposium on Hardware Oriented Security and Trust, HOST 2019, McLean, VA, USA, May 5-10, 2019*, pp. 41–50 (2019), https://doi.org/10.1109/HST.2019.8741029

[GYS15] N.F. Ghalaty, B. Yuce, P. Schaumont, Differential fault intensity analysis on PRESENT and LED block ciphers, in *Constructive Side-Channel Analysis and Secure Design - 6th International Workshop, COSADE 2015, Berlin, Germany, April 13-14, 2015*. Revised Selected Papers, pp. 174–188 (2015), https://doi.org/10.1007/978-3-319-21476-4%5C_12

[LRD+12] R. Lashermes, G. Reymond, J. Dutertre, J.J.A. Fournier, B. Robisson, A. Tria, A DFA on AES based on the entropy of error distributions, in *2012 Workshop on Fault Diagnosis and Tolerance in Cryptography, Leuven, Belgium, September 9,2012*, pp. 34–43 (2012), https://doi.org/10.1109/FDTC.2012.18

[FJL+13] T. Fuhr, É. Jaulmes, V. Lomné, A. Thillard, Fault attacks on AES with faulty ciphertexts only, in *2013 Workshop on Fault Diagnosis and Tolerance in Cryptography, Los Alamitos, CA, USA, August 20, 2013*, pp. 108–118 (2013), https://doi.org/10.1109/FDTC.2013.18

[LSG+10] Y. Li, K. Sakiyama, S. Gomisawa, T. Fukunaga, J. Takahashi, K. Ohta, Fault sensitivity analysis, in *Cryptographic Hardware and Embedded Systems, CHES 2010, 12th International Workshop, Santa Barbara, CA, USA, August 17-20, 2010. Proceedings*, pp. 320–334 (2010), https://doi.org/10.1007/978-3-642-15031-9_22

[GYT+14] N.F. Ghalaty, B. Yuce, M.M.I. Taha, P. Schaumont, Differential fault intensity analysis, in *2014 Workshop on Fault Diagnosis and Tolerance in Cryptography*, pp. 49–58 (2014) (page: 74)

[GPT19] M. Gruber, M. Probst, M. Tempelmeier, Statistical ineffective fault analysis of GIMLI. CoRR, abs/1911.03212 (2019), arXiv: 1911.03212

[ZLZ+18] F. Zhang, X. Lou, X. Zhao, S. Bhasin, W. He, R. Ding, S. Qureshi, K. Ren, Persistent fault analysis on block ciphers. IACR Trans. Cryptogr. Hardw. Embed. Syst. 150–172 (2018)

[CB19] A. Caforio, S. Banik, A study of persistent fault analysis, in *Security, Privacy, and Applied Cryptography Engineering - 9th International Conference, SPACE 2019, Gandhinagar, India, December 3-7, 2019, Proceedings*, pp. 13–33 (2019), https://doi.org/10.1007/978-3-030-35869-3

[ZZY+19] Y. Zhang, F. Zhang, B. Yang, G. Xu, B. Shao, X. Zhao, K. Ren, Persistent fault injection in fpga via bram modification, in *2019 IEEE Conference on Dependable and Secure Computing (DSC)*, pp. 1–6 (2019) (page: 75)

[BHB19] J. Breier, X. Hou, S. Bhasin (eds.) *Automated Methods in Cryptographic Fault Analysis*, 1st edn. (Springer, 2019). ISBN: 978-3-030-11332-2

[ZGZ+16] F. Zhang, S. Guo, X. Zhao, T. Wang, J. Yang, F. Standaert, D. Gu, A framework for the analysis and evaluation of algebraic fault attacks on lightweight block ciphers. IEEE Trans. Inf. Forens. Secur. **11**(5), 1039–1054 (2016), https://doi.org/10.1109/TIFS.2016.2516905

[KRH17] P. Khanna, C. Rebeiro, A. Hazra, Xfc: a framework for exploitable fault characterization in block ciphers, in *Design Automation Conference (DAC), 2017 54th ACM/EDAC/IEEE* (IEEE, 2017), pp. 1–6

[SKM+17] S. Saha, U. Kumar, D. Mukhopadhyay, P. Dasgupta, Differential fault analysis automation. IACR Cryptology ePrint Archive, 2017:673 (2017), http://eprint.iacr.org/2017/673

[SJP+19] S. Saha, D. Jap, S. Patranabis, D. Mukhopadhyay, S. Bhasin, P. Dasgupta, Automatic characterization of exploitable faults: A machine learning approach. IEEE Trans. Inf. Forens. Secur. **14**,(4), 954–968 (2019), https://doi.org/10.1109/TIFS.2018.2868245

[SMD18] S. Saha, D. Mukhopadhyay, P. Dasgupta, Expfault: an automated framework for exploitable fault characterization in block ciphers. IACR Trans. Cryptogr. Hardw. Embed. Syst. **2018**(2), 242–276 (2018), https://doi.org/10.13154/tches.v2018.i2.242-276

[SDA+17] S. Sarkar, P. Dey, A. Adhikari, S. Maitra, Probabilistic signature based generalized framework for differential fault analysis of stream ciphers. Cryptogr. Commun. **9**(4), 523–543 (2017), https://doi.org/10.1007/s12095-016-0197-2

[BSS+20] A. Baksi, S. Sarkar, A. Siddhanti, R. Anand, A. Chattopadhyay, Fault location identification by machine learning. Cryptology ePrint Archive, Report 2020/717 (2020), https://eprint.iacr.org/2020/717

[BHL18] J. Breier, X. Hou, Y. Liu, Fault attacks made easy: differential fault analysis automation on assembly code. IACR Trans. Cryptogr. Hardw. Embed. Syst. **2018**(2), 96–122 (2018)

[HBZ+19] X. Hou, J. Breier, F. Zhang, Y. Liu, Fully automated differential fault analysis on software implementations of block ciphers. IACR Trans. Cryptogr. Hardw. Embed. Syst. 1–29 (2019)

[BGE+17] J. Burchard, M. Gay, A.-S. M. Ekossono, J. Horcek, B. Becker, T. Schubert, M. Kreuzer, I. Polian, Autofault: towards automatic construction of algebraic fault attacks, in *2017 Workshop on Fault Diagnosis and Tolerance in Cryptography (FDTC)* (IEEE, 2017), pp. 65–72

[GPU+19] M. Gay, T. Paxian, D. Upadhyaya, B. Becker, I. Polian, Hardware-oriented algebraic fault attack framework with multiple fault injection support, in *2019 Workshop on Fault Diagnosis and Tolerance in Cryptography, FDTC 2019, Atlanta, GA, USA, August 24, 2019*, pp. 25–32 (2019), https://doi.org/10.1109/FDTC.2019.00012

[BCN+04] H. Bar-El, H. Choukri, D. Naccache, M. Tunstall, C. Whelan, The sorcerer's apprentice guide to fault attacks. IACR Cryptology ePrint Archive, 2004:100 (2004), http://dblp.uni-trier.de/db/journals/iacr/iacr2004.html#Bar-ElCNTW04

[ZDT+14] L. Zussa, A. Dehbaoui, K. Tobich, J.-M. Dutertre, P. Maurine, L. Guillaume-Sage, J. Clediere, A. Tria, Efficiency of a glitch detector against electromagnetic fault injection, in *Design, Automation and Test in Europe Conference and Exhibition (DATE), 2014* (IEEE, 2014), pp. 1–6

[HBB+16'2] W. He, J. Breier, S. Bhasin, N. Miura, M. Nagata, Ring oscillator under laser: potential of pll-based countermeasure against laser fault injection, in *2016 Workshop on Fault Diagnosis and Tolerance in Cryptography (FDTC)* (IEEE, 2016), pp. 102–113

[HBB16] W. He, J. Breier, S. Bhasin, Cheap and cheerful: a low-cost digital sensor for detecting laser fault injection attacks, in *Security, Privacy, and Applied Cryptography Engineering - 6th International Conference, SPACE 2016, Hyderabad, India, December 14-18, 2016, Proceedings*, pp. 27–46 (2016), https://doi.org/10.1007/978-3-319-49445-6_2

[BSS20] A. Baksi, D. Saha, S. Sarkar, To infect or not to infect: A critical analysis of infective countermeasures in fault attacks. J. Cryptogr. Eng. (2020), https://link.springer.com/article/10.1007%2Fs13389-020-00224-9

[SHS16] B. Selmke, J. Heyszl, G. Sigl, Attack on a DFA protected AES by simultaneous laser fault injections, in *2016 Workshop on Fault Diagnosis and Tolerance in Cryptography, FDTC 2016, Santa Barbara, CA, USA, August 16, 2016*, pp. 36–46 (2016), https://doi.org/10.1109/FDTC.2016.16

[BBK+10] A. Barenghi, L. Breveglieri, I. Koren, G. Pelosi, F. Regazzoni, Countermeasures against fault attacks on software implemented AES: effectiveness and cost, in *Proceedings of the 5th Workshop on Embedded Systems Security, WESS 2010, Scottsdale, AZ, USA, October 24, 2010*, p. 7 (2010), http://doi.acm.org/10.1145/1873548.1873555

[BJB16] J. Breier, D. Jap, S. Bhasin, The other side of the coin: analyzing software encoding schemes against fault injection attacks, in *2016 IEEE International Symposium on Hardware Oriented Security and Trust, HOST 2016, McLean, VA, USA, May 3-5, 2016*, pp. 209–216 (2016), https://doi.org/10.1109/HST.2016.7495584

[LRT12] V. Lomné, T. Roche, A. Thillard, On the need of randomness in fault attack countermeasures - application to AES, in *2012 Workshop on Fault Diagnosis and Tolerance in Cryptography, Leuven, Belgium, September 9, 2012*, pp. 85–94 (2012), https://doi.org/10.1109/FDTC.2012.19

[SMG16] T. Schneider, A. Moradi, T. Güneysu, Parti - towards combined hardware countermeasures against side-channel and fault-injection attacks, in *Advances in Cryptology CRYPTO2016: 36th Annual International Cryptology Conference, Santa Barbara, CA, USA, August 14-18, 2016, Proceedings, Part II* ed. by M. Robshawand J. Katz (Springer, Berlin, Heidelberg, 2016), pp. 302–332. ISBN: 978-3-662-53008-5, https://doi.org/10.1007/978-3-662-53008-5_11

[BH17] J. Breier, X. Hou, Feeding two cats with one bowl: on designing a fault and side-channel resistant software encoding scheme, in *Topics in Cryptology - CT-RSA 2017 - The Cryptographers' Track at the RSA Conference 2017, San Francisco, CA, USA, February 14-17, 2017, Proceedings*, pp. 77–94 (2017), https://doi.org/10.1007/978-3-319-52153-4%5C_5

[BHL19] J. Breier, X. Hou, Y. Liu, On evaluating fault resilient encoding schemes in software. IEEE Trans. Depend. Secur. Comput. (2019)

[KGK+18] B. Karp, M. Gay, O. Keren, I. Polian, Detection and correction of malicious and natural faults in cryptographic modules, in *PROOFS 2018, 7th International Workshop on Security Proofs for Embedded Systems, colocated with CHES 2018, Amsterdam, The Netherlands, September 13, 2018*, pp. 68–82 (2018), http://www.easychair.org/publications/paper/zMjh

[SBG+09] N. Selmane, S. Bhasin, S. Guilley, T. Graba, J. Danger, WDDL is protected against setup time violation attacks, in *Sixth International Workshop on Fault Diagnosis and Tolerance in Cryptography, FDTC 2009, Lausanne, Switzerland, 6 September 2009*, pp. 73–83 (2009), https://doi.org/10.1109/FDTC.2009.40

[TBM14] H. Tupsamudre, S. Bisht, D. Mukhopadhyay, Destroying fault invariant with randomization, in *International Workshop on Cryptographic Hardware and Embedded Systems* (Springer, 2014), pp. 93–111

[GSS+15] S. Ghosh, D. Saha, A. Sengupta, D.R. Chowdhury, Preventing fault attacks using fault randomization with a case study on AES, in *Information Security and Privacy - 20th Australasian Conference, ACISP 2015, Brisbane, QLD, Australia, June 29 - July 1, 2015, Proceedings*, pp. 343–355 (2015), https://doi.org/10.1007/978-3-319-19962-7_20

[GST12] B. Gierlichs, J. Schmidt, M. Tunstall, Infective computation and dummy rounds: fault protection for block ciphers without check-before-output, in *Progress in Cryptology - LATINCRYPT 2012 - 2nd International Conference on Cryptology and Information Security in Latin America, Santiago, Chile, October 7-10, 2012. Proceedings*, pp. 305–321 (2012), https://doi.org/10.1007/978-3-642-33481-8_17

[BG13] A. Battistello, C. Giraud, Fault analysis of infective AES computations, in *2013 Workshop on Fault Diagnosis and Tolerance in Cryptography, Los Alamitos, CA, USA, August 20, 2013*, pp. 101–107 (2013), https://doi.org/10.1109/FDTC.2013.12

[GSS+17] S. Ghosh, D. Saha, A. Sengupta, D.R. Chowdhury, Preventing fault attacks using fault randomisation with a case study on AES. ACT **3**(3), 225–235 (2017), https://doi.org/10.1504/IJACT.2017.10007295

[BB15] S. Banik, A. Bogdanov, Cryptanalysis of two fault countermeasure schemes, in *Progress in Cryptology - INDOCRYPT 2015 - 16th International Conference on Cryptology in India, Bangalore, India, December 6-9, 2015, Proceedings*, pp. 241–252 (2015), https://doi.org/10.1007/978-3-319-26617-6_13

[PCM15] S. Patranabis, A. Chakraborty, D. Mukhopadhyay, Fault tolerant infective countermeasure for AES, in *Security, Privacy, and Applied Cryptography Engineering - 5th International Conference, SPACE 2015, Jaipur, India, October 3-7, 2015, Proceedings*, pp. 190–209 (2015), https://doi.org/10.1007/978-3-319-24126-5_12

[FCL+20] J. Feng, H. Chen, Y. Li, Z. Jiao, W. Xi, A framework for evaluation and analysis on infection countermeasures against fault attacks. IEEE Trans. Inf. Forens. Secur. **15**, 391–406 (2020). ISSN: 1556-6021

[MHE+14] N. Moro, K. Heydemann, E. Encrenaz, B. Robisson, Formal verification of a software countermeasure against instruction skip attacks. J. Cryptogr. Eng. **4**(3), 145–156 (2014), https://doi.org/10.1007/s13389-014-0077-7

[PYG+16] C. Patrick, B. Yuce, N.F. Ghalaty, P. Schaumont, Lightweight fault attack resistance in software using intra-instruction redundancy, in *Selected Areas in Cryptography - SAC 2016 - 23rd International Conference, St. John's, NL, Canada, August 10-12, 2016*, Revised Selected Papers, pp. 231–244 (2016), https://doi.org/10.1007/978-3-319-69453-5%5C_13

[MSG+10] M. Medwed, F. Standaert, J. Grodl, F. Regazzoni, Fresh re-keying: security against side-channel and fault attacks for low-cost devices, in *Progress in Cryptology - AFRICACRYPT 2010, Third International Conference on Cryptology in Africa, Stellenbosch, South Africa, May 3-6, 2010. Proceedings*, pp. 279–296 (2010), https://doi.org/10.1007/978-3-642-12678-9_17

[MPR+11] M. Medwed, C. Petit, F. Regazzoni, M. Renauld, F. Standaert, Fresh re-keying II: securing multiple parties against side-channel and fault attacks, in *Smart Card Research and Advanced Applications - 10th IFIP WG 8.8/11.2 International Conference, CARDIS 2011, Leuven, Belgium, September 14-16, 2011*, Revised Selected Papers (2011), pp. 115–132, https://doi.org/10.1007/978-3-642-27257-8_8

[DEM+14] C. Dobraunig, M. Eichlseder, S. Mangard, F. Mendel, On the security of fresh re-keying to counteract side-channel and fault attacks, in *Smart Card Research and Advanced Applications - 13th International Conference, CARDIS 2014, Paris, France, November 5-7, 2014*. Revised Selected Papers (2014), pp. 233–244, https://doi.org/10.1007/978-3-319-16763-3_14

[DKM+15] C. Dobraunig, F. Koeune, S. Mangard, F. Mendel, F. Standaert, Towards fresh and hybrid re-keying schemes with beyond birthday security, in *Smart Card Research and Advanced Applications - 14th International Conference, CARDIS*

2015, Bochum, Germany, November 4-6, 2015. Revised Selected Papers (2015), pp. 225–241, https://doi.org/10.1007/978-3-319-31271-2_14

[PRM16] S. Patranabis, D.B. Roy, D. Mukhopadhyay, Using tweaks to design fault resistant ciphers, in *2016 29th International Conference on VLSI Design and 2016 15th International Conference on Embedded Systems (VLSID)* (IEEE, 2016), pp. 585–586

[GSD+10] S. Guilley, L. Sauvage, J. Danger, N. Selmane, Fault injection resilience, in *2010 Workshop on Fault Diagnosis and Tolerance in Cryptography, FDTC 2010, Santa Barbara, California, USA, 21 August 2010* (2010), pp. 51–65, https://doi.org/10.1109/FDTC.2010.15

[PCN+15] S. Patranabis, A. Chakraborty, P.H. Nguyen, D. Mukhopadhyay, A biased fault attack on the time redundancy countermeasure for AES, in *Constructive Side-Channel Analysis and Secure Design - 6th International Workshop, COSADE 2015, Berlin, Germany, April 13-14, 2015*. Revised Selected Papers (2015), pp. 189–203, https://doi.org/10.1007/978-3-319-21476-4%5C_13

[HBB+16'1] W. He, J. Breier, S. Bhasin, A. Chattopadhyay, Bypassing parity protected cryptography using laser fault injection in cyber-physical system, in *Proceedings of the 2nd ACM International Workshop on Cyber-Physical System Security* (ACM, 2016), pp. 15–21

[BJB17] J. Breier, D. Jap, S. Bhasin, A study on analyzing side-channel resistant encoding schemes with respect to fault attacks. J. Cryptogr. Eng. **7**(4), 311–320 (2017), https://doi.org/10.1007/s13389-017-0166-5

[SJB+18] S. Saha, D. Jap, J. Breier, S. Bhasin, D. Mukhopadhyay, P. Dasgupta, Breaking redundancy-based countermeasures with random faults and power side channel, in *2018 Workshop on Fault Diagnosis and Tolerance in Cryptography, FDTC 2018, Amsterdam, The Netherlands, September 13, 2018* (IEEE Computer Society, 2018), pp. 15-22. ISBN: 978-1-5386-8197-8. https://doi.org/10.1109/FDTC.2018.00011

[BG16] A. Battistello, C. Giraud, A note on the security of CHES 2014 symmetric infective countermeasure, in *Constructive Side-Channel Analysis and Secure Design - 7th International Workshop, COSADE 2016, Graz, Austria, April 14-15, 2016*, Revised Selected Papers, pp. 144–159 (2016), https://doi.org/10.1007/978-3-319-43283-0_9

[BKH+19] J. Breier, M. Khairallah, X. Hou, Y. Liu, A countermeasure against statistical ineffective fault analysis. Cryptology ePrint Archive, Report 2019/515 (2019), https://eprint.iacr.org/2019/515

[SJR+20] S. Saha, D. Jap, D.B. Roy, A. Chakraborty, S. Bhasin, D. Mukhopadhyay, A framework to counter statistical ineffective fault analysis of block ciphers using domain transformation and error correction. IEEE Trans. Inf. Forens. Secur. **15**, 1905–1919 (2020). https://doi.org/10.1109/TIFS.2019.2952262

[DDE+19] J. Daemen, C. Dobraunig, M. Eichlseder, H. Gross, F. Mendel, R. Primas, Protecting against statistical ineffective fault attacks. Cryptology ePrint Archive, Report 2019/536 (2019), https://eprint.iacr.org/2019/536

[BKK+20'2] A. Baksi, V.B. Kumar, B. Karmakar, S. Bhasin, D. Saha, A. Chattopadhyay, A novel duplication based countermeasure to statistical ineffective fault analysis, in *Australasian Conference on Information Security and Privacy* (2020), https://link.springer.com/chapter/10.1007/978-3-030-55304-3_27

[SRM19] A.R. Shahmirzadi, S. Rasoolzadeh, A. Moradi, Impeccable circuits ii. Cryptology ePrint Archive, Report 2019/1369 (2019), https://eprint.iacr.org/2019/1369

Chapter 4
Side Channel Attack

4.1 Introduction and Background

The *Side Channel Attack* or *Side Channel Analysis* (SCA) makes use of the relation between the physical measurements during the execution and actual state of the cipher that depends on the secret key. Sources of side channel leakages include power consumption/electromagnetic (EM) leakage, timing, acoustics, etc.

Power and Electromagnetic Leakage

Modern electronic devices are based on Complementary Metal Oxide Semiconductor (CMOS) technology. In CMOS, the power consumption and electromagnetic emanation are two viable sources of leakage. The variation in such sources while executing a cryptographic operation has enabled the attacker to gain information on the secret key.

Power consumption in CMOS devices can be classified in three categories, namely *static*, *dynamic*, and *short-circuit*. Static power refers to the situation when the transistor does not switch its logic state. The amount of such power consumption is typically small. Short-circuit power consumption happens during the short period of time when the transistor switches its logic state. Dynamic power consumption, on the other hand, is the situation when the transistor switches its state from 0 to 1 or vice-versa. When this happens, electricity flows due to charging and discharging of capacitors. Dynamic power consumption is of particular interest in cryptography, as it directly relates between a device's internal status with the power consumption (which is easily observable from outside). A CMOS transistor consumes almost negligible power when it is static (holds a 1 or 0), but only consumes power (higher current) when there is a change in its state (either $1 \rightarrow 0$ or $0 \rightarrow 1$). This provides an intuition that the power consumption of the CMOS device is directly correlated

A. Baksi, *Classical and Physical Security of Symmetric Key Cryptographic Algorithms*, Computer Architecture and Design Methodologies,
https://doi.org/10.1007/978-981-16-6522-6_4

with the data (more precisely, some form of Hamming weight of the data) being processed in the device. For example, the power consumption reaches maximum during a $0 \rightarrow 1$ logic state transition.

Power consumption of a CMOS device is usually measured by measuring the voltage difference across a small resistor placed in the power path or the ground path. The current flowing through the ground path (through the resistor) is proportional to the voltage measured across it and thus provides information about the power consumption of the device. On the other hand, electromagnetic emanation results from current flows within the circuit. The current loops (V_{DD}—ground) in the circuit act as antennas and generates time varying electric and magnetic fields which are correlated with the data being computed inside the device. These EM emanations can be coupled to a near-field EM probe which can further be used to indirectly measure the current/power consumption of the target device.

There has been a large body of work done with respect to a number of aspects of side channel analysis such as development of theoretically sound techniques and countermeasures [BCO04, CZ06], development of holistic side channel evaluation strategies [BCD+13], demonstration of attack on real-world devices [OP11], and development of auxiliary techniques to improve side channel attacks [DSN+11]. It is important to note with respect to power side channel attacks is that the relation between the operated data and the power consumption is characterized using device-specific leakage models. The leakage model for any given device enables the attacker to model its power consumption provided the data being computed inside the device is known. We will focus on microcontrollers for the rest of this chapter as it remains our Device under Test (DuT).

There are multiple active units such as Arithmetic Logic Unit (ALU), decoder, bus controller among others during any operation occurring inside a microcontroller. It is well-known that the bus reads and writes leak the most among all the active units and thus activity on the microcontroller bus can be easily observed on the power traces. Whenever data on the bus changes from say value A to value B, the bus in state A undergoes a pre-charging process where all the bus lines are pulled to 1 before the bus lines change to state B. So, what can be seen on the power traces is the change of the bus lines from state A to all 1's state (Activity A) and back to state B (Activity B). Thus, each of the activities can be observed on the power traces and we can see that the power consumption/EM radiation corresponding to each activity (P_A, P_B) is directly correlated to the Hamming weight (HW) of A and B respectively. This can be denoted as $P_A \propto \mathrm{HW}(A)$ and $P_B \propto \mathrm{HW}(B)$. Thus, it is said that the microcontroller leaks according to the Hamming weight model.

Power-based side channel attacks can be broadly classified into two categories—

(i) attacks that require the knowledge of the leakage model,
(ii) attacks that work without the leakage model.

Among the attacks that are based on the knowledge of the leakage model, one of the most commonly used attacks is the *Correlation Power Analysis/Correlation Power Attack* (CPA) [BCO04]/Sect. 4.2.4. We utilize CPA to demonstrate key recovery attacks from XOR operations of the CHASKEY MAC in Sect. 4.3.

The analysis technique for the electromagnetic leakage is basically same as that of power analysis [Lom10, Chap. 2.2]. Hence, we use as the umbrella term to indicate both the power and EM analysis techniques.

Other Sources of Leakage

Other sources of side channel information leakage include:

- Acoustics
- Temperature
- Light emission
- Network traffic
- Data usage statistics
- Cache memory [OST05].

4.2 Power Analysis

As mentioned earlier, power consumption is among the most commonly used source for SCA. Here we present an idea on some of the common techniques to analyse the power traces. These techniques can be classified in to two heads, *non-profiling* and *profiling*. If the attacker has physical access to the DuT (an identical copy or temporarily prior to the attack), then she can employ profiling attacks (where she makes use of her knowledge of the device). Non-profiling attacks, on the other hand, can work directly on the DuT without any prior knowledge of the micro-architecture.

To see how the power analysis techniques work, let us consider a simplified case: $S(p \oplus k)$; S is typically an SBox, p is the plaintext and k is the secret key she wants to find out. Note that k is fixed, but she can access p (she can even choose p). The working principle can be described as follows:

1. Collect n traces t_i corresponding to each plaintext p_i for $i = 0, \ldots, n-1$.
2. Choose an appropriate model H, such as the Hamming weight model.
3. Compute the key hypotheses corresponding to each key guess:
$$h(\hat{k}, i) = H(S(p_i \oplus \hat{k})).$$
4. Determine a statistical distinguisher, D, which takes pairs of n-bit vectors and outputs a real number.
5. Compute the distinguisher for every key hypotheses:
$$D(\hat{k}) = D((t_0, h(\hat{k}, 0)), \ldots, (t_{n-1}, h(\hat{k}, n-1))).$$
6. Combine all values of $D(\hat{k})$ and deduce the best key guess.

Due to pre-charge logic, the power leakage model in software (H) is generally taken as Hamming weight. In hardware, the model is generally taken as the Hamming distance of state at consecutive clocks, i.e., the leakage is assumed proportional to the number of bits which change $1 \rightarrow 0$ or $0 \rightarrow 1$.

4.2.1 Simple Power Analysis

The maiden work [KJJ99] on power analysis introduces the concept of *Simple Power Analysis* (SPA). In this very basic technique, one power trace (or, a few) is used to deduce information on (say) one particular location of the state at a certain round. With advanced analysis methods at disposal, it basically functions as a proof-of-concept (e.g., in [SJB20, Sect. 5.5]) than a full-fledged attack.

4.2.2 Differential Power Analysis (DPA)

The authors in [KJJ99] introduce the concept of *Differential Power Analysis* (DPA). Under this model, the attacker chooses a proper attack model depending on the underlying architecture and implementation details. Then she makes hypotheses for the key in a divide-and-conquer fashion. That means, she breaks the key in to multiple sub-keys and tries brute-force search for these sub-keys. For each sub-key, she computes a graph. If her hypothesis is incorrect, then this graph looks more like random; whereas, for correct sub-key guess, she will observe a peak. This can be more clearly visualized in graphical form, where a band indicates incorrect sub-key guess, and a correct guess is indicated by a spike. She repeats the same procedure for each sub-key one by one, until she has sufficient information on the entire secret key.

DPA Contest

Launched in CHES 2008, the *DPA contest*[1] aims at a fair comparison among SCA techniques. Several winning criteria were defined (e.g., execution time, number of traces, memory footprint), and winners were declared. The first 3 versions of the contest are concluded, and at present the version 4 is officially open to the best of our knowledge.

[1] http://www.dpacontest.org/home/.

4.2.3 Template Attack

Template Attack [CRR02] is considered a powerful SCA technique, since it can theoretically work with only one trace. Once the attacker has a copy of the target device, she builds templates for different values for the key, where she characterizes the multivariate normal distribution associated with each key. Once this is done, she measures traces on the target device to match with the templates she built using the concept of maximum likelihood. However, the template building is a huge time and memory consuming process, hence some pre-processing on the raw traces is generally recommended.

4.2.4 Correlation Power Analysis (CPA)

Correlation Power Analysis (CPA) is an advanced version of DPA, which uses correlation coefficient and hence are easier to implement. As mentioned, the idea is first proposed in [BCO04] and is one of the most common attacks.

CPA works by constructing simulated power traces for known input data and various hypotheses of the key based on the knowledge of the leakage model. The attacker then uses a statistical distinguisher, Pearson's correlation coefficient in this case to correlate the traces (for the various hypotheses of each of the sub-keys, separately) with the actual traces observed by power analysis setup. This analysis returns a high (absolute) value for correlation at that point of the power trace where the target operation is taking place. The attacker thus essentially gains information about both the key and the location of the target operation within the power trace.

4.2.5 Countermeasures

The SCA countermeasures can be broadly classified into two categories, *hiding* and *masking*. Hiding countermeasures attempt to decrease the Signal to Noise Ratio (SNR). Here, the underlying algorithmic description of the cipher practically remains untouched. Masking countermeasures work by introducing randomized structures, which cancel each other at the end. Masking greatly depends on the cipher description. Table 4.1 shows the common schemes under each of the categories. For detailed discussion, one may refer to [MOP07], among other sources. To the best of our knowledge, most countermeasures can be broken if the attacker is able to take more side channel traces.

Table 4.1 Common schemes under hiding and masking

 Shuffling instructions Dummy operations Random jitters Altered clock frequency Hiding	 Boolean masking Arithmetic masking Masking

4.3 Case Study: Side Channel Analysis of CHASKEY

With the background information on SCA discussed already, here we show how such an attack is carried out in practice. For this purpose, we take the message authentication code (MAC) CHASKEY (see Sect. 2.4.4 for its description). Based on power and EM leakage[2] from the reference C implementation by the designers, we show how the initial XOR operation (where the message is XORed with the key) can be recovered. It is relatively easy to retrieve secret information if the initial or the final round is targeted (for an intermediate round, the computational cost for the same would be more). For each 8-bit sub-keys, we are able to narrow down the search space to 2 candidates only using CPA (Sect. 4.2.4), from which each sub-key can be uniquely retrieved by exhaustive search.

4.3.1 Practical Attack Setups

Power Analysis Setup

We demonstrate our power analysis attack over CHASKEY implemented in an ATMega328P, 8-bit microcontroller DuT housed in a custom-made power analysis evaluation board. We port the reference implementation of CHASKEY to our DuT and it is operated at a clock frequency of 20 MHz. The power measurements are collected using a Tektronix oscilloscope (TDS 2012C) at a sampling rate of 1 mega samples/second. The trace-acquisition process is synchronized with the operation of the DuT using a custom-made software running on a host computer. We collect 2500 sample points per trace.

---

[2] Visit https://bitbucket.org/anubhab001/chaskey-side-channel for our traces.

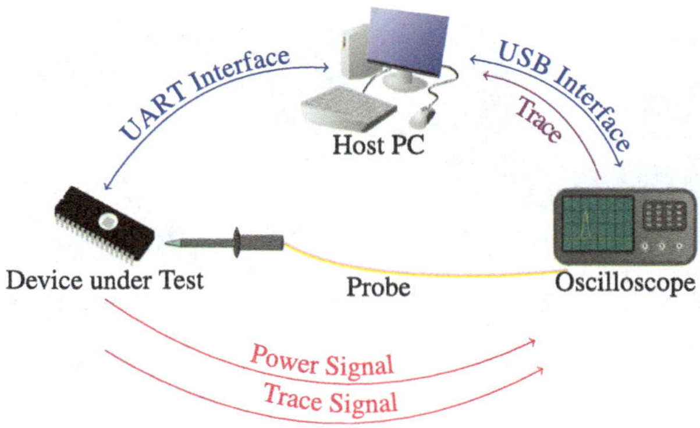

Fig. 4.1 Schematic of our power/electromagnetic trace collection setup

Electromagnetic Analysis Setup

We additionally mount our attack using EM leakage analysis, and our device under target in this case is the ARM Cortex-M4F, a 32-bit microcontroller. Similarly, we port the reference implementation to the STM32M4F07 discovery board (that houses the ARM microcontroller) running at a clock frequency of 24 MHz. The EM measurements are captured using a near-field probe from Riscure[3] and collected using the HDO6104A Lecroy oscilloscope at a sampling rate of 1.25 Giga samples/second. Here again, the trace-acquisition process is synchronized with the DuT using a custom trace-acquisition script running on a host computer. Here we sample each trace at 1250 discrete points. It may be noted that both the power and EM attacks work in a chosen message setting wherein the attacker provides message inputs of her choice to the MAC whose tags are to be generated using a single secret key.

In Fig. 4.1, we present the schematic diagram of the setups. Figure 4.2 shows the power analysis setup we used, and Fig. 4.3 shows the electromagnetic attack setup. For the sake of better clarity, we present a long and a close-up views of the equipment used.

4.3.2 Experimental Results

Since the reference C code works with 32-bits (`uint32_t` data type), the operations in the microcontrollers are carried out differently. Depending on this, the success of the key recovery varies. For the power attack setup, we can recover the first key byte. One such example is given in Fig. 4.4, where we plot the number of traces analysed versus the absolute correlation value. Also, we highlight the corresponding absolute correlation coefficient values for those two candidates, for which we consistently

[3] https://www.riscure.com/.

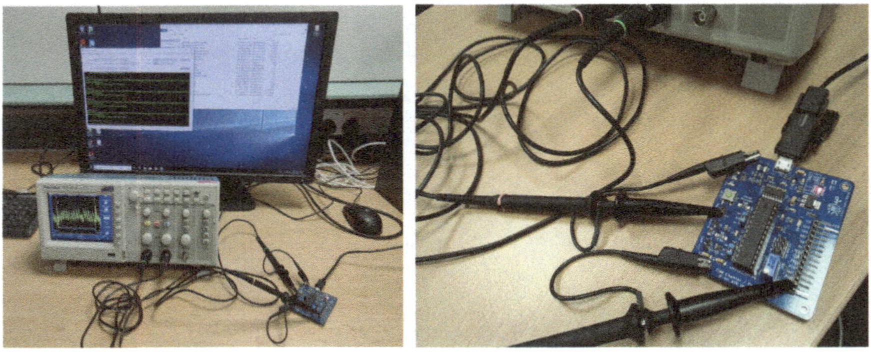

(a) Long shot (b) Close-up shot

Fig. 4.2 Power leakage analysis setup with ATmega board

(a) Long shot (b) Close-up shot

Fig. 4.3 Electromagnetic leakage analysis setup with ARM board

observe maximum absolute correlation for a large number of traces. Here, the possible candidates for the first sub-key (8-bits) are `83` and `7c`, as both have the same absolute correlation coefficient.

As we use a 32-bit microcontroller in the EM attack setup, the XOR operations that we target are carried out consecutively, from where we directly get the power traces. This enables us to recover all bytes one by one. For example, one may refer to Fig. 4.5. Here, the third and fourth key byte recovery are shown (Fig. 4.5a, b, respectively). As again, the two sub-key candidates, namely, `34` and `cb` (respectively, `33` and `cc`) have maximum absolute correlation coefficient for the third (respectively, fourth) byte of the key. We choose two slightly different ranges for the number of traces for these two bytes to show the overall nature of the maximum absolute correlation coefficient curve.

The justification for the ambiguity of the sub-keys under XOR can be found in [YO19]. It is described in [YO19] that, in a noise-free scenario, if x has correlation of ρ then $\vec{1} \oplus x$ has correlation of $-\rho$, for all x, where $\vec{1}$ is the Boolean vector (of suitable length) with all bits set at logic high.[4]

---

[4] The observation in [JB17] that the other candidate is random appears to be incorrect.

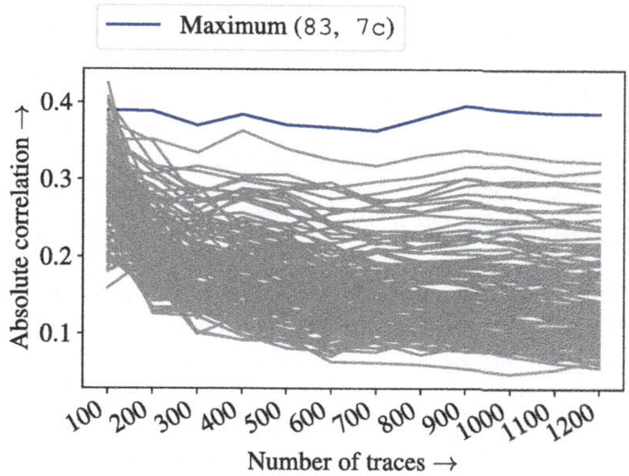

Fig. 4.4 CPA results for power leakage: First key byte

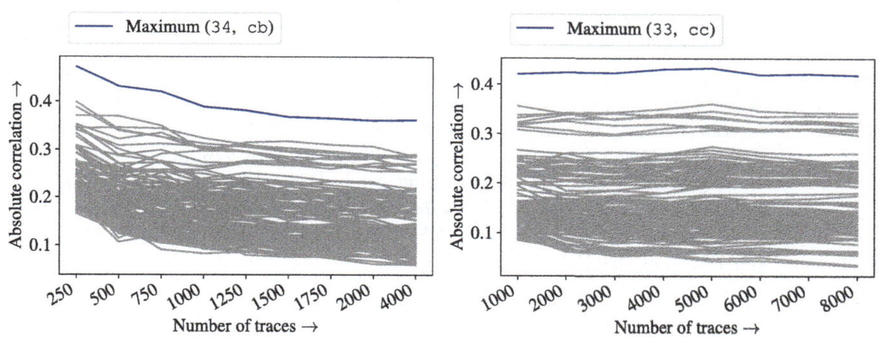

(a) Third key byte

(b) Fourth key byte

Fig. 4.5 CPA results for electromagnetic leakage

References

[BCO04] E. Brier, C. Clavier, F. Olivier, Correlation power analysis with a leakage model, in *Cryptographic Hardware and Embedded Systems - CHES 2004: 6th International Workshop Cambridge, MA, USA, August 11-13, 2004. Proceedings*, pp. 16–29 (2004). https://doi.org/10.1007/978-3-540-28632-5_2

[CZ06] Z. Chen, Y. Zhou, Dual-rail random switching logic: a countermeasure to reduce side channel leakage, in *International Workshop on Cryptographic Hardware and Embedded Systems* (Springer, 2006), pp. 242–254

[BCD+13] G. Becker, J. Cooper, E. DeMulder, G. Goodwill, J. Jaffe, G. Kenworthy, T. Kouzminov, A. Leiserson, M. Marson, P. Rohatgi, et al. Test vector leakage assessment (tvla) methodology in practice, in *International Cryptographic Module Conference*, vol. 1001, p. 13 (2013)

[OP11] D. Oswald, C. Paar, Breaking mifare desfire mf3icd40: power analysis and templates in the real world, in *International Workshop on Cryptographic Hardware and Embedded Systems*, (Springer, 2011), pp. 207–222

[DSN+11] N. Debande, Y. Souissi, M. Nassar, S. Guilley, T.-H. Le, J.-L. Danger, "Resynchronization by moments": an efficient solution to align side-channel traces, in *2011 IEEE*

International Workshop on Information Forensics and Security (WIFS) (IEEE, 2011), pp. 1–6

[Lom10] V. Lomne, Power and Electro-Magnetic Side-Channel Attacks: Threats and Countermeasures. Ph.D. thesis, Universite Montpellier II (2010), https://sites.google.com/site/victorlomne/research. Accessed on 2017-06-15

[OST05] D.A. Osvik, A. Shamir, E. Tromer, Cache attacks and countermeasures: the case of AES. IACR Cryptology ePrint Archive, 2005:271 (2005), http://eprint.iacr.org/2005/271

[KJJ99] P. C. Kocher, J. Jaffe, B. Jun, Differential power analysis, in *Advances in Cryptology - CRYPTO '99, 19th Annual International Cryptology Conference, Santa Barbara, California, USA, August 15-19, 1999, Proceedings*, pp. 388–3971 (1999). https://doi.org/10.1007/3-540-48405-1_25

[SJB20] S.M. Sim, D. Jap, S. Bhasin, Dapa: differential analysis aided power attack on (non-)linear feedback shift registers (extended version). Cryptology ePrint Archive, Report 2020/1241 (2020), https://eprint.iacr.org/2020/1241

[CRR02] S. Chari, J. R. Rao, P. Rohatgi, Template attacks, in *Cryptographic Hardware and Embedded Systems - CHES 2002, 4th International Workshop, Redwood Shores, CA, USA, August 13-15, 2002, Revised Papers*, pp. 13–28 (2002), https://doi.org/10.1007/3-540-36400-5_3

[MOP07] S. Mangard, E. Oswald, T. Popp, *Power Analysis Attacks - Revealing the Secrets of Smart Cards* (Springer, 2007). ISBN: 978-0-387-30857-9

[YO19] Y. Yan, E. Oswald, Examining the practical side channel resilience of arx-boxes. Cryptology ePrint Archive, Report 2019/335 (2019), https://eprint.iacr.org/2019/335

[JB17] B. Jungk, S. Bhasin. Don't fall into a trap: physical side-channel analysis of chacha20-poly1305, in *Design, Automation & Test in Europe Conference & Exhibition, DATE 2017, Lausanne, Switzerland, March 27-31, 2017*, pp. 1110–1115 (2017), https://doi.org/10.23919/DATE.2017.7927155

Chapter 5
New Insights on Differential and Linear Bounds Using Mixed Integer Linear Programming

Symmetric key cryptography is one of the cornerstones of security in the modern era of electronic communication. The symmetric key algorithms, known as the ciphers, are to satisfy certain requirements in order to be considered secure, which are broadly classified as Classical Attack and Physical Attack. We show new results in context of both the classical and physical attacks to advance the state of the art.

In classical attack, we first show an issue related to a common modelling using Mixed Integer Linear Programming (MILP). We provide a new MILP modelling that overcomes this issue and explore heuristic options to reduce the solution time taken by the MILP solver. Our analysis shows that the solution time can be improved nearly ten folds by using a proper heuristic. Second, we show how Machine Learning (ML) can be used as a generic tool in the analysis of the symmetric key ciphers. In the process, we demonstrate how the existing security notions (that do not use ML) underestimate the vulnerability of the ciphers. To the best of our knowledge, this is the first generic application of ML in this field.

In physical attack, we start with new mathematical results related to the Differential Fault Attack (DFA) from the point of view of the cipher designer. Next, we make use of these results to propose a cipher named DeFault, which has an in-built resistance against DFA. While all other methods to thwart DFA rely on some form of duplication, DeFault has an inherent protection against DFA that does not use duplication and hence is the first-of-its-kind. Third, we analyse the so-called Infective Countermeasure that is used as a duplication-based DFA countermeasure in more depth. We construct new schemes, show weakness of an existing scheme, and propose a simple patch to another scheme to fix its weakness, among other results. Last, we propose a low-cost countermeasure to a newly proposed fault model, named Statistical Ineffective Fault analysis (SIFA). In contrast to the existing SIFA countermeasures that rely on some form of triplication at its core, our countermeasure uses duplication and thus is the most cost-effective.

A. Baksi, *Classical and Physical Security of Symmetric Key Cryptographic Algorithms*, Computer Architecture and Design Methodologies,
https://doi.org/10.1007/978-981-16-6522-6_5

5.1 Introduction

Mixed Integer Linear Programming (MILP) is among the most frequently used tool in symmetric key cryptography, as evident from a large volume of research works [MWG+11, SHW+13, SHW+14, ZDY18, LWZ+19, JZD19, AST+17, ZZD+19]. More particularly, MILP is used to determine the bounds for differential cryptanalysis [BS90] and linear cryptanalysis [Mat93] for a lot of modern ciphers. MILP aided techniques generally give an opportunity to cover more rounds with more precision. Since the differential and linear cryptanalytic methods are essential for any cipher design, MILP aided techniques are frequently used in cipher design and further cryptanalysis, such as design of `GIFT` [BPP+17] (see Sect. 2.4.3 for description of `GIFT-128`) or the improved differential bounds on `GIFT` in [ZDY18]

In the process, the problem of differential/linear bounds for a (reduced round) cipher[1] is converted to an MILP instance, which is then solved using a standard solver. It is to be noted that the modelling from a differential/linear bound to an MILP instance can be done in various ways. As the efficiency of such a solver greatly depends on the formulated MILP instance, the research community has been active to find out optimal way for modelling. For example, the authors of [LWZ+19] experiment with heuristic techniques that can make the MILP solver can solve those instances in less time.

Apart from this, the MILP aided techniques are used in context of the classical impossible differential cryptanalysis [BGN05]. The MILP modelling for impossible differential is directly taken from that of the differential case [CJF+16, ST17'2]. Hence, any improvement in the modelling for the differential case will likely improve the impossible differential too.

Our Contributions

The authors of [SHW+13] have introduced a MILP modelling by incorporating the concept of *Convex Hull* (CH), which becomes quite popular in the literature [SHW+14, ZDY18, ZZ18, ST17'1]. It follows-up from Mouha et al.'s model [MWG+11].

We observe a problem related to this modelling which is not reported so far in the literature. We show the convex hull modelling only works when the SBox meets certain condition. Thus, applying the convex hull method to an arbitrary SBox may lead to incorrect results . The problem arises due to a property of the convex hull. In particular, it may happen that a point may be inside the hyper-volume of the convex hull while the convex hull is generated excluding that point. Variations of the CH model, such as [SHW+14], are also affected by this problem. Eventually, this

[1] Note that we only need to analyse the un-keyed permutation (which is described in Sect. 2.2.1), as the round key addition (and hence the key schedule) and the round constants addition (if applicable) do not play any role.

problem does not occur in common SBoxes, like `GIFT (1A4C6F392DB7508E)` [BPP+17] or `PRESENT (C56B90AD3EF84712)` [BKL+07].

In case the problem occurs, one may need to look for an alternate modelling, since apparently no variant of the convex hull modelling can properly work with such an SBox. To address this issue, i.e., to work with all SBoxes, we propose a different MILP modelling. This modelling works for both differential and linear cases and can give the exact bound (similar to [SHW+14]). Further, this model can be fine-tuned to use with impossible differential cases (such as [ST17'2]), or to find the minimum number of active SBoxes (used in the design of the `GIFT` family of block ciphers [BPP+17]), or iterative trails (like [ZDY18]).

On top of this, we propose heuristic methods to reduce the execution time. This is inspired from [LWZ+19], where the authors experiment with various heuristics to improve run time of the MILP instances. These heuristics create a new MILP instance, but do not alter the actual MILP problem (thus, the optimal bound remains unchanged). Using suitable heuristic may help the MILP solver to return the optimal solution faster.

We experiment with three main heuristics, along with its combinations. We show how one heuristic applies better to the differential cases, whereas the other applies better to the linear cases. We indeed show a few folds reduction of run time compared to the corresponding no heuristic cases. Interestingly, the idea for two of our main heuristics is actually taken from the convex hull modelling of [SHW+13]. The third idea makes use of the optimal bounds already available for smaller rounds.

For benchmark, we take the lightweight cipher `GIFT-128` [BPP+17], and run the MILP differential and linear instances for reduced rounds (Sect. 5.4.3). As mentioned, the problem associated with convex hull modelling does not appear for the `GIFT` SBox. Therefore, we are able to verify our results with existing MILP bounds on `GIFT` reported in the literature, namely, [SHW+14, LWZ+19, JZD19, ZZD+19]; and also with [LLL+19]. We also test with `GIFT-64` [BPP+17] and `PRESENT` [BKL+07] (a brief description of `PRESENT` can be found in Sect. 2.4.2); and report the optimal linear bounds for `GIFT-128` for 11 and 12 rounds for the first time in the literature.

5.2 Background

As previously mentioned, the problem of differential or linear bounds is converted to an MILP instance through a proper modelling. After this, this MILP instance is solved by some state-of-the-art solver, like Gurobi (see Sect. 1.4). For the sake of completeness, we briefly describe the relevant MILP models in the subsequent part.

This idea of using MILP is introduced by Mouha et al. [MWG+11]. The next major contribution comes from Sun et al. [SHW+13], where the convex hull (which is actually a concept in computation geometry [dBCvK+08, Mou16]) is used to form the linear constraints. This is quite useful as the code to find convex hull is implemented in the open source tool Sage (see Sect. 1.4). The authors of [SHW+13] also propose a heuristic algorithm to reduce the number of constraints returned by the convex hull, so that the MILP instance thus constructed becomes faster to solve (but it does not change the underlying problem of differential/linear bound). However, this claim is apparently not backed-up by any experimental result.

It may be noted that we actually model the Difference Distribution Table (DDT) [Sti06, Sect. 3.4]/Sect. 2.1.2 for the differential case, and the Linear Approximation Table (LAT) [Sti06, Sect. 3.3]/Sect. 2.1.2 for the linear case.[2] The convex hull method treats a given DDT/LAT as distinct points in a hyper-cube. Then, the hyper-dimensional convex hull of the desired points in the DDT/LAT is computed. Since a convex hull is represented by a set of linear inequalities, those can be directly used for the constraints for the MILP instance.[3] As a side note, it can be mentioned that the sign of the elements in the LAT does not play any role in finding the linear bound, hence the absolute LAT can be used for simplification purpose.

5.2.1 *Branch Number to Model SBox (Inscrypt'11)*

As mentioned, Mouha et al. first propose MILP modelling for differential and linear cases [MWG+11]. The corresponding bounds are found by finding the minimum number of active SBoxes.

Differential

The modelling for the differential case is described here. The first type of constraints comes from modelling an XOR operation (refer to Sect. 5.6.1 for an in-depth description). Denoting x_0, x_1 as the inputs to the XOR and y as the output bit, the following linear constraints model this operation (d is a dummy variable of type binary):

$$\begin{aligned} x_0 + x_1 - 2d &\geq 0; \\ d - x_i &\geq 0 \text{ for } i = 0, 1; \\ d - y &\geq 0. \end{aligned}$$

Next the constraints for the linear layer, L. For simplicity, we consider it in the binary non-singular matrix form, say, of dimension b. Assume, as before, $(x_0, x_1, \ldots, x_{b-1})$ as the input variables denoting the difference (i.e., if the ith bit of L is active then $x_i = 1$), and $(y_0, y_1, \ldots, y_{b-1})$ as the output variables denoting

---

[2] Any two SBoxes with identical DDT (respectively, absolute LAT) behave identically with respect to the differential (respectively, linear) attack, at least with respect to the currently existing attack models.

[3] The inequalities are not strict, i.e., of the type $\leq$ or $\geq$ (but not of the type $<$ or $>$). The MILP solvers generally cannot handle strict inequalities, hence the inequalities representing CH suits well for forming the constraints of MILP instances. See Sect. 2.8.2.

the difference. Given the DBN is D, a new dummy variable of type binary d is created. Finally the following linear constraints represent L:

$$\sum_{i=0}^{b-1} x_i + \sum_{i=0}^{b-1} y_i - D \cdot d \geq 0;$$

$$d - x_i \geq 0 \text{ for } i = 0, 1, \ldots, b-1;$$

$$d - y_i \geq 0 \text{ for } i = 0, 1, \ldots, b-1.$$

It is to be noted that the variable d in both cases indicates whether the corresponding operation is active. The objective function of the MILP problem is set to minimize the arithmetic sum of the d-variables which indicate the SBoxes.

Once the minimum number of active SBoxes is found, the bound for differential probability can be computed by assuming the transition in each SBox takes maximum probability. This generally amounts for a lower bound for the differential probability, since it is possible for the transition in several SBoxes to take some lower probability.

Linear

The modelling for the linear bound is similar to that of the differential. Therefore only the modifications are mentioned here for brevity: No constraint is needed for the XOR operation, and DBN is replaced by LBN.

5.2.2 *Convex Hull to Model SBox—Active SBox Count (Eprint'13)*

Convex Hull-Based Modelling

The authors of [SHW+13] propose to use CH, which converts the input difference—output difference relations of the DDT into linear constraints. This modelling is particularly useful for bit-oriented ciphers like `GIFT` [BPP+17]. Since this model, similar to that of [MWG+11], uses the number of active SBoxes; it is not possible to get the exact bound, as the maximum transition probability for each active SBox is assumed when computing the bound. Although not explored, a similar modelling would also work for the linear case.

Consider a $w \times v$ SBox where $(x_0, x_1, \ldots, x_{w-1})$ denotes the (non-zero) input difference vector, and $(y_0, y_1, \ldots, y_{v-1})$ denotes the (non-zero) output difference vector. A dummy variable A of type binary is created, which indicates whether the SBox is active ($A = 1$) or not ($A = 0$). The following constraints capture this property:

$$A - x_i \geq 0 \text{ for } i = 0, 1, \ldots, w - 1;$$
$$\sum_{i=0}^{w-1} x_i - A \geq 0.$$

Now the augmented vector $(x_0, x_1, \ldots, x_{w-1}, y_0, y_1, \ldots, y_{v-1})$ denote a (non-zero) input difference—(non-zero) output difference pattern in the DDT. Each of the vectors denotes a point in the hyper-dimension. Then the concept of CH is applied on the set of all of the augmented vectors to convert the set of hyper-dimensional points to non-strict linear inequalities.

Generally, the number of constraints returned by the mathematical tools[4] is in a few hundred for a 4×4 SBox. For example, the number of such constraints for the GIFT SBox is 237 [ZDY18, Section 3.3].

In order to reduce the number of linear constraints, which in turn is expected to reduce the execution time for the MILP solver; the authors of [SHW+13] also discuss a greedy algorithm. For example, the number of constraints for the GIFT SBox can be reduced to 21 [ZDY18, Section 4.2].

The basic idea of the greedy algorithm is to remove some of the linear constraints from the set of all linear constraints for the CH. To see how it works, consider two constraints l_0 and l_1. Among all the 2^{w+v} points in the hyper-dimension, suppose n_i of those are not satisfied by l_i, respectively, $i = 0, 1$. Those hyper-dimensional points which do not satisfy at least one constraint correspond to the 0 transitions of the DDT. Suppose, $n_0 > n_1$, then l_0 is given preference to be in the set of chosen constraints. This is done over all constraints.

In addition to the convex hull constraints, other constraints are also considered, as discussed here. If the SBox is bijective ($v = w$), the following two constraints can be inserted into the MILP instance:

$$w \sum_{i=0}^{w-1} y_i - \sum_{i=0}^{w-1} x_i \geq 0,$$
$$w \sum_{i=0}^{w-1} x_i - \sum_{i=0}^{w-1} y_i \geq 0.$$

The branch number-based constraints from [MWG+11] are used only if the branch number is > 2. The authors mention that the branch number-based constraints are redundant and hence are skipped if the branch number $= 2$ [SHW+13, Sect. 2.2].

[4] The only tool used in the literature so far is Sage, to the best of our knowledge.

Logical Condition Modelling

Additionally, the authors of [SHW+13] also mention another type of modelling, termed as the Logical Condition Modelling [SHW+13, Sect. 3]. While the authors report a relatively small number of constraints obtained by this modelling for the PRESENT SBox (C56B90AD3EF84712) [BKL+07] compared to that of the convex hull, this modelling is not applicable to a variety of SBoxes. As such, this modelling is not pursued in the follow-up research works, to the best of our knowledge. We omit the details of this modelling for simplicity.

5.2.3 Convex Hull to Model SBox—Exact Bound (Eprint'14)

Extending the works of [SHW+13], the authors of [SHW+14] incorporate the individual transitions into the hyper-dimensional points (by increasing the dimension). After this, the convex hull is computed as in [SHW+13]. As the individual transitions of DDT are modelled, now it is possible to find the exact bound, thereby improving the modelling from [SHW+13]. The greedy algorithm proposed in [SHW+13] is used to reduce the number of linear constraints in the CH.

To see how this modelling works, we adopt (from [ZDY18, Sect. 4.2]) the example of the 4×4 SBox used in GIFT, 1A4C6F392DB7508E. There are 4 non-zero transitions in the DDT, namely, (16, 6, 4, 2). Each of those non-zero transitions are modelled respectively by a three-dimensional point. In particular, the 16 transition (corresponds to $2^{\log_2 16/16} = 1$-probability) is modelled by (0, 0, 0); the 6 transition (corresponds to $2^{\log_2 6/16} \approx 2^{-1.415}$-probability) is modelled by (0, 0, 1); the 4 transition (corresponds to $2^{\log_2 4/16} = 2^{-2}$-probability) is modelled by (0, 1, 0), and finally the 2 transition (corresponds to $2^{\log_2 2/16} = 2^{-3}$-probability) is modelled by (1, 0, 0). Now the eight-dimensional vectors $(x_0, x_1, x_2, x_3, y_0, y_1, y_2, y_3)$, which indicate the (non-zero) input difference—(non-zero) output difference relations of this SBox, are augmented with the corresponding three-dimensional vectors indicate the individual transitions. Thus now we have points over the binary eleven-dimensional space, on which the convex hull is computed. Finally, the objective function is set to minimize $\sum_{i=0,(p_0,p_1,p_2)=(2,4,6)}^{2} \left(\log_2 \frac{p_i}{16} \times q_i\right)$, where (q_0, q_1, q_2) is the augmented vector (denoting the individual transitions). The exact complexity for differential distinguisher is calculated by raising the result of the MILP solver to the power of 2.

5.2.4 *Redundant Constraints to Reduce Solution Time (Eprint'19)*

On top of the MILP modelling proposed in [SHW+14], the authors of [LWZ+19] experiment by altering the following parameters (in such a way that the solution to the MILP problem remains unchanged), as given next.

1. **Number of constraints.** Insert redundant constraint (i.e., this constraint is satisfied given other constraints) to the MILP instance. This will not change the solution but will likely influence the solver's run time.
2. **Ordering of constraints.** The ordering of the constraints given by the code for convex hull can be altered. Similar to the previous case, the solution will not change.
3. **Ordering of variables.** The linear constraints can be written by changing the variables. For example, the constraint $a_0x_0 + a_1x_1 + \cdots + a_{n-1}x_{n-1} \geq b$ can be written as, $a_1x_1 + \cdots + a_{n-1}x_{n-1} + a_0x_0 \geq b$.

The authors show that all three types of heuristics influence the search strategy of Gurobi. Thus, by carefully studying the effect of those heuristics on run time of Gurobi, it is possible to create a new instance of the MILP problem which is faster to solve (but has the same optimal solution).

5.3 Problem with Convex Hull Modelling

In this part, we describe our finding on the convex hull modelling used to find differential and linear bounds. For simplicity, we only consider the modelling from [SHW+13], i.e., without probability encoding for individual transitions.

As noted already, the convex hull method solves the problem of converting a DDT/LAT to MILP-compatible format. It maps a set of binary hyper-dimensional points to a system of non-strict linear inequalities with real coefficients. Thus, all the points will satisfy the corresponding linear constraints which describe the convex hull.

However, there is no check to ascertain a point not in the set of hyper-dimensional points does not satisfy all the linear constraints. As a convex hull has certain other properties, it may happen that a particular point which is not within the set of points (which is used to create the convex hull) still satisfies all the inequalities that govern the convex hull. In other words, the hyper-volume created by some other points includes this point. Thus, the convex hull model will take this point as a valid point (i.e., as if this point belongs to the set of points based on which the convex hull is generated). If used in the MILP instance, the solver will consider the point (which is outside the set of points for which the CH is generated, but inside the hyper-volume of the CH) as valid. This can lead to wrong results. Indeed, our experiments confirm that this actually occurs to a number of SBoxes.

For a simple example, consider the 4 × 4 trivial SBox: 0123456789ABCDEF. There are 16 non-zero transitions in its DDT. Each being 16, are at the diagonal: (0, 0), (1, 1), (2, 2), (3, 3), (4, 4), (5, 5), (6, 6), (7, 7), (8, 8), (9, 9), (a, a), (b, b), (c, c), (d, d), (e, e), (f, f). Thus the convex hull is generated for the 16 eight-dimensional points. This is given by the following eight linear inequalities with dummy variables $z_0, z_1, \ldots, z_7$ (each of type binary), as returned by Sage:

$$- z_4 + 1 \geq 0, -z_5 + 1 \geq 0, -z_6 + 1 \geq 0, -z_7 + 1 \geq 0,$$
$$+ z_4 \geq 0, + z_5 \geq 0, + z_6 \geq 0, + z_7 \geq 0.$$

Since the corner points of the eight-dimensional cube are used to create the CH, it inherently contains all other points, i.e., all 256 points $\in \mathbb{F}_2^8$ satisfy all the inequalities of the CH. This is due to the following property: For any two points **u** and **v** in convex hull, any linear combination of **u** and **v** is also in the convex hull.

Out of those 256 points which satisfy the generated CH, 16 are of the form: $(0, 0, 0, 0, y_0, y_1, y_2, y_3)$ or $(x_0, x_1, x_2, x_3, 0, 0, 0, 0)$. Hence those points can be caught by proper modelling; such as $\sum_{i=0}^{3} x_i - 1 \geq 0$, $\sum_{i=0}^{3} y_i - 1 \geq 0$. The rest 240 points (none of which is at the diagonal of the DDT), correspond to zero transitions of the DDT; yet the convex hull model cannot capture it. As a result, the MILP instance will consider those points as non-zero transitions.

Note that, the modelling in [SHW+13] does not explicitly check whether any zero transition satisfies all the inequalities of the convex hull. Thus, the problem will go unnoticed, if an extra cross-check is not used. So far, this step seems missing in the literature.

Since those points with zero transitions satisfy all the inequalities for the CH, the same will hold after few inequalities are removed (as per the greedy algorithm in [SHW+13]/Sect. 5.2.2). The greedy approach starts with the set of the zero transitions, and then checks which inequality removes the maximum number of zero transitions. However, since all zero transitions satisfy all the inequalities, the set of points which do not satisfy at least of the inequalities is null, and hence, this step will be trivially skipped.

Here we present a few typical SBoxes with the same undesirable CH property: Some zero transitions in DDT or LAT lie within the hyper-volume of the convex hull (which is generated by points with non-zero transitions). We take the representatives of each of the 302 AE (see Sect. 2.1.2) classes from [DeC07, Chapter 5.4.2]. The results are summarized in Table 5.1 (Table 5.1(a) for differential and Table 5.1(b) for linear). Out of the 302 SBoxes tested, 49 show undesired property for differential and 13 for linear. For instance, with the AE representative #245 of [DeC07] (40132567E8A9CDBF), the non-zero transitions in the DDT are 16, 6, 4, 2. See Table 5.2 for its DDT. The number of points with non-zero transitions in its DDT is

Table 5.1 Typical SBoxes with undesired zero transitions in convex hull

(a) Differential

AE # representative	# Points for CH	# CH inequalities	# Zero transitions in CH	AE # representative	# Points for CH	# CH inequalities	# Zero transitions in CH	AE # representative	# Points for CH	# CH inequalities	# Zero transitions in CH	AE # representative	# Points for CH	# CH inequalities	# Zero transitions in CH
245	86	59	80	261	75	67	69	278	70	58	65	292	55	22	150
246	82	54	77	262	75	59	68	279	64	78	59	294	58	48	56
247	84	49	78	263	75	57	69	280	72	73	66	295	58	60	53
250	80	56	74	264	80	65	74	281	70	79	64	296	52	72	47
251	80	52	74	265	74	72	68	282	71	61	66	297	46	22	126
252	80	49	76	266	68	80	62	283	71	46	68	298	46	22	125
253	73	42	70	267	68	52	64	284	68	52	61	299	42	22	114
254	73	76	69	268	68	98	63	286	67	63	61	300	30	12	189
255	73	56	69	269	82	68	75	287	58	28	53	301	28	12	176
256	73	48	68	270	82	39	77	288	49	18	135	302	16	8	210
257	73	54	66	272	72	80	68	289	58	16	159				
258	79	71	70	276	82	55	75	290	48	46	44				
260	80	59	74	277	76	72	69	291	50	22	137				

(b) Linear

AE # representative	# Points for CH	# CH inequalities	# Zero transitions in CH
258	28	34	24
290	40	36	26
292	32	16	26
293	44	93	38
294	28	22	26
295	28	53	24
296	40	36	34
297	24	12	62
298	24	14	20
299	32	16	26
300	16	8	154
301	16	8	186
302	16	8	210

86, the corresponding convex hull is given by 59 linear inequalities. Out of the 170 zero transitions in the DDT (ignoring the cases where either the input difference or the output difference is 0), 80 of those satisfy all the convex hull inequalities. Thus, for this SBox, the usual MILP modelling given in [SHW+13] could lead to incorrect results for the differential case.

Illustration for Differential Case with SBox `40132567E8A9CDBF`

For completeness, we present the complete system of 59 inequalities that represent the convex hull with the dummy variables $z_0, z_1, z_2, z_3, z_4, z_5, z_6, z_7$ (each of type binary) for the 4×4 SBox 40132567E8A9CDBF; and highlight the associated problem.

This convex hull is created for the non-zero transitions in the DDT, namely with these 86 trails: (0, 0), (1, 2), (1, 3), (1, 6), (1, 7), (2, 1), (2, 3), (2, 5), (2, 7), (3, 1), (3, 2), (3, 5), (3, 6), (4, 1), (4, 2), (4, 4), (4, 7), (5, 1), (5, 3), (5, 4), (5, 6), (6, 2), (6, 3), (6, 4), (6, 5), (7, 4), (7, 5), (7, 6), (7, 7), (8, b), (8, d), (8, e), (9, 8), (9, e), (9, f), (a, b), (a, e), (a, f), (b, 9), (b, a), (b, e), (c, 8), (c, a), (c, e), (d, 8), (d, 9), (d, a), (d, b), (d, c), (d, d), (d, e), (d, f), (e, c), (e, d), (e, e), (f, 9), (f, c), (f, e), (1, 1), (1, 4), (2, 2), (2, 4), (3, 3), (3, 7), (4, 5), (4, 6), (5, 2), (5, 5), (6, 6), (6, 7), (7, 1), (7, 3), (8, a), (9, c), (a, 9), (b, b), (c, d), (e, 8), (f, f), (8, 8), (9, 9), (a, a), (b, d), (c, c), (e, f), (f, b).

The following 80 trails corresponding to the zero transition in the DDT (refer to Table 5.2) satisfy all the inequalities for the convex hull: (1, 8), (1, 9), (1, c), (1, e), (1, f), (2, 9), (2, a), (2, b), (2, e), (2, f), (3, 9), (3, a), (3, b), (3, d), (3, e), (4, 8), (4, a), (4, c), (4, d), (4, e), (5, 8), (5, 9), (5, a), (5, b), (5, c), (5, d), (5, e), (5, f), (6, 8), (6, c), (6, d), (6, e), (6, f), (7, 9), (7, b), (7, c), (7, e), (7, f), (9, 1), (9, 2), (9, 3), (9, 4), (9, 6), (9, 7), (a, 1), (a, 2), (a, 3), (a, 4), (a, 5), (a, 7), (b, 1), (b, 2), (b, 3), (b, 5), (b, 6), (b, 7), (c, 1), (c, 2), (c, 4), (c, 5), (c, 6), (c, 7), (d, 1), (d, 2), (d, 3), (d, 4), (d, 5), (d, 6), (e, 2), (e, 3), (e, 4), (e, 5), (e, 6), (e, 7), (f, 1), (f, 3), (f, 4), (f, 5), (f, 6), (f, 7).

Table 5.2 DDT for SBox 40132567E8A9CDBF

	0	1	2	3	4	5	6	7	8	9	a	b	c	d	e	f
0	16	0	0	0	0	0	0	0	0	0	0	0	0	0	0	0
1	0	4	2	2	4	0	2	2	0	0	0	0	0	0	0	0
2	0	2	4	2	4	2	0	2	0	0	0	0	0	0	0	0
3	0	2	2	4	0	2	2	4	0	0	0	0	0	0	0	0
4	0	2	2	0	2	4	4	2	0	0	0	0	0	0	0	0
5	0	2	4	2	2	4	2	0	0	0	0	0	0	0	0	0
6	0	0	2	2	2	2	4	4	0	0	0	0	0	0	0	0
7	0	4	0	4	2	2	2	2	0	0	0	0	0	0	0	0
8	0	0	0	0	0	0	0	0	6	0	4	2	0	2	2	0
9	0	0	0	0	0	0	0	0	2	6	0	0	4	0	2	2
a	0	0	0	0	0	0	0	0	0	4	6	2	0	0	2	2
b	0	0	0	0	0	0	0	0	0	2	2	4	0	6	2	0
c	0	0	0	0	0	0	0	0	2	0	2	0	6	4	2	0
d	0	0	0	0	0	0	0	0	2	2	2	2	2	2	2	2
e	0	0	0	0	0	0	0	0	4	0	0	0	2	2	2	6
f	0	0	0	0	0	0	0	0	0	2	0	6	2	0	2	4

Therefore, the convex hull model will consider these 80 trails as if these correspond to non-zero transitions, if no further cross-check in conjunction to the model of [SHW+13] is used.

The convex hull inequalities are given by

$$
\begin{array}{llll}
-z_1+1 & \geq 0, & +z_1 & \geq 0,\\
-z_4+1 & \geq 0, & +2z_1+2z_2+z_3+2z_4-z_5+z_6-z_7 & \geq 0,\\
-z_2+1 & \geq 0, & +z_1+2z_2+2z_3+2z_4+z_5-z_6-z_7 & \geq 0,\\
-z_3+1 & \geq 0, & +z_1+z_2+z_3+z_4-z_5 & \geq 0,\\
-z_5+1 & \geq 0, & +z_7 & \geq 0,\\
-z_6+1 & \geq 0, & -z_3+z_4+z_5+z_6+z_7 & \geq 0,\\
-z_7+1 & \geq 0, & +z_1+z_2-z_3-z_5+z_6-z_7+2 & \geq 0,\\
+z_1-2z_2-z_3+z_4+2z_5+3z_6+3z_7 & \geq 0, & +z_1-z_2+z_3-z_4-z_5+z_6+2 & \geq 0,\\
-z_1-z_2-z_4+z_5-z_6+z_7+3 & \geq 0, & -z_1+z_3-z_4+z_5-z_7+2 & \geq 0,\\
-z_1-z_2-z_3+z_5+z_7+2 & \geq 0, & +z_1+z_3+z_4-z_5-z_6+z_7+1 & \geq 0,\\
+3z_1+3z_2+3z_3+2z_4-z_5-z_6-z_7 & \geq 0, & +z_1-z_2-z_4+z_6+z_7+1 & \geq 0,\\
+z_6 & \geq 0, & +z_1-z_2-z_3+z_6+z_7+1 & \geq 0,\\
+z_1-2z_2-z_3+z_5+2z_6+2z_7+1 & \geq 0, & +z_2 & \geq 0,\\
-z_1-z_2+z_3+z_5+z_6-z_7+2 & \geq 0, & -z_1+z_2+z_3-z_4-z_6-z_7+3 & \geq 0,\\
+2z_1-z_2+z_3-2z_4-z_5+2z_6+z_7+2 & \geq 0, & -z_1+z_2+z_3+z_5-z_6-z_7+2 & \geq 0,\\
-z_1+z_2-z_3+z_4-z_5-z_6-z_7+4 & \geq 0, & +z_1-z_2+z_5+z_6+z_7 & \geq 0,\\
+z_1+z_2-z_3-z_4+z_5-z_6+2 & \geq 0, & -z_2-z_3+z_4+2z_5+2z_6+2z_7 & \geq 0,\\
+z_5 & \geq 0, & -z_1-z_2+z_3-z_4+z_5-z_6+3 & \geq 0,\\
+z_1-z_2-z_3+z_4+z_5+2z_6+2z_7 & \geq 0, & -z_1-z_2-z_3+2z_4+3z_5+2z_6+3z_7 & \geq 0,\\
+z_3 & \geq 0, & +z_4 & \geq 0,\\
+z_1+z_2+z_3-z_5-z_6-z_7+2 & \geq 0, & +z_2+z_3-z_4-z_5-z_6-z_7+3 & \geq 0,\\
+z_1+z_2+z_3+z_4-z_6 & \geq 0, & -z_1+z_4+z_5+z_6+z_7 & \geq 0,\\
+z_2+z_3-z_4+z_5+z_6-z_7+1 & \geq 0, & -z_1-z_2+z_3+2z_4+2z_5+2z_6+z_7 & \geq 0,\\
+z_1+z_2+z_3-z_5+z_6+z_7 & \geq 0, & -z_1-z_2+z_3+z_4+z_5+z_6+1 & \geq 0,\\
+z_1+z_2+z_3+z_5+z_6-z_7 & \geq 0, & -z_2+z_4+z_5+z_6+z_7 & \geq 0,\\
+z_1+z_2+z_4-z_5+z_6-z_7+1 & \geq 0, & -2z_1+z_2+2z_3-z_4+z_5-z_6-2z_7+4 & \geq 0,\\
+z_1+z_2+z_3+z_4-z_7 & \geq 0, & +z_1+z_3-z_4-z_5+z_6+z_7+1 & \geq 0,\\
+z_2+z_3+z_4+z_5-z_6-z_7+1 & \geq 0, & +z_1-z_2-z_3-z_4-z_5-z_6-z_7+5 & \geq 0,\\
-z_2-z_3+z_5+z_6+z_7+1 & \geq 0, & -z_1-z_2-z_3-z_4-z_5+z_6-z_7+5 & \geq 0.\\
+2z_1+z_2+2z_3+2z_4-z_5-z_6+z_7 & \geq 0, & &
\end{array}
$$

Effect of Reduced Set of Inequalities

As noted earlier, this problem is not implicitly resolved by the greedy algorithm to reduce the number of linear inequalities. The undesired points satisfy all the convex hull inequalities, whereas the greedy algorithm only looks for the points which do not satisfy at least one of the inequalities. Thus, by picking the inequalities one by one, and checking how many points do not satisfy this particular inequality will not reveal those points. Stated in other words, since those points satisfy the set of all the inequalities of the convex hull, all inequalities of any subset (of the set of all inequalities) will be satisfied by those points. However, one may incorporate a simple check by testing whether any point with zero transitions satisfies all the inequalities.

Separate Convex Hulls for Individual Transitions

It may be noted that the undesired property may also hold if each transition in DDT/LAT is modelled by separate convex hulls. For example, with the same SBox as before (`40132567E8A9CDBF`), the convex hull for the 6 transition in the DDT is given by 7 eight-dimensional points. The corresponding convex hull is given by 7 linear inequalities. Out of 170 zero transitions in the DDT, 12 satisfy all the inequalities.

Relation to Affine Equivalence

It appears that this characteristic of CH does not follow the affine equivalence property. The case for the SBoxes `1032456789ABCDEF` (representative of AE #301 from [DeC07]) and `126CDE39F58BA047` (belongs to the same AE class) can be taken as an example. For the former SBox, the CH problem appears for both the differential and linear cases, but the latter SBox is free from it for both the cases. Therefore there are likely more SBoxes with this problem than presented in Table 5.1, finding which is left open for the future research.

Equality Constraints

Together with the inequality constraints, there is a Sage API which returns the equality constraints.[5] The problem mentioned here will likely not appear if both the equality and the inequality constraints are used, although it is not verified explicitly. Nonetheless, this is not mentioned in the literature to the best of our knowledge.

5.4 Automated Bounds with MILP: Our Proposal

In case the problem (described in Sect. 5.3) is observed for the given SBox, a new modelling different from that of [SHW+13] may be of interest since no variant of the CH-based model can circumvent this problem, to the best of our knowledge. In this regard, we devise a new strategy for a Substitution–Permutation Network (SPN) permutation. We describe our modelling for 4×4 SBox only for simplicity, though

[5] https://doc.sagemath.org/html/en/reference/discrete_geometry/sage/geometry/polyhedron/base.html#sage.geometry.polyhedron.base.Polyhedron_base.equations_list.

it can be generalized if needed. We denote the state size and number of rounds as b (counting from 0) and η (counting from 1), respectively.

Our modelling is inspired from the MILP modelling proposed in [AST+17] and the concept of indicator constraint (see Sect. 2.8.2) used in linear programming where the large constant, M, is chosen. In our case, it is sufficient to choose $M =$ twice the SBox size ($= 8$). Unlike [AST+17], however, we do not rely on any Boolean function-based optimization; the constraints are directly fed to the MILP instance instead.

To get the optimal differential probability, the result from the MILP solver is negated (assuming the maximization variation, see Sect. 5.4.1), and raised to the power of 2. So, if ϵ_d is the result from the MILP solver, the attacker will need at least 2^{ϵ_d+1} chosen inputs, following [Sti06, Sect. 3.4]. For the linear case, if the result from the MILP solver (for the maximization variation) is ϵ_l, the attacker would need at least $2^{2\epsilon_l}$ known inputs [Sti06, Sect. 3.3].

5.4.1 Modelling

In this part, we describe the MILP modelling for the differential case. The MILP formulation for the linear case is much alike, hence we skip the details for conciseness. For completeness, the main differences in the linear case are as follow. The absolute values for the biases are considered. Since $\pm\frac{1}{2}$ linear bias is equivalent to 1 differential probability, each $Q_{i,j}^p$ in the linear case is multiplied by 2 (the notations are described later).

Assume each of the p transitions (corresponds to $2^{\log_2 p/16}$ probability transition, $1 \geq p > 0$) has q_p frequency in the DDT. For example, there are fifty-seven 2 transitions (each corresponds to 2^{-3} probability transition) for the SBox `40132567E8A9CDBF`, as can be seen from its DDT in Table 5.2; hence $q_2 = 57$.

First, for the ith SBox ($i = 0, 1, \ldots, b/4 - 1$) at the jth round ($j = 1, \ldots, \eta$); we create the binary variables:

$Q_{i,j}$	to indicate if it is active;
$Q_{i,j}^p$	to indicate if it takes a p transition;
$Q_{i,j,l}^p$, for $l = 0(1)q_p - 1$	to indicate which among the q_p trails is chosen;
$\mathbf{x}_{i,j} = (x_{i,j}^0, x_{i,j}^1, x_{i,j}^2, x_{i,j}^3)$	to indicate the input difference;
$\mathbf{y}_{i,j} = (y_{i,j}^0, y_{i,j}^1, y_{i,j}^2, y_{i,j}^3)$	to indicate the output difference.

Next, we set the constraints for each SBoxes:

$M Q_{i,j} \geq \sum_{l=0}^{3} x_{i,j}^l + \sum_{l=0}^{3} y_{i,j}^l$	to check if it is active;
$Q_{i,j} = \sum_p Q_{i,j}^p$	to ensure if active, it will take exactly one of the q_p trails;
For each p transition, do:	
$Q_{i,j,l}^p = \sum_{l=0}^{q_p-1} Q_{i,j}^l$	to check which trail among all q_p-trails is chosen.

After this, each $Q^p_{i,j,l}$ for $l = 0(1)q_p - 1$ and for each p is used to model respective transitions. For example, the (6, 7) trail which is a 4 transition in the DDT for the SBox `40132567E8A9CDBF` (see Table 5.2 for its DDT) is modelled as $MQ^4_{i,j} \geq (x^0_{i,j}) + (1 - x^1_{i,j}) + (1 - x^2_{i,j}) + (x^3_{i,j}) + (y^0_{i,j}) + (1 - y^1_{i,j}) + (1 - y^2_{i,j}) + (1 - y^3_{i,j})$. Basically, each negative literal is taken as is, and each positive literal is subtracted from 1; then added together.

Also, we have to give the initial input difference to at least one variable at the beginning ($j = 1$), i.e., $\sum_{i=0}^{b/4-1} \sum_{l=0}^{3} x^l_{i,1} \geq 1$.

There will be additional constraints representing the linear layer. For a bit permutation-based cipher like `GIFT-128`, 128 equality constraints are inserted for each round from 2 to η. For example, the second entry in the permutation ($1 \rightarrow 33$) is modelled as $x^1_{8,j} = y^1_{0,j-1}$. If required, this can be generalized to other type of linear layers.

Finally, the objective function is fit: Minimize $\sum_{i=0}^{b/4-1} \sum_{j=1}^{\eta} \sum_{p<1} \log_2 \frac{p}{16} \times Q^p_{i,j}$. For better readability, we express it as Maximize $\sum_{i=0}^{b/4-1} \sum_{j=1}^{\eta} \sum_{p<1} (-\log_2 \frac{p}{16}) \times Q^p_{i,j}$.

An abridged example with the differential bound for 1-round `GIFT-128` is given in Sect. 5.6.2 for more clarity.

5.4.2 *Optimizations*

Using the idea described earlier (in Sect. 5.4.1), we construct the MILP problems and attempt to solve them using the Gurobi solver. Being inspired from [LWZ+19] (since we are using the same solver), we put redundant constraints in the MILP problem. Using redundant constraints together with the usual constraints does not change the optimal result but could make the execution faster.

As for the choice of the heuristics, we reuse the idea of CH [SHW+13]/Sect. 5.2.2. Therefore, we put additional constraints in the MILP problem together with the usual constraints (described in Sect. 5.4.1). We basically employ three main heuristics and combinations of those heuristics. The main heuristics are termed as *All Convex Hull*, *Chosen Convex Hull*, and *Previous Solutions* and are described next.

All Convex Hull (All-CH)

In this heuristic, we use all the inequalities generated for the convex hull. The convex hull is generated with all points in the DDT/LAT correspond to non-zero transitions.

The choice of this heuristic is motivated by the observation that the claim made in [SHW+13] (i.e., having all the inequalities from the convex hull in the MILP instance will make the solver taking more time) is not apparently supported by any experimental result. Follow-up works, such as [SHW+14, ZDY18], seem to accept this claim without any apparent experimental result too. In fact, it is suggested in

[ST17'1] that increasing the size of the constraints that describe the DDT of an SBox may indeed reduce the run time.

Chosen Convex Hull (Chosen-CH)

In this heuristic, we reduce the number of inequalities for the convex hull (which is generated with points corresponding to non-zero transitions of the DDT/LAT) by applying a randomized version of the greedy algorithm (the greedy algorithm is proposed in [SHW+13]). The randomization is applied in tie-breaking: When multiple inequalities are not satisfied by same number of zero transitions, we break time tie uniformly. This way, we run the algorithm few times to get the smallest number of constraints.

Therefore, this heuristic can be thought as an improvement over that of [SHW+13], as it is non-deterministic in nature and chooses the smallest system of inequalities after a few runs.

Previous Solutions (Prev-Sol)

Suppose, we want to know the optimal bound for round η and we already have optimal bounds (possibly by solving MILP instances) for rounds $0, 1, \ldots, \eta - 1$.[6] In this case, the solutions for the previous rounds can be fed to the MILP instance. Since we know the optimal bounds for smaller rounds, the objective functions can be assigned to those optimal bounds, thus creating new constraints. Refer to Sect. 5.6.3 for an illustration.

For example, suppose we have the optimal solutions up to the 3rd round: s_1, s_2, s_3, respectively; and want the optimal solution for the 4th (so, $\eta = 4$) round. We create the 1-round objective function for each round $\{(1), (2), (3), (4)\}$, and assign with s_1 to create 4 constraints. Next, we create the 2-round objective function by adding the 1-round objective functions for two consecutive rounds $\{(1, 2), (2, 3), (3, 4)\}$ and assign each with s_2 to create 3 constraints. Finally, we create the 3-round objective function by adding the 1-round objective functions for three consecutive rounds $\{(1, 2, 3), (2, 3, 4)\}$ and assign each with s_3 to create 2 constraints.

More specifically, assume the optimal bound corresponding to the jth ($j \in \{1, \ldots, \eta - 1\}$) round is s_j, we insert the following inequalities (taking the maximization variant of the objective function, which is described in Sect. 5.4.1):

[6] Note that, this assumption is practical. As the run time for higher rounds takes significantly longer than the smaller rounds, generally the solutions for the smaller rounds are available.

$$\left.\begin{array}{r}
\sum_{i=0}^{b/4-1}\sum_{p<1}\left(-\log_2\frac{p}{16}\right)\times Q_{i,1}^{p}\le s_1,\\
\sum_{i=0}^{b/4-1}\sum_{p<1}\left(-\log_2\frac{p}{16}\right)\times Q_{i,2}^{p}\le s_1,\\
\vdots\\
\sum_{i=0}^{b/4-1}\sum_{p<1}\left(-\log_2\frac{p}{16}\right)\times Q_{i,\eta-1}^{p}\le s_1,\\
\sum_{i=0}^{b/4-1}\sum_{p<1}\left(-\log_2\frac{p}{16}\right)\times Q_{i,\eta}^{p}\le s_1,
\end{array}\right\}\eta\text{ constraints}$$

$$\left.\begin{array}{r}
\sum_{i=0}^{b/4-1}\sum_{p<1}\left(-\log_2\frac{p}{16}\right)\times Q_{i,1}^{p}+\sum_{i=0}^{b/4-1}\sum_{p<1}\left(-\log_2\frac{p}{16}\right)\times Q_{i,2}^{p}\le s_2,\\
\sum_{i=0}^{b/4-1}\sum_{p<1}\left(-\log_2\frac{p}{16}\right)\times Q_{i,2}^{p}+\sum_{i=0}^{b/4-1}\sum_{p<1}\left(-\log_2\frac{p}{16}\right)\times Q_{i,3}^{p}\le s_2,\\
\vdots\\
\sum_{i=0}^{b/4-1}\sum_{p<1}\left(-\log_2\frac{p}{16}\right)\times Q_{i,\eta-2}^{p}+\sum_{i=0}^{b/4-1}\sum_{p<1}\left(-\log_2\frac{p}{16}\right)\times Q_{i,\eta-1}^{p}\le s_2,\\
\sum_{i=0}^{b/4-1}\sum_{p<1}\left(-\log_2\frac{p}{16}\right)\times Q_{i,\eta-1}^{p}+\sum_{i=0}^{b/4-1}\sum_{p<1}\left(-\log_2\frac{p}{16}\right)\times Q_{i,\eta}^{p}\le s_2,
\end{array}\right\}\eta-1\text{ constraints}$$

$$\vdots$$

$$\left.\begin{array}{r}
\sum_{i=0}^{b/4-1}\sum_{j=1}^{\eta-1}\sum_{p<1}\left(-\log_2\frac{p}{16}\right)\times Q_{i,j}^{p}\le s_{\eta-1},\\
\sum_{i=0}^{b/4-1}\sum_{j=2}^{\eta}\sum_{p<1}\left(-\log_2\frac{p}{16}\right)\times Q_{i,j}^{p}\le s_{\eta-1}.
\end{array}\right\}2\text{ constraints}$$

It may be stated that this heuristic can still be used if the solution for some round $j \in \{1, \ldots, \eta - 1\}$ is not available. In this case, the constraints with s_j at the right-hand side will be skipped. A variant, where not all available constraints are not used, can be considered in the future scope.

Note that, the all-CH modelling itself is sufficient to construct the MILP instances to find the exact bound. The same argument applies to the chosen-CH modelling (which is indeed used in [SHW+13] and its follow-up works, such as [SHW+14, ZDY18]) as well. In other words, it is possible not to use the usual constraints in our model (generated as per Sect. 5.4.1), but only use either of the all-CH or chosen-CH model to obtain the optimal bounds. Here we term these two models are heuristics, as these are redundantly used with the usual constraints in a bid to make the execution faster. The previous solutions based approach is not self-sufficient and will likely not give correct results if the usual constraints are not used together with it.

Although we use the convex hull inequalities (namely, all-CH and chosen-CH) here, the problem mentioned earlier (in Sect. 5.3) does not apply (even if the SBox has undesired zero transitions in its DDT/LAT). This is because those inequalities are used as redundant constraints, on top the usual constraints (described in Sect. 5.4.1). Those usual constraints are sufficient to ensure no unwanted point is included. Also, all our models (regardless of whether a heuristic is used or not) are designed for finding the optimal bounds.

5.4.3 Results

Here we present our experimental results for the differential and linear cases for GIFT-128, GIFT-64 and PRESENT for reduced rounds. Results are obtained from a workstation with 16× Intel Xeon E7-8880 physical processors (shared among multiple users) running Gurobi 8.1 on 64-bit Ubuntu 18.04. It remains to see how those heuristics perform with a different solver and/or environment.

As the Gurobi solver is deterministic,[7] only extraneous factor that can affect the execution time is the system load. To mitigate this factor, we run each MILP instances for a few times and take the average for benchmarking with round-reduced GIFT-128.

The experimental results for run time are summarized in Table 5.3. Here we present the average run time (in seconds) for the differential and linear MILP instances for reduced round (1–8) GIFT-128 corresponding to the cases where no heuristic is applied (only the usual constraints are used); usual constraints with all-CH constraints are used, usual constraints with chosen-CH constraints are used, usual constraints with previous solutions are used, usual constraints with previous solutions and all-CH constraints are used, usual constraints with previous solutions and chosen-CH constraints are used. The MILP instances with previous solutions as heuristic appear to slow down the solver for the differential case, particularly round 5 onward.

The relative (1.0×) run times for each of the heuristics, with respect to the cases where no heuristic is used; can be seen from Fig. 5.1. The differential case is shown in Fig. 5.1a (a zoomed in version till the 5th round is also shown), and the linear case is given in Fig. 5.1b. As evident from the experimental results, it is generally difficult to find a general trend. Still, one may notice significant improvement in run time, generally by a few folds. For example, the all-CH heuristic is more suitable for the differential case; whereas the both the previous solutions with all-CH and the previous solutions with chosen-CH are more suitable for the linear case. Also, it appears that all three of the previous solutions based heuristics model perform somewhat similarly for the differential case, and in general are slower than that of no heuristic. However, the same three heuristics perform faster than no heuristic for the linear case.

[7] https://support.gurobi.com/hc/en-us/articles/360031636051-Is-Gurobi-Optimizer-deterministic-.

Table 5.3 Average run time for MILP instances for `GIFT-128` with various heuristics

Round(s)		1	2	3	4	5	6	7	8
Differential	–	4.29	16.68	16.06	504.66	6698.07	914.91	1142.62	3142.78
	All-CH	0.25	3.47	15.05	63.15	931.07	607.34	754.12	2708.20
	Chosen-CH	0.69	6.23	25.56	173.39	1949.50	1148.18	1239.64	4671.95
	Prev-Sol	2.49	0.55	11.00	2097.36	12754.98	70514.22	122874.35	123434.43
	Prev-Sol + All-CH	0.89	1.21	12.94	2034.16	9841.02	47055.10	127680.92	200204.50
	Prev-Sol + Chosen-CH	0.38	1.06	10.51	1943.44	10867.38	40964.27	103575.79	229649.26
Linear	–	0.33	0.62	2.71	80.98	4546.46	2108.19	6814.77	38826.96
	All-CH	0.31	1.67	4.18	28.84	80.36	2698.96	3374.94	15166.05
	Chosen-CH	0.38	0.73	2.24	10.69	3205.41	1241.32	2649.87	13120.24
	Prev-Sol	0.15	0.49	1.40	32.87	2435.58	1110.57	2639.03	20523.93
	Prev-Sol + All-CH	0.22	0.61	2.01	19.46	52.88	1546.41	1750.24	9199.43
	Prev-Sol + Chosen-CH	0.21	1.35	2.56	16.42	50.77	1764.47	1992.35	7065.79

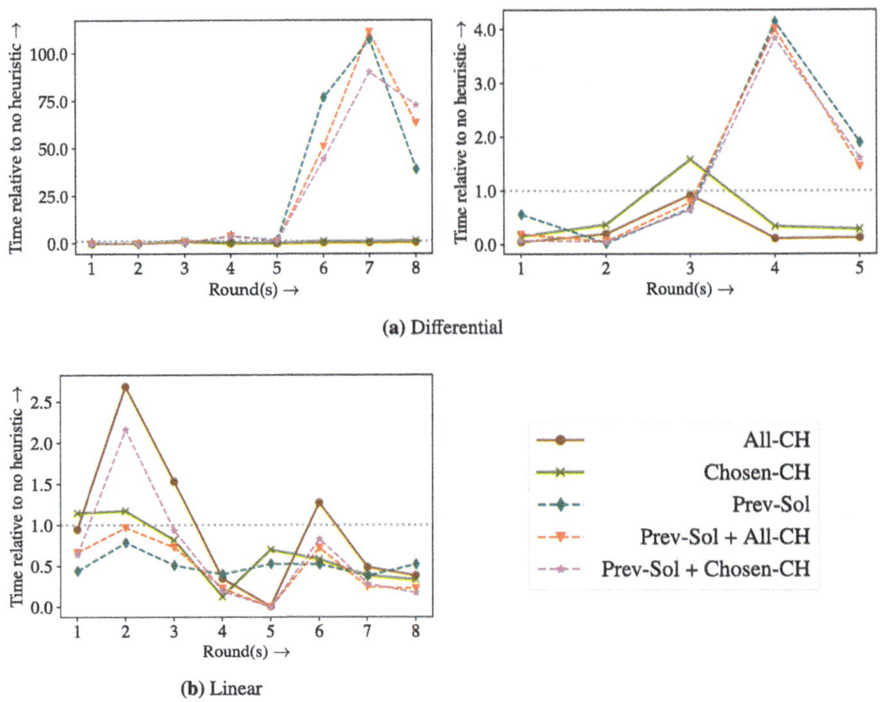

Fig. 5.1 Relative performance of MILP heuristics for `GIFT-128`

Table 5.4 shows the optimal bounds till round 12 for `GIFT-128`, `GIFT-64` and `PRESENT`. We do not put the average run time corresponding to rounds 9 onward as those cases are not run sufficient times (as each of such cases takes a long time to run). These results are consistent with the existing literature [LWZ+19, JZD19, LLL+19, SHW+14, ZZD+19].

Moreover, we present the optimal linear bounds of `GIFT-128` for 11th and 12th rounds. The input masks are shown in Table reft:linearspsgift128 (Table 5.5(a) for 11 rounds and Table 5.5(b) for 12 rounds). Here weight counts the sum of $-\log_2(|p|/8)$ over the state, where p is a transition in the LAT for the corresponding input mask. The inactive SBoxes are shown by ·, and the 1- and 2-weights are shown, respectively, by the markers $\underline{\gamma}$ and γ. The total of the weights is 31 and 36 for 11th and 12th rounds, respectively; which are also indicated in Table 5.4. Further, we show the (input mask, output mask) for 11 rounds in Fig. 5.2, and for 12 rounds in Fig. 5.3 for each SBox. The active SBoxes with weight 1 (i.e., ±4 transitions in the LAT) are shown by the marker $\gamma\Gamma$, and the active SBoxes with weight 2 (i.e., ±2 transitions in the LAT) are shown by the marker $\gamma\Gamma$ —γ represents the input mask and Γ represents the output mask. The LAT of the `GIFT` SBox is not given here for conciseness, an interested reader may find it in [BPP+17, Table 19].

Table 5.4 Optimal bounds for GIFT-128, GIFT-64 and PRESENT

Round(s)		1	2	3	4	5	6	7	8	9	10	11	12
Differential	GIFT-128	1.415	3.415	7.000	11.415	17.000	22.415	28.415	39.000	45.415	49.415	54.415	60.415
	GIFT-64	1.415	3.415	7.000	11.415	17.000	22.415	28.415	38.000	42.000	48.000	52.000	58.000
	PRESENT	2.000	4.000	8.000	12.000	20.000	24.000	28.000	32.000	36.000	41.000	46.000	52.000
Linear	GIFT-128	1.000	2.000	3.000	5.000	7.000	10.000	13.000	17.000	22.000	26.000	31.000	36.000
	GIFT-64	1.000	2.000	3.000	5.000	7.000	10.000	13.000	16.000	20.000	25.000	29.000	31.000
	PRESENT	1.000	2.000	4.000	6.000	8.000	10.000	12.000	14.000	16.000	18.000	20.000	22.000

Table 5.5 Optimal linear bounds for `GIFT-128`

(a) 11-Round

Round	Input Mask	Weight
1	d·d·········58··················	4
2	1··6···········1··9·············	5
3	·························c··c···	2
4	······11························	3
5	·c······························	1
6	2·······························	1
7	1······2························	3
8	·························8·8····	2
9	······5················5········	4
10	·4··4···························	2
11	22·····44·······················	4
	Total	31

(b) 12-Round

Round	Input Mask	Weight
1	·······a···············b········	2
2	················2··2············	2
3	····22······················11··	5
4	·1·······6·············c·1······	5
5	···············8·8··28··········	5
6	····56··············59··········	7
7	················c··c············	2
8	····························11··	3
9	························c·······	1
10	············1···················	1
11	··················8·············	1
12	···········1················4···	2
	Total	36

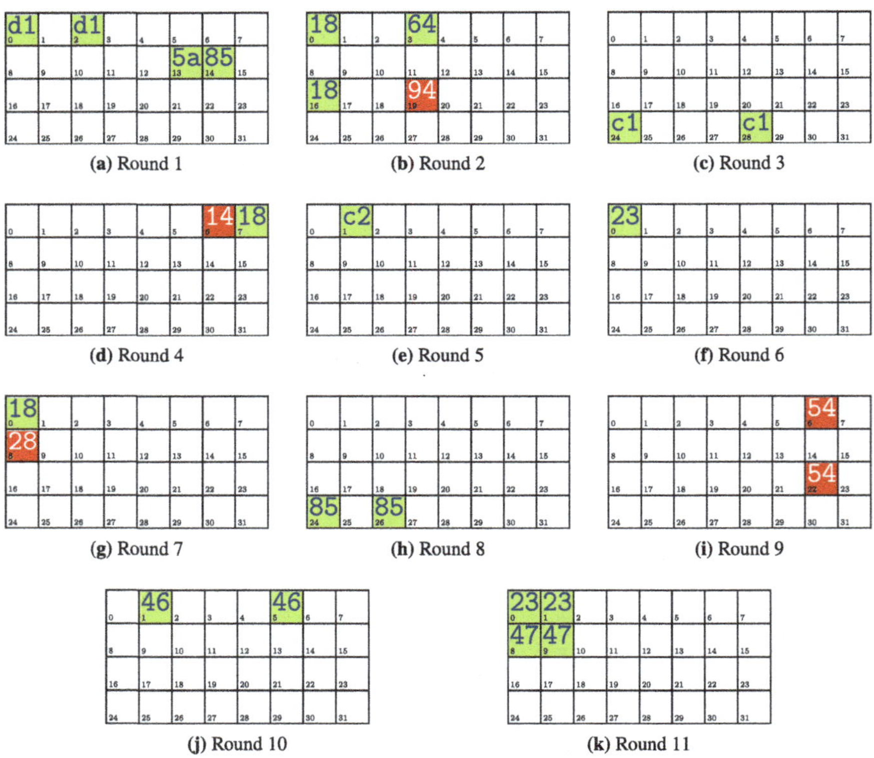

Fig. 5.2 Input and output masks for an 11-round optimal linear bound of `GIFT-128`

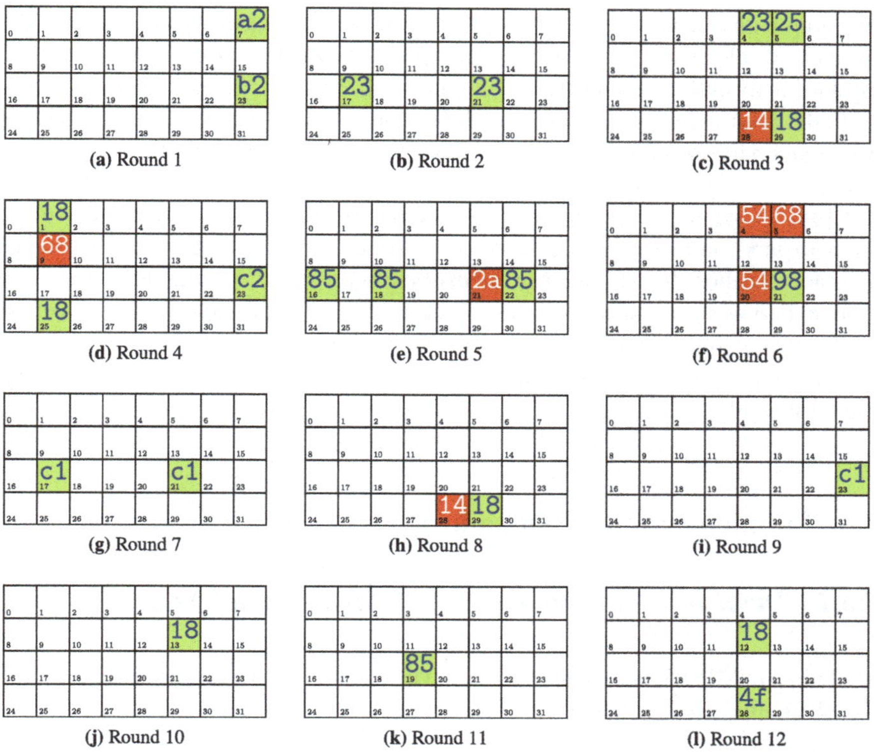

Fig. 5.3 Input and output masks for a 12-round optimal linear bound of `GIFT-128`

5.5 Conclusion

We attempt to study the problem of modelling differential and linear bounds using MILP in depth. In the process, we revisit the modelling proposed in [SHW+13] and explore a related shortcoming. I t may happen for an SBox that the hyper-volume of convex hull will contain some undesired points. Although this probably does not happen with the commonly used SBoxes, it can still be inferred that this model is not generic as it depends on specific properties of the SBox.

Therefore, we propose our new MILP modelling which works for any SBox and is partly inspired from [AST+17]. Our modelling is simpler, and it does not require specialized library or Boolean logic minimization.

At the same time, we also follow [LWZ+19], where the authors observe that the number of constraints can influence the solution time taken by the MILP solver Gurobi. Being motivated by their research, we experiment with redundant constraints. The redundant constraints are inserted along with the usual constraints (which are enough to specify the MILP instance). Those constraints do not change the optimal solution but can improve the run time. With our experiment, we observe significant speed-up with the redundant constraints, compared to the case with only usual constraints. We also show that one needs a separate heuristic depending on the MILP instance is for differential or linear bound.

In the future scope, one may extend the search for heuristics. As we are always using our own constraints (described in Sect. 5.4.1), any other modelling can be used as a heuristic. This includes the branch number-based model [MWG+11] or that of [AST+17]. One may also try to model the zero transitions (using an analogous modelling to ours), and use those as redundant constraints in a way that the MILP instance does not take those transitions. The effect of the heuristics is not straightforward, and more experiments are needed in this direction. Next, the problem with convex hull modelling can be more formally studied, and the SBoxes can be characterized with respect to this problem. Lastly, the effect of a different solver and/or environment can be studied.

5.6 Supplementary Discussion

5.6.1 Detailed Description on MILP Modelling of XOR

Modelling to MILP is somewhat non-trivial in general. For better clarity, we describe the MILP modelling of one XOR operation from [MWG+11]/Sect. 5.2.1.

An MILP problem requires all the constraints to be linear. If an "if-else" clause could be allowed in MILP constraints, we could have formulated $y = \text{XOR}(x_0, x_1)$ as

$$\begin{aligned}\text{if } x_0 = x_1 : &\\ &y = 0\\ \text{else} : &\\ &y = 1\end{aligned}$$

Since this is not allowed, we need to find a way to represent "if-else" as linear constraints. This is where the dummy variable, d, comes in. We use it in such a way that

- The objective function forces d to be 0 whenever possible, or equivalently does not make $d = 1$ unless it has a choice to make it 0. Thus, $d = 1$ only if $d = 0$ makes the system unsatisfiable.
- The linear constraints are carefully chosen such that
  - For a valid situation, d can be either 0 or 1.
  - For an invalid situation, d has to be 1.

Combining the constraints with the objective function makes sure that only the valid situations are taken by the solver. Ultimately, this leads to the MILP instance (it can be verified that the invalid situations like $(x_0, x_1, y, d) = (1, 1, 1, 1)$ or $(1, 1, 0, 1)$ are filtered):

$$\text{Minimize } d$$
$$\text{Subject to} \begin{cases} x_0 + x_1 - 2d \geq 0 \\ d - x_0 \geq 0 \\ d - x_1 \geq 0 \\ d - y \geq 0 \end{cases}$$
$$x_0, x_1, y, d \text{ are binary}$$

That said, as noted in [BKD+21], a modern MILP solver like Gurobi can actually work directly with XOR clauses as the constraints (hence formulation to linear constraints is not strictly needed). Not clear why, but this feature is not used in any formulation of differential or linear bound in the literature.

5.6.2 *Illustration with MILP Model for 1-Round Differential Bound for* `GIFT-128`

An abridged (and partly annotated) version of the MILP instance corresponding to the 1-round differential bound for `GIFT-128` (without any heuristic) in LP format[8] is shown in Implementation 5.4. The unabridged LP file contains 3297 linear constraints with 3520 binary variables (i.e., 110 variables per SBox). Here in the abridged version, only the constraints corresponding to the 0th SBox is shown in Lines 14–118 (i.e., 103 constraints). In the unabridged version, a similar pattern would be repeated for the rest 31 SBoxes (totaling in 3296 constraints). The additional constraint (Line 10) comes from the initial non-zero input difference. The objective function is given in Line 4. There is no constraint for the linear layer (mentioned in Line 122), those will come from round 2 onward. Marked in Line 126, all the variables are declared as type binary.

5.6.3 *Illustration with MILP Model of Previous Constraints for 4-Round Differential Bound for* `GIFT-128`

Similar to Sect. 5.6.2, here we present an example on the previous constraints for 4-round differential bound for `GIFT-128` in LP format in Implementation 5.8

[8] LP is a human-readable format of expressing MILP instances. Here all the variables are on the left-hand side (the right-hand side only has the constant). See https://www.gurobi.com/documentation/9.1/refman/lp_format.html for more details.

```
\ Differential modelling for 1-round GIFT-128

\ Objective function
Minimize 3 Q02_00_00 + 3 Q02_01_00 + 3 Q02_02_00 + 3 Q02_03_00 + 3 Q02_04_00 + 3 Q02_05_00 + 3 Q02_06_00 + 3 Q02_07_00 + 3
↪ Q02_08_00 + 3 Q02_09_00 + 3 Q02_10_00 + 3 Q02_11_00 + 3 Q02_12_00 + 3 Q02_13_00 + 3 Q02_14_00 + 3 Q02_15_00 + 3 Q02_16_00 + 3
↪ Q02_17_00 + 3 Q02_18_00 + 3 Q02_19_00 + 3 Q02_20_00 + 3 Q02_21_00 + 3 Q02_22_00 + 3 Q02_23_00 + 3 Q02_24_00 + 3 Q02_25_00 + 3
↪ Q02_26_00 + 3 Q02_27_00 + 3 Q02_28_00 + 3 Q02_29_00 + 3 Q02_30_00 + 3 Q02_31_00 + 2 Q04_00_00 + 2 Q04_01_00 + 2 Q04_02_00 + 2
↪ Q04_03_00 + 2 Q04_04_00 + 2 Q04_05_00 + 2 Q04_06_00 + 2 Q04_07_00 + 2 Q04_08_00 + 2 Q04_09_00 + 2 Q04_10_00 + 2 Q04_11_00 + 2
↪ Q04_12_00 + 2 Q04_13_00 + 2 Q04_14_00 + 2 Q04_15_00 + 2 Q04_16_00 + 2 Q04_17_00 + 2 Q04_18_00 + 2 Q04_19_00 + 2 Q04_20_00 + 2
↪ Q04_21_00 + 2 Q04_22_00 + 2 Q04_23_00 + 2 Q04_24_00 + 2 Q04_25_00 + 2 Q04_26_00 + 2 Q04_27_00 + 2 Q04_28_00 + 2 Q04_29_00 + 2
↪ Q04_30_00 + 2 Q04_31_00 + 1.4150374992788437 Q06_00_00 + 1.4150374992788437 Q06_01_00 + 1.4150374992788437 Q06_02_00 +
↪ 1.4150374992788437 Q06_03_00 + 1.4150374992788437 Q06_04_00 + 1.4150374992788437 Q06_05_00 + 1.4150374992788437 Q06_06_00 +
↪ 1.4150374992788437 Q06_07_00 + 1.4150374992788437 Q06_08_00 + 1.4150374992788437 Q06_09_00 + 1.4150374992788437 Q06_10_00 +
↪ 1.4150374992788437 Q06_11_00 + 1.4150374992788437 Q06_12_00 + 1.4150374992788437 Q06_13_00 + 1.4150374992788437 Q06_14_00 +
↪ 1.4150374992788437 Q06_15_00 + 1.4150374992788437 Q06_16_00 + 1.4150374992788437 Q06_17_00 + 1.4150374992788437 Q06_18_00 +
↪ 1.4150374992788437 Q06_19_00 + 1.4150374992788437 Q06_20_00 + 1.4150374992788437 Q06_21_00 + 1.4150374992788437 Q06_22_00 +
↪ 1.4150374992788437 Q06_23_00 + 1.4150374992788437 Q06_24_00 + 1.4150374992788437 Q06_25_00 + 1.4150374992788437 Q06_26_00 +
↪ 1.4150374992788437 Q06_27_00 + 1.4150374992788437 Q06_28_00 + 1.4150374992788437 Q06_29_00 + 1.4150374992788437 Q06_30_00 +
↪ 1.4150374992788437 Q06_31_00

\ Linear constraints
Subject To

\ Non-zero input difference
x0_00_00 + x0_01_00 + x0_02_00 + x0_03_00 + x0_04_00 + x0_05_00 + x0_06_00 + x0_07_00 + x0_08_00 + x0_09_00 + x0_10_00 + x0_11_00 +
↪ x0_12_00 + x0_13_00 + x0_14_00 + x0_15_00 + x0_16_00 + x0_17_00 + x0_18_00 + x0_19_00 + x0_20_00 + x0_21_00 + x0_22_00 +
↪ x0_23_00 + x0_24_00 + x0_25_00 + x0_26_00 + x0_27_00 + x0_28_00 + x0_29_00 + x0_30_00 + x0_31_00 + x1_00_00 + x1_01_00 +
↪ x1_02_00 + x1_03_00 + x1_04_00 + x1_05_00 + x1_06_00 + x1_07_00 + x1_08_00 + x1_09_00 + x1_10_00 + x1_11_00 + x1_12_00 +
↪ x1_13_00 + x1_14_00 + x1_15_00 + x1_16_00 + x1_17_00 + x1_18_00 + x1_19_00 + x1_20_00 + x1_21_00 + x1_22_00 + x1_23_00 +
↪ x1_24_00 + x1_25_00 + x1_26_00 + x1_27_00 + x1_28_00 + x1_29_00 + x1_30_00 + x1_31_00 + x2_00_00 + x2_01_00 + x2_02_00 +
↪ x2_03_00 + x2_04_00 + x2_05_00 + x2_06_00 + x2_07_00 + x2_08_00 + x2_09_00 + x2_10_00 + x2_11_00 + x2_12_00 + x2_13_00 +
↪ x2_14_00 + x2_15_00 + x2_16_00 + x2_17_00 + x2_18_00 + x2_19_00 + x2_20_00 + x2_21_00 + x2_22_00 + x2_23_00 + x2_24_00 +
↪ x2_25_00 + x2_26_00 + x2_27_00 + x2_28_00 + x2_29_00 + x2_30_00 + x2_31_00 + x3_00_00 + x3_01_00 + x3_02_00 + x3_03_00 +
↪ x3_04_00 + x3_05_00 + x3_06_00 + x3_07_00 + x3_08_00 + x3_09_00 + x3_10_00 + x3_11_00 + x3_12_00 + x3_13_00 + x3_14_00 +
↪ x3_15_00 + x3_16_00 + x3_17_00 + x3_18_00 + x3_19_00 + x3_20_00 + x3_21_00 + x3_22_00 + x3_23_00 + x3_24_00 + x3_25_00 +
↪ x3_26_00 + x3_27_00 + x3_28_00 + x3_29_00 + x3_30_00 + x3_31_00 >= 1

\ SBox constraints
\ SBox : 0
- 8 Q_00_00 + x0_00_00 + x1_00_00 + x2_00_00 + x3_00_00 + y0_00_00 + y1_00_00 + y2_00_00 + y3_00_00 <= 0
Q02_00_00 + Q04_00_00 + Q06_00_00 - Q_00_00 = 0
Q02_00_00 - Q02_00_00_00 - Q02_01_00_00 - Q02_02_00_00 - Q02_03_00_00 - Q02_04_00_00 - Q02_05_00_00 - Q02_06_00_00 - Q02_07_00_00 -
↪ Q02_08_00_00 - Q02_09_00_00 - Q02_10_00_00 - Q02_11_00_00 - Q02_12_00_00 - Q02_13_00_00 - Q02_14_00_00 - Q02_15_00_00 -
↪ Q02_16_00_00 - Q02_17_00_00 - Q02_18_00_00 - Q02_19_00_00 - Q02_20_00_00 - Q02_21_00_00 - Q02_22_00_00 - Q02_23_00_00 -
↪ Q02_24_00_00 - Q02_25_00_00 - Q02_26_00_00 - Q02_27_00_00 - Q02_28_00_00 - Q02_29_00_00 - Q02_30_00_00 - Q02_31_00_00 -
↪ Q02_32_00_00 - Q02_33_00_00 - Q02_34_00_00 - Q02_35_00_00 - Q02_36_00_00 - Q02_37_00_00 - Q02_38_00_00 - Q02_39_00_00 -
↪ Q02_40_00_00 - Q02_41_00_00 - Q02_42_00_00 - Q02_43_00_00 - Q02_44_00_00 - Q02_45_00_00 - Q02_46_00_00 - Q02_47_00_00 -
↪ Q02_48_00_00 - Q02_49_00_00 - Q02_50_00_00 - Q02_51_00_00 - Q02_52_00_00 - Q02_53_00_00 - Q02_54_00_00 - Q02_55_00_00 -
↪ Q02_56_00_00 - Q02_57_00_00 - Q02_58_00_00 - Q02_59_00_00 - Q02_60_00_00 - Q02_61_00_00 - Q02_62_00_00 - Q02_63_00_00 -
↪ Q02_64_00_00 - Q02_65_00_00 - Q02_66_00_00 - Q02_67_00_00 - Q02_68_00_00 - Q02_69_00_00 - Q02_70_00_00 - Q02_71_00_00 -
↪ Q02_72_00_00 - Q02_73_00_00 - Q02_74_00_00 - Q02_75_00_00 - Q02_76_00_00 - Q02_77_00_00 = 0
- 8 Q02_00_00_00 + x0_00_00 - x1_00_00 - x2_00_00 - x3_00_00 + y0_00_00 - y1_00_00 + y2_00_00 - y3_00_00 >= - 5
- 8 Q02_01_00_00 + x0_00_00 - x1_00_00 - x2_00_00 - x3_00_00 - y0_00_00 + y1_00_00 + y2_00_00 - y3_00_00 >= - 5
- 8 Q02_02_00_00 + x0_00_00 - x1_00_00 - x2_00_00 - x3_00_00 - y0_00_00 - y1_00_00 - y2_00_00 + y3_00_00 >= - 6
- 8 Q02_03_00_00 + x0_00_00 - x1_00_00 - x2_00_00 - x3_00_00 + y0_00_00 - y1_00_00 - y2_00_00 + y3_00_00 >= - 5
- 8 Q02_04_00_00 + x0_00_00 - x1_00_00 - x2_00_00 - x3_00_00 - y0_00_00 + y1_00_00 - y2_00_00 + y3_00_00 >= - 5
- 8 Q02_05_00_00 + x0_00_00 - x1_00_00 - x2_00_00 - x3_00_00 + y0_00_00 + y1_00_00 - y2_00_00 + y3_00_00 >= - 4
- 8 Q02_06_00_00 + x0_00_00 - x1_00_00 - x2_00_00 - x3_00_00 - y0_00_00 - y1_00_00 + y2_00_00 + y3_00_00 >= - 5
- 8 Q02_07_00_00 + x0_00_00 - x1_00_00 - x2_00_00 - x3_00_00 + y0_00_00 + y1_00_00 + y2_00_00 + y3_00_00 >= - 3
- 8 Q02_08_00_00 - x0_00_00 + x1_00_00 - x2_00_00 - x3_00_00 + y0_00_00 - y1_00_00 - y2_00_00 + y3_00_00 >= - 5
- 8 Q02_09_00_00 - x0_00_00 + x1_00_00 - x2_00_00 - x3_00_00 - y0_00_00 + y1_00_00 - y2_00_00 + y3_00_00 >= - 5
- 8 Q02_10_00_00 - x0_00_00 + x1_00_00 - x2_00_00 - x3_00_00 + y0_00_00 - y1_00_00 + y2_00_00 + y3_00_00 >= - 4
- 8 Q02_11_00_00 - x0_00_00 + x1_00_00 - x2_00_00 - x3_00_00 - y0_00_00 + y1_00_00 + y2_00_00 + y3_00_00 >= - 4
- 8 Q02_12_00_00 + x0_00_00 + x1_00_00 - x2_00_00 - x3_00_00 + y0_00_00 - y1_00_00 + y2_00_00 - y3_00_00 >= - 4
- 8 Q02_13_00_00 + x0_00_00 + x1_00_00 - x2_00_00 - x3_00_00 - y0_00_00 + y1_00_00 + y2_00_00 - y3_00_00 >= - 4
- 8 Q02_14_00_00 + x0_00_00 + x1_00_00 - x2_00_00 - x3_00_00 - y0_00_00 - y1_00_00 - y2_00_00 + y3_00_00 >= - 5
- 8 Q02_15_00_00 + x0_00_00 + x1_00_00 - x2_00_00 - x3_00_00 + y0_00_00 + y1_00_00 - y2_00_00 + y3_00_00 >= - 3
- 8 Q02_16_00_00 + x0_00_00 + x1_00_00 - x2_00_00 - x3_00_00 - y0_00_00 - y1_00_00 + y2_00_00 + y3_00_00 >= - 4
- 8 Q02_17_00_00 + x0_00_00 + x1_00_00 - x2_00_00 - x3_00_00 + y0_00_00 - y1_00_00 + y2_00_00 + y3_00_00 >= - 3
- 8 Q02_18_00_00 + x0_00_00 + x1_00_00 - x2_00_00 - x3_00_00 - y0_00_00 + y1_00_00 + y2_00_00 + y3_00_00 >= - 3
- 8 Q02_19_00_00 + x0_00_00 + x1_00_00 - x2_00_00 - x3_00_00 + y0_00_00 + y1_00_00 + y2_00_00 + y3_00_00 >= - 2
- 8 Q02_20_00_00 - x0_00_00 - x1_00_00 + x2_00_00 - x3_00_00 + y0_00_00 + y1_00_00 - y2_00_00 - y3_00_00 >= - 5
- 8 Q02_21_00_00 - x0_00_00 - x1_00_00 + x2_00_00 - x3_00_00 + y0_00_00 - y1_00_00 - y2_00_00 + y3_00_00 >= - 5
- 8 Q02_22_00_00 - x0_00_00 - x1_00_00 + x2_00_00 - x3_00_00 + y0_00_00 - y1_00_00 + y2_00_00 + y3_00_00 >= - 4
- 8 Q02_23_00_00 + x0_00_00 - x1_00_00 + x2_00_00 - x3_00_00 - y0_00_00 + y1_00_00 - y2_00_00 - y3_00_00 >= - 5
- 8 Q02_24_00_00 + x0_00_00 - x1_00_00 + x2_00_00 - x3_00_00 + y0_00_00 - y1_00_00 + y2_00_00 - y3_00_00 >= - 4
- 8 Q02_25_00_00 + x0_00_00 - x1_00_00 + x2_00_00 - x3_00_00 - y0_00_00 - y1_00_00 - y2_00_00 + y3_00_00 >= - 5
- 8 Q02_26_00_00 + x0_00_00 - x1_00_00 + x2_00_00 - x3_00_00 - y0_00_00 - y1_00_00 + y2_00_00 + y3_00_00 >= - 4
- 8 Q02_27_00_00 + x0_00_00 - x1_00_00 + x2_00_00 - x3_00_00 + y0_00_00 - y1_00_00 + y2_00_00 + y3_00_00 >= - 3
- 8 Q02_28_00_00 + x0_00_00 - x1_00_00 + x2_00_00 - x3_00_00 - y0_00_00 + y1_00_00 + y2_00_00 + y3_00_00 >= - 3
- 8 Q02_29_00_00 - x0_00_00 + x1_00_00 + x2_00_00 - x3_00_00 + y0_00_00 + y1_00_00 + y2_00_00 - y3_00_00 >= - 3
- 8 Q02_30_00_00 - x0_00_00 + x1_00_00 + x2_00_00 - x3_00_00 - y0_00_00 + y1_00_00 - y2_00_00 + y3_00_00 >= - 4
- 8 Q02_31_00_00 - x0_00_00 + x1_00_00 + x2_00_00 - x3_00_00 - y0_00_00 + y1_00_00 + y2_00_00 + y3_00_00 >= - 3
```

Fig. 5.4 GIFT-128 1-round differential bound in LP format (part I)

```
- 8 Q02_32_00_00 + x0_00_00 + x1_00_00 + x2_00_00 - x3_00_00 - y0_00_00 + y1_00_00 - y2_00_00 - y3_00_00 >= - 4
- 8 Q02_33_00_00 + x0_00_00 + x1_00_00 + x2_00_00 - x3_00_00 + y0_00_00 - y1_00_00 + y2_00_00 - y3_00_00 >= - 3
- 8 Q02_34_00_00 + x0_00_00 + x1_00_00 + x2_00_00 - x3_00_00 - y0_00_00 - y1_00_00 - y2_00_00 + y3_00_00 >= - 4
- 8 Q02_35_00_00 + x0_00_00 + x1_00_00 + x2_00_00 - x3_00_00 + y0_00_00 - y1_00_00 - y2_00_00 + y3_00_00 >= - 3
- 8 Q02_36_00_00 + x0_00_00 + x1_00_00 + x2_00_00 - x3_00_00 - y0_00_00 + y1_00_00 - y2_00_00 + y3_00_00 >= - 3
- 8 Q02_37_00_00 + x0_00_00 + x1_00_00 + x2_00_00 - x3_00_00 - y0_00_00 - y1_00_00 + y2_00_00 + y3_00_00 >= - 3
- 8 Q02_38_00_00 + x0_00_00 - x1_00_00 - x2_00_00 + x3_00_00 + y0_00_00 - y1_00_00 - y2_00_00 - y3_00_00 >= - 5
- 8 Q02_39_00_00 + x0_00_00 - x1_00_00 - x2_00_00 + x3_00_00 + y0_00_00 + y1_00_00 - y2_00_00 - y3_00_00 >= - 4
- 8 Q02_40_00_00 + x0_00_00 - x1_00_00 - x2_00_00 + x3_00_00 - y0_00_00 + y1_00_00 + y2_00_00 - y3_00_00 >= - 4
- 8 Q02_41_00_00 + x0_00_00 - x1_00_00 - x2_00_00 + x3_00_00 + y0_00_00 + y1_00_00 + y2_00_00 - y3_00_00 >= - 3
- 8 Q02_42_00_00 + x0_00_00 - x1_00_00 - x2_00_00 + x3_00_00 - y0_00_00 - y1_00_00 - y2_00_00 + y3_00_00 >= - 5
- 8 Q02_43_00_00 + x0_00_00 - x1_00_00 - x2_00_00 + x3_00_00 - y0_00_00 + y1_00_00 - y2_00_00 + y3_00_00 >= - 4
- 8 Q02_44_00_00 + x0_00_00 - x1_00_00 - x2_00_00 + x3_00_00 - y0_00_00 - y1_00_00 + y2_00_00 + y3_00_00 >= - 4
- 8 Q02_45_00_00 + x0_00_00 - x1_00_00 - x2_00_00 + x3_00_00 + y0_00_00 - y1_00_00 + y2_00_00 + y3_00_00 >= - 3
- 8 Q02_46_00_00 - x0_00_00 + x1_00_00 - x2_00_00 + x3_00_00 + y0_00_00 - y1_00_00 - y2_00_00 + y3_00_00 >= - 4
- 8 Q02_47_00_00 - x0_00_00 + x1_00_00 - x2_00_00 + x3_00_00 - y0_00_00 + y1_00_00 - y2_00_00 + y3_00_00 >= - 4
- 8 Q02_48_00_00 - x0_00_00 + x1_00_00 - x2_00_00 + x3_00_00 + y0_00_00 - y1_00_00 + y2_00_00 + y3_00_00 >= - 3
- 8 Q02_49_00_00 - x0_00_00 + x1_00_00 - x2_00_00 + x3_00_00 - y0_00_00 + y1_00_00 + y2_00_00 + y3_00_00 >= - 3
- 8 Q02_50_00_00 + x0_00_00 + x1_00_00 - x2_00_00 + x3_00_00 + y0_00_00 - y1_00_00 - y2_00_00 - y3_00_00 >= - 4
- 8 Q02_51_00_00 + x0_00_00 + x1_00_00 - x2_00_00 + x3_00_00 + y0_00_00 + y1_00_00 - y2_00_00 - y3_00_00 >= - 3
- 8 Q02_52_00_00 + x0_00_00 + x1_00_00 - x2_00_00 + x3_00_00 - y0_00_00 + y1_00_00 + y2_00_00 - y3_00_00 >= - 3
- 8 Q02_53_00_00 + x0_00_00 + x1_00_00 - x2_00_00 + x3_00_00 + y0_00_00 + y1_00_00 + y2_00_00 - y3_00_00 >= - 2
- 8 Q02_54_00_00 + x0_00_00 + x1_00_00 - x2_00_00 + x3_00_00 - y0_00_00 - y1_00_00 - y2_00_00 + y3_00_00 >= - 4
- 8 Q02_55_00_00 + x0_00_00 + x1_00_00 - x2_00_00 + x3_00_00 + y0_00_00 - y1_00_00 - y2_00_00 + y3_00_00 >= - 3
- 8 Q02_56_00_00 + x0_00_00 + x1_00_00 - x2_00_00 + x3_00_00 - y0_00_00 - y1_00_00 + y2_00_00 + y3_00_00 >= - 3
- 8 Q02_57_00_00 + x0_00_00 + x1_00_00 - x2_00_00 + x3_00_00 - y0_00_00 + y1_00_00 + y2_00_00 + y3_00_00 >= - 2
- 8 Q02_58_00_00 - x0_00_00 - x1_00_00 + x2_00_00 + x3_00_00 - y0_00_00 - y1_00_00 - y2_00_00 + y3_00_00 >= - 5
- 8 Q02_59_00_00 - x0_00_00 - x1_00_00 + x2_00_00 + x3_00_00 - y0_00_00 + y1_00_00 - y2_00_00 + y3_00_00 >= - 4
- 8 Q02_60_00_00 - x0_00_00 - x1_00_00 + x2_00_00 + x3_00_00 - y0_00_00 - y1_00_00 + y2_00_00 + y3_00_00 >= - 4
- 8 Q02_61_00_00 - x0_00_00 - x1_00_00 + x2_00_00 + x3_00_00 - y0_00_00 + y1_00_00 + y2_00_00 + y3_00_00 >= - 3
- 8 Q02_62_00_00 + x0_00_00 - x1_00_00 + x2_00_00 + x3_00_00 + y0_00_00 - y1_00_00 - y2_00_00 - y3_00_00 >= - 4
- 8 Q02_63_00_00 + x0_00_00 - x1_00_00 + x2_00_00 + x3_00_00 - y0_00_00 + y1_00_00 - y2_00_00 - y3_00_00 >= - 4
- 8 Q02_64_00_00 + x0_00_00 - x1_00_00 + x2_00_00 + x3_00_00 - y0_00_00 + y1_00_00 - y2_00_00 + y3_00_00 >= - 3
- 8 Q02_65_00_00 + x0_00_00 - x1_00_00 + x2_00_00 + x3_00_00 + y0_00_00 + y1_00_00 - y2_00_00 + y3_00_00 >= - 2
- 8 Q02_66_00_00 + x0_00_00 - x1_00_00 + x2_00_00 + x3_00_00 + y0_00_00 - y1_00_00 + y2_00_00 + y3_00_00 >= - 2
- 8 Q02_67_00_00 + x0_00_00 - x1_00_00 + x2_00_00 + x3_00_00 + y0_00_00 + y1_00_00 + y2_00_00 + y3_00_00 >= - 1
- 8 Q02_68_00_00 - x0_00_00 + x1_00_00 + x2_00_00 + x3_00_00 - y0_00_00 - y1_00_00 - y2_00_00 + y3_00_00 >= - 4
- 8 Q02_69_00_00 - x0_00_00 + x1_00_00 + x2_00_00 + x3_00_00 + y0_00_00 - y1_00_00 - y2_00_00 + y3_00_00 >= - 3
- 8 Q02_70_00_00 - x0_00_00 + x1_00_00 + x2_00_00 + x3_00_00 - y0_00_00 - y1_00_00 + y2_00_00 + y3_00_00 >= - 3
- 8 Q02_71_00_00 - x0_00_00 + x1_00_00 + x2_00_00 + x3_00_00 + y0_00_00 - y1_00_00 + y2_00_00 + y3_00_00 >= - 2
- 8 Q02_72_00_00 + x0_00_00 + x1_00_00 + x2_00_00 + x3_00_00 + y0_00_00 - y1_00_00 - y2_00_00 - y3_00_00 >= - 3
- 8 Q02_73_00_00 + x0_00_00 + x1_00_00 + x2_00_00 + x3_00_00 - y0_00_00 + y1_00_00 - y2_00_00 - y3_00_00 >= - 3
- 8 Q02_74_00_00 + x0_00_00 + x1_00_00 + x2_00_00 + x3_00_00 + y0_00_00 - y1_00_00 - y2_00_00 + y3_00_00 >= - 2
- 8 Q02_75_00_00 + x0_00_00 + x1_00_00 + x2_00_00 + x3_00_00 + y0_00_00 + y1_00_00 - y2_00_00 + y3_00_00 >= - 1
- 8 Q02_76_00_00 + x0_00_00 + x1_00_00 + x2_00_00 + x3_00_00 - y0_00_00 + y1_00_00 + y2_00_00 + y3_00_00 >= - 1
- 8 Q02_77_00_00 + x0_00_00 + x1_00_00 + x2_00_00 + x3_00_00 + y0_00_00 + y1_00_00 + y2_00_00 + y3_00_00 >= 0
Q04_00_00 - Q04_00_00_00 - Q04_01_00_00 - Q04_02_00_00 - Q04_03_00_00 - Q04_04_00_00 - Q04_05_00_00 - Q04_06_00_00 - Q04_07_00_00 -
 ↪  Q04_08_00_00 - Q04_09_00_00 - Q04_10_00_00 - Q04_11_00_00 - Q04_12_00_00 - Q04_13_00_00 - Q04_14_00_00 - Q04_15_00_00 -
 ↪  Q04_16_00_00 - Q04_17_00_00 = 0
- 8 Q04_00_00_00 - x0_00_00 + x1_00_00 - x2_00_00 - x3_00_00 + y0_00_00 - y1_00_00 + y2_00_00 - y3_00_00 >= - 5
- 8 Q04_01_00_00 - x0_00_00 + x1_00_00 - x2_00_00 - x3_00_00 - y0_00_00 + y1_00_00 + y2_00_00 - y3_00_00 >= - 5
- 8 Q04_02_00_00 - x0_00_00 - x1_00_00 + x2_00_00 - x3_00_00 + y0_00_00 - y1_00_00 + y2_00_00 - y3_00_00 >= - 5
- 8 Q04_03_00_00 + x0_00_00 - x1_00_00 + x2_00_00 - x3_00_00 + y0_00_00 + y1_00_00 + y2_00_00 + y3_00_00 >= - 2
- 8 Q04_04_00_00 - x0_00_00 + x1_00_00 + x2_00_00 - x3_00_00 - y0_00_00 + y1_00_00 - y2_00_00 - y3_00_00 >= - 5
- 8 Q04_05_00_00 + x0_00_00 + x1_00_00 + x2_00_00 - x3_00_00 + y0_00_00 + y1_00_00 - y2_00_00 + y3_00_00 >= - 2
- 8 Q04_06_00_00 - x0_00_00 - x1_00_00 - x2_00_00 + x3_00_00 + y0_00_00 + y1_00_00 - y2_00_00 - y3_00_00 >= - 5
- 8 Q04_07_00_00 - x0_00_00 - x1_00_00 - x2_00_00 + x3_00_00 + y0_00_00 + y1_00_00 + y2_00_00 - y3_00_00 >= - 4
- 8 Q04_08_00_00 - x0_00_00 - x1_00_00 - x2_00_00 + x3_00_00 + y0_00_00 + y1_00_00 - y2_00_00 + y3_00_00 >= - 4
- 8 Q04_09_00_00 - x0_00_00 - x1_00_00 - x2_00_00 + x3_00_00 + y0_00_00 + y1_00_00 + y2_00_00 + y3_00_00 >= - 3
- 8 Q04_10_00_00 - x0_00_00 + x1_00_00 - x2_00_00 + x3_00_00 + y0_00_00 - y1_00_00 - y2_00_00 - y3_00_00 >= - 5
- 8 Q04_11_00_00 - x0_00_00 + x1_00_00 - x2_00_00 + x3_00_00 - y0_00_00 + y1_00_00 + y2_00_00 - y3_00_00 >= - 4
- 8 Q04_12_00_00 - x0_00_00 - x1_00_00 + x2_00_00 + x3_00_00 - y0_00_00 + y1_00_00 - y2_00_00 - y3_00_00 >= - 5
- 8 Q04_13_00_00 - x0_00_00 - x1_00_00 + x2_00_00 + x3_00_00 - y0_00_00 - y1_00_00 + y2_00_00 - y3_00_00 >= - 5
- 8 Q04_14_00_00 + x0_00_00 - x1_00_00 + x2_00_00 + x3_00_00 - y0_00_00 - y1_00_00 + y2_00_00 - y3_00_00 >= - 4
- 8 Q04_15_00_00 - x0_00_00 + x1_00_00 + x2_00_00 + x3_00_00 + y0_00_00 - y1_00_00 - y2_00_00 - y3_00_00 >= - 4
- 8 Q04_16_00_00 - x0_00_00 + x1_00_00 + x2_00_00 + x3_00_00 - y0_00_00 - y1_00_00 + y2_00_00 - y3_00_00 >= - 4
- 8 Q04_17_00_00 + x0_00_00 + x1_00_00 + x2_00_00 + x3_00_00 - y0_00_00 - y1_00_00 + y2_00_00 - y3_00_00 >= - 3
Q06_00_00 - Q06_00_00_00 - Q06_01_00_00 = 0
- 8 Q06_00_00_00 - x0_00_00 - x1_00_00 + x2_00_00 - x3_00_00 + y0_00_00 + y1_00_00 + y2_00_00 - y3_00_00 >= - 4
- 8 Q06_01_00_00 - x0_00_00 + x1_00_00 + x2_00_00 - x3_00_00 + y0_00_00 + y1_00_00 - y2_00_00 - y3_00_00 >= - 4
- 8 Q_01_00 + x0_01_00 + x1_01_00 + x2_01_00 + x3_01_00 + y0_01_00 + y1_01_00 + y2_01_00 + y3_01_00 <= 0
Q02_01_00 + Q04_01_00 + Q06_01_00 - Q_01_00 = 0

\ The same is repeated for the rest 31 SBoxes

\ Linear layer constraints : Empty

\ Variable type declaration
Binary
\ All 3520 variables are binary, omitted here for brevity

End
```

Fig. 5.5 GIFT-128 1-round differential bound in LP format (part II)

```
\ Constraints based on round 1
3 Q02_00_00 + 3 Q02_01_00 + 3 Q02_02_00 + 3 Q02_03_00 + 3 Q02_04_00 + 3 Q02_05_00 + 3 Q02_06_00 + 3 Q02_07_00 + 3 Q02_08_00 + 3
  ↪ Q02_09_00 + 3 Q02_10_00 + 3 Q02_11_00 + 3 Q02_12_00 + 3 Q02_13_00 + 3 Q02_14_00 + 3 Q02_15_00 + 3 Q02_16_00 + 3 Q02_17_00 + 3
  ↪ Q02_18_00 + 3 Q02_19_00 + 3 Q02_20_00 + 3 Q02_21_00 + 3 Q02_22_00 + 3 Q02_23_00 + 3 Q02_24_00 + 3 Q02_25_00 + 3 Q02_26_00 + 3
  ↪ Q02_27_00 + 3 Q02_28_00 + 3 Q02_29_00 + 3 Q02_30_00 + 3 Q02_31_00 + 2 Q04_00_00 + 2 Q04_01_00 + 2 Q04_02_00 + 2 Q04_03_00 + 2
  ↪ Q04_04_00 + 2 Q04_05_00 + 2 Q04_06_00 + 2 Q04_07_00 + 2 Q04_08_00 + 2 Q04_09_00 + 2 Q04_10_00 + 2 Q04_11_00 + 2 Q04_12_00 + 2
  ↪ Q04_13_00 + 2 Q04_14_00 + 2 Q04_15_00 + 2 Q04_16_00 + 2 Q04_17_00 + 2 Q04_18_00 + 2 Q04_19_00 + 2 Q04_20_00 + 2 Q04_21_00 + 2
  ↪ Q04_22_00 + 2 Q04_23_00 + 2 Q04_24_00 + 2 Q04_25_00 + 2 Q04_26_00 + 2 Q04_27_00 + 2 Q04_28_00 + 2 Q04_29_00 + 2 Q04_30_00 + 2
  ↪ Q04_31_00 + 1.415037 Q06_00_00 + 1.415037 Q06_01_00 + 1.415037 Q06_02_00 + 1.415037 Q06_03_00 + 1.415037 Q06_04_00 + 1.415037
  ↪ Q06_05_00 + 1.415037 Q06_06_00 + 1.415037 Q06_07_00 + 1.415037 Q06_08_00 + 1.415037 Q06_09_00 + 1.415037 Q06_10_00 + 1.415037
  ↪ Q06_11_00 + 1.415037 Q06_12_00 + 1.415037 Q06_13_00 + 1.415037 Q06_14_00 + 1.415037 Q06_15_00 + 1.415037 Q06_16_00 + 1.415037
  ↪ Q06_17_00 + 1.415037 Q06_18_00 + 1.415037 Q06_19_00 + 1.415037 Q06_20_00 + 1.415037 Q06_21_00 + 1.415037 Q06_22_00 + 1.415037
  ↪ Q06_23_00 + 1.415037 Q06_24_00 + 1.415037 Q06_25_00 + 1.415037 Q06_26_00 + 1.415037 Q06_27_00 + 1.415037 Q06_28_00 + 1.415037
  ↪ Q06_29_00 + 1.415037 Q06_30_00 + 1.415037 Q06_31_00 >= 1.415037
3 Q02_00_01 + 3 Q02_01_01 + 3 Q02_02_01 + 3 Q02_03_01 + 3 Q02_04_01 + 3 Q02_05_01 + 3 Q02_06_01 + 3 Q02_07_01 + 3 Q02_08_01 + 3
  ↪ Q02_09_01 + 3 Q02_10_01 + 3 Q02_11_01 + 3 Q02_12_01 + 3 Q02_13_01 + 3 Q02_14_01 + 3 Q02_15_01 + 3 Q02_16_01 + 3 Q02_17_01 + 3
  ↪ Q02_18_01 + 3 Q02_19_01 + 3 Q02_20_01 + 3 Q02_21_01 + 3 Q02_22_01 + 3 Q02_23_01 + 3 Q02_24_01 + 3 Q02_25_01 + 3 Q02_26_01 + 3
  ↪ Q02_27_01 + 3 Q02_28_01 + 3 Q02_29_01 + 3 Q02_30_01 + 3 Q02_31_01 + 2 Q04_00_01 + 2 Q04_01_01 + 2 Q04_02_01 + 2 Q04_03_01 + 2
  ↪ Q04_04_01 + 2 Q04_05_01 + 2 Q04_06_01 + 2 Q04_07_01 + 2 Q04_08_01 + 2 Q04_09_01 + 2 Q04_10_01 + 2 Q04_11_01 + 2 Q04_12_01 + 2
  ↪ Q04_13_01 + 2 Q04_14_01 + 2 Q04_15_01 + 2 Q04_16_01 + 2 Q04_17_01 + 2 Q04_18_01 + 2 Q04_19_01 + 2 Q04_20_01 + 2 Q04_21_01 + 2
  ↪ Q04_22_01 + 2 Q04_23_01 + 2 Q04_24_01 + 2 Q04_25_01 + 2 Q04_26_01 + 2 Q04_27_01 + 2 Q04_28_01 + 2 Q04_29_01 + 2 Q04_30_01 + 2
  ↪ Q04_31_01 + 1.415037 Q06_00_01 + 1.415037 Q06_01_01 + 1.415037 Q06_02_01 + 1.415037 Q06_03_01 + 1.415037 Q06_04_01 + 1.415037
  ↪ Q06_05_01 + 1.415037 Q06_06_01 + 1.415037 Q06_07_01 + 1.415037 Q06_08_01 + 1.415037 Q06_09_01 + 1.415037 Q06_10_01 + 1.415037
  ↪ Q06_11_01 + 1.415037 Q06_12_01 + 1.415037 Q06_13_01 + 1.415037 Q06_14_01 + 1.415037 Q06_15_01 + 1.415037 Q06_16_01 + 1.415037
  ↪ Q06_17_01 + 1.415037 Q06_18_01 + 1.415037 Q06_19_01 + 1.415037 Q06_20_01 + 1.415037 Q06_21_01 + 1.415037 Q06_22_01 + 1.415037
  ↪ Q06_23_01 + 1.415037 Q06_24_01 + 1.415037 Q06_25_01 + 1.415037 Q06_26_01 + 1.415037 Q06_27_01 + 1.415037 Q06_28_01 + 1.415037
  ↪ Q06_29_01 + 1.415037 Q06_30_01 + 1.415037 Q06_31_01 >= 1.415037
3 Q02_00_02 + 3 Q02_01_02 + 3 Q02_02_02 + 3 Q02_03_02 + 3 Q02_04_02 + 3 Q02_05_02 + 3 Q02_06_02 + 3 Q02_07_02 + 3 Q02_08_02 + 3
  ↪ Q02_09_02 + 3 Q02_10_02 + 3 Q02_11_02 + 3 Q02_12_02 + 3 Q02_13_02 + 3 Q02_14_02 + 3 Q02_15_02 + 3 Q02_16_02 + 3 Q02_17_02 + 3
  ↪ Q02_18_02 + 3 Q02_19_02 + 3 Q02_20_02 + 3 Q02_21_02 + 3 Q02_22_02 + 3 Q02_23_02 + 3 Q02_24_02 + 3 Q02_25_02 + 3 Q02_26_02 + 3
  ↪ Q02_27_02 + 3 Q02_28_02 + 3 Q02_29_02 + 3 Q02_30_02 + 3 Q02_31_02 + 2 Q04_00_02 + 2 Q04_01_02 + 2 Q04_02_02 + 2 Q04_03_02 + 2
  ↪ Q04_04_02 + 2 Q04_05_02 + 2 Q04_06_02 + 2 Q04_07_02 + 2 Q04_08_02 + 2 Q04_09_02 + 2 Q04_10_02 + 2 Q04_11_02 + 2 Q04_12_02 + 2
  ↪ Q04_13_02 + 2 Q04_14_02 + 2 Q04_15_02 + 2 Q04_16_02 + 2 Q04_17_02 + 2 Q04_18_02 + 2 Q04_19_02 + 2 Q04_20_02 + 2 Q04_21_02 + 2
  ↪ Q04_22_02 + 2 Q04_23_02 + 2 Q04_24_02 + 2 Q04_25_02 + 2 Q04_26_02 + 2 Q04_27_02 + 2 Q04_28_02 + 2 Q04_29_02 + 2 Q04_30_02 + 2
  ↪ Q04_31_02 + 1.415037 Q06_00_02 + 1.415037 Q06_01_02 + 1.415037 Q06_02_02 + 1.415037 Q06_03_02 + 1.415037 Q06_04_02 + 1.415037
  ↪ Q06_05_02 + 1.415037 Q06_06_02 + 1.415037 Q06_07_02 + 1.415037 Q06_08_02 + 1.415037 Q06_09_02 + 1.415037 Q06_10_02 + 1.415037
  ↪ Q06_11_02 + 1.415037 Q06_12_02 + 1.415037 Q06_13_02 + 1.415037 Q06_14_02 + 1.415037 Q06_15_02 + 1.415037 Q06_16_02 + 1.415037
  ↪ Q06_17_02 + 1.415037 Q06_18_02 + 1.415037 Q06_19_02 + 1.415037 Q06_20_02 + 1.415037 Q06_21_02 + 1.415037 Q06_22_02 + 1.415037
  ↪ Q06_23_02 + 1.415037 Q06_24_02 + 1.415037 Q06_25_02 + 1.415037 Q06_26_02 + 1.415037 Q06_27_02 + 1.415037 Q06_28_02 + 1.415037
  ↪ Q06_29_02 + 1.415037 Q06_30_02 + 1.415037 Q06_31_02 >= 1.415037
3 Q02_00_03 + 3 Q02_01_03 + 3 Q02_02_03 + 3 Q02_03_03 + 3 Q02_04_03 + 3 Q02_05_03 + 3 Q02_06_03 + 3 Q02_07_03 + 3 Q02_08_03 + 3
  ↪ Q02_09_03 + 3 Q02_10_03 + 3 Q02_11_03 + 3 Q02_12_03 + 3 Q02_13_03 + 3 Q02_14_03 + 3 Q02_15_03 + 3 Q02_16_03 + 3 Q02_17_03 + 3
  ↪ Q02_18_03 + 3 Q02_19_03 + 3 Q02_20_03 + 3 Q02_21_03 + 3 Q02_22_03 + 3 Q02_23_03 + 3 Q02_24_03 + 3 Q02_25_03 + 3 Q02_26_03 + 3
  ↪ Q02_27_03 + 3 Q02_28_03 + 3 Q02_29_03 + 3 Q02_30_03 + 3 Q02_31_03 + 2 Q04_00_03 + 2 Q04_01_03 + 2 Q04_02_03 + 2 Q04_03_03 + 2
  ↪ Q04_04_03 + 2 Q04_05_03 + 2 Q04_06_03 + 2 Q04_07_03 + 2 Q04_08_03 + 2 Q04_09_03 + 2 Q04_10_03 + 2 Q04_11_03 + 2 Q04_12_03 + 2
  ↪ Q04_13_03 + 2 Q04_14_03 + 2 Q04_15_03 + 2 Q04_16_03 + 2 Q04_17_03 + 2 Q04_18_03 + 2 Q04_19_03 + 2 Q04_20_03 + 2 Q04_21_03 + 2
  ↪ Q04_22_03 + 2 Q04_23_03 + 2 Q04_24_03 + 2 Q04_25_03 + 2 Q04_26_03 + 2 Q04_27_03 + 2 Q04_28_03 + 2 Q04_29_03 + 2 Q04_30_03 + 2
  ↪ Q04_31_03 + 1.415037 Q06_00_03 + 1.415037 Q06_01_03 + 1.415037 Q06_02_03 + 1.415037 Q06_03_03 + 1.415037 Q06_04_03 + 1.415037
  ↪ Q06_05_03 + 1.415037 Q06_06_03 + 1.415037 Q06_07_03 + 1.415037 Q06_08_03 + 1.415037 Q06_09_03 + 1.415037 Q06_10_03 + 1.415037
  ↪ Q06_11_03 + 1.415037 Q06_12_03 + 1.415037 Q06_13_03 + 1.415037 Q06_14_03 + 1.415037 Q06_15_03 + 1.415037 Q06_16_03 + 1.415037
  ↪ Q06_17_03 + 1.415037 Q06_18_03 + 1.415037 Q06_19_03 + 1.415037 Q06_20_03 + 1.415037 Q06_21_03 + 1.415037 Q06_22_03 + 1.415037
  ↪ Q06_23_03 + 1.415037 Q06_24_03 + 1.415037 Q06_25_03 + 1.415037 Q06_26_03 + 1.415037 Q06_27_03 + 1.415037 Q06_28_03 + 1.415037
  ↪ Q06_29_03 + 1.415037 Q06_30_03 + 1.415037 Q06_31_03 >= 1.415037

\ Constraints based on round 2
3 Q02_00_00 + 3 Q02_00_01 + 3 Q02_01_00 + 3 Q02_01_01 + 3 Q02_02_00 + 3 Q02_02_01 + 3 Q02_03_00 + 3 Q02_03_01 + 3 Q02_04_00 + 3
  ↪ Q02_04_01 + 3 Q02_05_00 + 3 Q02_05_01 + 3 Q02_06_00 + 3 Q02_06_01 + 3 Q02_07_00 + 3 Q02_07_01 + 3 Q02_08_00 + 3 Q02_08_01 + 3
  ↪ Q02_09_00 + 3 Q02_09_01 + 3 Q02_10_00 + 3 Q02_10_01 + 3 Q02_11_00 + 3 Q02_11_01 + 3 Q02_12_00 + 3 Q02_12_01 + 3 Q02_13_00 + 3
  ↪ Q02_13_01 + 3 Q02_14_00 + 3 Q02_14_01 + 3 Q02_15_00 + 3 Q02_15_01 + 3 Q02_16_00 + 3 Q02_16_01 + 3 Q02_17_00 + 3 Q02_17_01 + 3
  ↪ Q02_18_00 + 3 Q02_18_01 + 3 Q02_19_00 + 3 Q02_19_01 + 3 Q02_20_00 + 3 Q02_20_01 + 3 Q02_21_00 + 3 Q02_21_01 + 3 Q02_22_00 + 3
  ↪ Q02_22_01 + 3 Q02_23_00 + 3 Q02_23_01 + 3 Q02_24_00 + 3 Q02_24_01 + 3 Q02_25_00 + 3 Q02_25_01 + 3 Q02_26_00 + 3 Q02_26_01 + 3
  ↪ Q02_27_00 + 3 Q02_27_01 + 3 Q02_28_00 + 3 Q02_28_01 + 3 Q02_29_00 + 3 Q02_29_01 + 3 Q02_30_00 + 3 Q02_30_01 + 3 Q02_31_00 + 3
  ↪ Q02_31_01 + 2 Q04_00_00 + 2 Q04_00_01 + 2 Q04_01_00 + 2 Q04_01_01 + 2 Q04_02_00 + 2 Q04_02_01 + 2 Q04_03_00 + 2 Q04_03_01 + 2
  ↪ Q04_04_00 + 2 Q04_04_01 + 2 Q04_05_00 + 2 Q04_05_01 + 2 Q04_06_00 + 2 Q04_06_01 + 2 Q04_07_00 + 2 Q04_07_01 + 2 Q04_08_00 + 2
  ↪ Q04_08_01 + 2 Q04_09_00 + 2 Q04_09_01 + 2 Q04_10_00 + 2 Q04_10_01 + 2 Q04_11_00 + 2 Q04_11_01 + 2 Q04_12_00 + 2 Q04_12_01 + 2
  ↪ Q04_13_00 + 2 Q04_13_01 + 2 Q04_14_00 + 2 Q04_14_01 + 2 Q04_15_00 + 2 Q04_15_01 + 2 Q04_16_00 + 2 Q04_16_01 + 2 Q04_17_00 + 2
  ↪ Q04_17_01 + 2 Q04_18_00 + 2 Q04_18_01 + 2 Q04_19_00 + 2 Q04_19_01 + 2 Q04_20_00 + 2 Q04_20_01 + 2 Q04_21_00 + 2 Q04_21_01 + 2
  ↪ Q04_22_00 + 2 Q04_22_01 + 2 Q04_23_00 + 2 Q04_23_01 + 2 Q04_24_00 + 2 Q04_24_01 + 2 Q04_25_00 + 2 Q04_25_01 + 2 Q04_26_00 + 2
  ↪ Q04_26_01 + 2 Q04_27_00 + 2 Q04_27_01 + 2 Q04_28_00 + 2 Q04_28_01 + 2 Q04_29_00 + 2 Q04_29_01 + 2 Q04_30_00 + 2 Q04_30_01 + 2
  ↪ Q04_31_00 + 2 Q04_31_01 + 1.415037 Q06_00_00 + 1.415037 Q06_00_01 + 1.415037 Q06_01_00 + 1.415037 Q06_01_01 + 1.415037
  ↪ Q06_02_00 + 1.415037 Q06_02_01 + 1.415037 Q06_03_00 + 1.415037 Q06_03_01 + 1.415037 Q06_04_00 + 1.415037 Q06_04_01 + 1.415037
  ↪ Q06_05_00 + 1.415037 Q06_05_01 + 1.415037 Q06_06_00 + 1.415037 Q06_06_01 + 1.415037 Q06_07_00 + 1.415037 Q06_07_01 + 1.415037
  ↪ Q06_08_00 + 1.415037 Q06_08_01 + 1.415037 Q06_09_00 + 1.415037 Q06_09_01 + 1.415037 Q06_10_00 + 1.415037 Q06_10_01 + 1.415037
  ↪ Q06_11_00 + 1.415037 Q06_11_01 + 1.415037 Q06_12_00 + 1.415037 Q06_12_01 + 1.415037 Q06_13_00 + 1.415037 Q06_13_01 + 1.415037
  ↪ Q06_14_00 + 1.415037 Q06_14_01 + 1.415037 Q06_15_00 + 1.415037 Q06_15_01 + 1.415037 Q06_16_00 + 1.415037 Q06_16_01 + 1.415037
  ↪ Q06_17_00 + 1.415037 Q06_17_01 + 1.415037 Q06_18_00 + 1.415037 Q06_18_01 + 1.415037 Q06_19_00 + 1.415037 Q06_19_01 + 1.415037
  ↪ Q06_20_00 + 1.415037 Q06_20_01 + 1.415037 Q06_21_00 + 1.415037 Q06_21_01 + 1.415037 Q06_22_00 + 1.415037 Q06_22_01 + 1.415037
  ↪ Q06_23_00 + 1.415037 Q06_23_01 + 1.415037 Q06_24_00 + 1.415037 Q06_24_01 + 1.415037 Q06_25_00 + 1.415037 Q06_25_01 + 1.415037
  ↪ Q06_26_00 + 1.415037 Q06_26_01 + 1.415037 Q06_27_00 + 1.415037 Q06_27_01 + 1.415037 Q06_28_00 + 1.415037 Q06_28_01 + 1.415037
  ↪ Q06_29_00 + 1.415037 Q06_29_01 + 1.415037 Q06_30_00 + 1.415037 Q06_30_01 + 1.415037 Q06_31_00 + 1.415037 Q06_31_01 >= 3.415037
```

Fig. 5.6 Previous constraints for GIFT-128 4-round differential bound in LP format (part I)

```
9   3 Q02_00_01 + 3 Q02_00_02 + 3 Q02_01_01 + 3 Q02_01_02 + 3 Q02_02_01 + 3 Q02_02_02 + 3 Q02_03_01 + 3 Q02_03_02 + 3 Q02_04_01 + 3
    ↪ Q02_04_02 + 3 Q02_05_01 + 3 Q02_05_02 + 3 Q02_06_01 + 3 Q02_06_02 + 3 Q02_07_01 + 3 Q02_07_02 + 3 Q02_08_01 + 3 Q02_08_02 + 3
    ↪ Q02_09_01 + 3 Q02_09_02 + 3 Q02_10_01 + 3 Q02_10_02 + 3 Q02_11_01 + 3 Q02_11_02 + 3 Q02_12_01 + 3 Q02_12_02 + 3 Q02_13_01 + 3
    ↪ Q02_13_02 + 3 Q02_14_01 + 3 Q02_14_02 + 3 Q02_15_01 + 3 Q02_15_02 + 3 Q02_16_01 + 3 Q02_16_02 + 3 Q02_17_01 + 3 Q02_17_02 + 3
    ↪ Q02_18_01 + 3 Q02_18_02 + 3 Q02_19_01 + 3 Q02_19_02 + 3 Q02_20_01 + 3 Q02_20_02 + 3 Q02_21_01 + 3 Q02_21_02 + 3 Q02_22_01 + 3
    ↪ Q02_22_02 + 3 Q02_23_01 + 3 Q02_23_02 + 3 Q02_24_01 + 3 Q02_24_02 + 3 Q02_25_01 + 3 Q02_25_02 + 3 Q02_26_01 + 3 Q02_26_02 + 3
    ↪ Q02_27_01 + 3 Q02_27_02 + 3 Q02_28_01 + 3 Q02_28_02 + 3 Q02_29_01 + 3 Q02_29_02 + 3 Q02_30_01 + 3 Q02_30_02 + 3 Q02_31_01 + 3
    ↪ Q02_31_02 + 2 Q04_00_01 + 2 Q04_00_02 + 2 Q04_01_01 + 2 Q04_01_02 + 2 Q04_02_01 + 2 Q04_02_02 + 2 Q04_03_01 + 2 Q04_03_02 + 2
    ↪ Q04_04_01 + 2 Q04_04_02 + 2 Q04_05_01 + 2 Q04_05_02 + 2 Q04_06_01 + 2 Q04_06_02 + 2 Q04_07_01 + 2 Q04_07_02 + 2 Q04_08_01 + 2
    ↪ Q04_08_02 + 2 Q04_09_01 + 2 Q04_09_02 + 2 Q04_10_01 + 2 Q04_10_02 + 2 Q04_11_01 + 2 Q04_11_02 + 2 Q04_12_01 + 2 Q04_12_02 + 2
    ↪ Q04_13_01 + 2 Q04_13_02 + 2 Q04_14_01 + 2 Q04_14_02 + 2 Q04_15_01 + 2 Q04_15_02 + 2 Q04_16_01 + 2 Q04_16_02 + 2 Q04_17_01 + 2
    ↪ Q04_17_02 + 2 Q04_18_01 + 2 Q04_18_02 + 2 Q04_19_01 + 2 Q04_19_02 + 2 Q04_20_01 + 2 Q04_20_02 + 2 Q04_21_01 + 2 Q04_21_02 + 2
    ↪ Q04_22_01 + 2 Q04_22_02 + 2 Q04_23_01 + 2 Q04_23_02 + 2 Q04_24_01 + 2 Q04_24_02 + 2 Q04_25_01 + 2 Q04_25_02 + 2 Q04_26_01 + 2
    ↪ Q04_26_02 + 2 Q04_27_01 + 2 Q04_27_02 + 2 Q04_28_01 + 2 Q04_28_02 + 2 Q04_29_01 + 2 Q04_29_02 + 2 Q04_30_01 + 2 Q04_30_02 + 2
    ↪ Q04_31_01 + 2 Q04_31_02 + 1.415037 Q06_00_01 + 1.415037 Q06_00_02 + 1.415037 Q06_01_01 + 1.415037 Q06_01_02 + 1.415037
    ↪ Q06_02_01 + 1.415037 Q06_02_02 + 1.415037 Q06_03_01 + 1.415037 Q06_03_02 + 1.415037 Q06_04_01 + 1.415037 Q06_04_02 + 1.415037
    ↪ Q06_05_01 + 1.415037 Q06_05_02 + 1.415037 Q06_06_01 + 1.415037 Q06_06_02 + 1.415037 Q06_07_01 + 1.415037 Q06_07_02 + 1.415037
    ↪ Q06_08_01 + 1.415037 Q06_08_02 + 1.415037 Q06_09_01 + 1.415037 Q06_09_02 + 1.415037 Q06_10_01 + 1.415037 Q06_10_02 + 1.415037
    ↪ Q06_11_01 + 1.415037 Q06_11_02 + 1.415037 Q06_12_01 + 1.415037 Q06_12_02 + 1.415037 Q06_13_01 + 1.415037 Q06_13_02 + 1.415037
    ↪ Q06_14_01 + 1.415037 Q06_14_02 + 1.415037 Q06_15_01 + 1.415037 Q06_15_02 + 1.415037 Q06_16_01 + 1.415037 Q06_16_02 + 1.415037
    ↪ Q06_17_01 + 1.415037 Q06_17_02 + 1.415037 Q06_18_01 + 1.415037 Q06_18_02 + 1.415037 Q06_19_01 + 1.415037 Q06_19_02 + 1.415037
    ↪ Q06_20_01 + 1.415037 Q06_20_02 + 1.415037 Q06_21_01 + 1.415037 Q06_21_02 + 1.415037 Q06_22_01 + 1.415037 Q06_22_02 + 1.415037
    ↪ Q06_23_01 + 1.415037 Q06_23_02 + 1.415037 Q06_24_01 + 1.415037 Q06_24_02 + 1.415037 Q06_25_01 + 1.415037 Q06_25_02 + 1.415037
    ↪ Q06_26_01 + 1.415037 Q06_26_02 + 1.415037 Q06_27_01 + 1.415037 Q06_27_02 + 1.415037 Q06_28_01 + 1.415037 Q06_28_02 + 1.415037
    ↪ Q06_29_01 + 1.415037 Q06_29_02 + 1.415037 Q06_30_01 + 1.415037 Q06_30_02 + 1.415037 Q06_31_01 + 1.415037 Q06_31_02 >= 3.415037
10  3 Q02_00_02 + 3 Q02_00_03 + 3 Q02_01_02 + 3 Q02_01_03 + 3 Q02_02_02 + 3 Q02_02_03 + 3 Q02_03_02 + 3 Q02_03_03 + 3 Q02_04_02 + 3
    ↪ Q02_04_03 + 3 Q02_05_02 + 3 Q02_05_03 + 3 Q02_06_02 + 3 Q02_06_03 + 3 Q02_07_02 + 3 Q02_07_03 + 3 Q02_08_02 + 3 Q02_08_03 + 3
    ↪ Q02_09_02 + 3 Q02_09_03 + 3 Q02_10_02 + 3 Q02_10_03 + 3 Q02_11_02 + 3 Q02_11_03 + 3 Q02_12_02 + 3 Q02_12_03 + 3 Q02_13_02 + 3
    ↪ Q02_13_03 + 3 Q02_14_02 + 3 Q02_14_03 + 3 Q02_15_02 + 3 Q02_15_03 + 3 Q02_16_02 + 3 Q02_16_03 + 3 Q02_17_02 + 3 Q02_17_03 + 3
    ↪ Q02_18_02 + 3 Q02_18_03 + 3 Q02_19_02 + 3 Q02_19_03 + 3 Q02_20_02 + 3 Q02_20_03 + 3 Q02_21_02 + 3 Q02_21_03 + 3 Q02_22_02 + 3
    ↪ Q02_22_03 + 3 Q02_23_02 + 3 Q02_23_03 + 3 Q02_24_02 + 3 Q02_24_03 + 3 Q02_25_02 + 3 Q02_25_03 + 3 Q02_26_02 + 3 Q02_26_03 + 3
    ↪ Q02_27_02 + 3 Q02_27_03 + 3 Q02_28_02 + 3 Q02_28_03 + 3 Q02_29_02 + 3 Q02_29_03 + 3 Q02_30_02 + 3 Q02_30_03 + 3 Q02_31_02 + 3
    ↪ Q02_31_03 + 2 Q04_00_02 + 2 Q04_00_03 + 2 Q04_01_02 + 2 Q04_01_03 + 2 Q04_02_02 + 2 Q04_02_03 + 2 Q04_03_02 + 2 Q04_03_03 + 2
    ↪ Q04_04_02 + 2 Q04_04_03 + 2 Q04_05_02 + 2 Q04_05_03 + 2 Q04_06_02 + 2 Q04_06_03 + 2 Q04_07_02 + 2 Q04_07_03 + 2 Q04_08_02 + 2
    ↪ Q04_08_03 + 2 Q04_09_02 + 2 Q04_09_03 + 2 Q04_10_02 + 2 Q04_10_03 + 2 Q04_11_02 + 2 Q04_11_03 + 2 Q04_12_02 + 2 Q04_12_03 + 2
    ↪ Q04_13_02 + 2 Q04_13_03 + 2 Q04_14_02 + 2 Q04_14_03 + 2 Q04_15_02 + 2 Q04_15_03 + 2 Q04_16_02 + 2 Q04_16_03 + 2 Q04_17_02 + 2
    ↪ Q04_17_03 + 2 Q04_18_02 + 2 Q04_18_03 + 2 Q04_19_02 + 2 Q04_19_03 + 2 Q04_20_02 + 2 Q04_20_03 + 2 Q04_21_02 + 2 Q04_21_03 + 2
    ↪ Q04_22_02 + 2 Q04_22_03 + 2 Q04_23_02 + 2 Q04_23_03 + 2 Q04_24_02 + 2 Q04_24_03 + 2 Q04_25_02 + 2 Q04_25_03 + 2 Q04_26_02 + 2
    ↪ Q04_26_03 + 2 Q04_27_02 + 2 Q04_27_03 + 2 Q04_28_02 + 2 Q04_28_03 + 2 Q04_29_02 + 2 Q04_29_03 + 2 Q04_30_02 + 2 Q04_30_03 + 2
    ↪ Q04_31_02 + 2 Q04_31_03 + 1.415037 Q06_00_02 + 1.415037 Q06_00_03 + 1.415037 Q06_01_02 + 1.415037 Q06_01_03 + 1.415037
    ↪ Q06_02_02 + 1.415037 Q06_02_03 + 1.415037 Q06_03_02 + 1.415037 Q06_03_03 + 1.415037 Q06_04_02 + 1.415037 Q06_04_03 + 1.415037
    ↪ Q06_05_02 + 1.415037 Q06_05_03 + 1.415037 Q06_06_02 + 1.415037 Q06_06_03 + 1.415037 Q06_07_02 + 1.415037 Q06_07_03 + 1.415037
    ↪ Q06_08_02 + 1.415037 Q06_08_03 + 1.415037 Q06_09_02 + 1.415037 Q06_09_03 + 1.415037 Q06_10_02 + 1.415037 Q06_10_03 + 1.415037
    ↪ Q06_11_02 + 1.415037 Q06_11_03 + 1.415037 Q06_12_02 + 1.415037 Q06_12_03 + 1.415037 Q06_13_02 + 1.415037 Q06_13_03 + 1.415037
    ↪ Q06_14_02 + 1.415037 Q06_14_03 + 1.415037 Q06_15_02 + 1.415037 Q06_15_03 + 1.415037 Q06_16_02 + 1.415037 Q06_16_03 + 1.415037
    ↪ Q06_17_02 + 1.415037 Q06_17_03 + 1.415037 Q06_18_02 + 1.415037 Q06_18_03 + 1.415037 Q06_19_02 + 1.415037 Q06_19_03 + 1.415037
    ↪ Q06_20_02 + 1.415037 Q06_20_03 + 1.415037 Q06_21_02 + 1.415037 Q06_21_03 + 1.415037 Q06_22_02 + 1.415037 Q06_22_03 + 1.415037
    ↪ Q06_23_02 + 1.415037 Q06_23_03 + 1.415037 Q06_24_02 + 1.415037 Q06_24_03 + 1.415037 Q06_25_02 + 1.415037 Q06_25_03 + 1.415037
    ↪ Q06_26_02 + 1.415037 Q06_26_03 + 1.415037 Q06_27_02 + 1.415037 Q06_27_03 + 1.415037 Q06_28_02 + 1.415037 Q06_28_03 + 1.415037
    ↪ Q06_29_02 + 1.415037 Q06_29_03 + 1.415037 Q06_30_02 + 1.415037 Q06_30_03 + 1.415037 Q06_31_02 + 1.415037 Q06_31_03 >= 3.415037
11
12  \ Constraints based on round 3
```

Fig. 5.7 Previous constraints for GIFT-128 4-round differential bound in LP format (part II)

(only the previous constraints are given for compactness). Notice from Table 5.4 that the (approximated) bounds for GIFT-128 differential are 1.415, 3.415, and 7.000, respectively, for first three rounds. The 1-round bound is used 4 times (Lines 2–5), the 2-round bound is used 3 times (Lines 8–10), and the 3-round bound is used 2 times (Lines 13, 14).

```
3 Q02_00_00 + 3 Q02_00_01 + 3 Q02_00_02 + 3 Q02_01_00 + 3 Q02_01_01 + 3 Q02_01_02 + 3 Q02_02_00 + 3 Q02_02_01 + 3 Q02_02_02 + 3
↪ Q02_03_00 + 3 Q02_03_01 + 3 Q02_03_02 + 3 Q02_04_00 + 3 Q02_04_01 + 3 Q02_04_02 + 3 Q02_05_00 + 3 Q02_05_01 + 3 Q02_05_02 + 3
↪ Q02_06_00 + 3 Q02_06_01 + 3 Q02_06_02 + 3 Q02_07_00 + 3 Q02_07_01 + 3 Q02_07_02 + 3 Q02_08_00 + 3 Q02_08_01 + 3 Q02_08_02 + 3
↪ Q02_09_00 + 3 Q02_09_01 + 3 Q02_09_02 + 3 Q02_10_00 + 3 Q02_10_01 + 3 Q02_10_02 + 3 Q02_11_00 + 3 Q02_11_01 + 3 Q02_11_02 + 3
↪ Q02_12_00 + 3 Q02_12_01 + 3 Q02_12_02 + 3 Q02_13_00 + 3 Q02_13_01 + 3 Q02_13_02 + 3 Q02_14_00 + 3 Q02_14_01 + 3 Q02_14_02 + 3
↪ Q02_15_00 + 3 Q02_15_01 + 3 Q02_15_02 + 3 Q02_16_00 + 3 Q02_16_01 + 3 Q02_16_02 + 3 Q02_17_00 + 3 Q02_17_01 + 3 Q02_17_02 + 3
↪ Q02_18_00 + 3 Q02_18_01 + 3 Q02_18_02 + 3 Q02_19_00 + 3 Q02_19_01 + 3 Q02_19_02 + 3 Q02_20_00 + 3 Q02_20_01 + 3 Q02_20_02 + 3
↪ Q02_21_00 + 3 Q02_21_01 + 3 Q02_21_02 + 3 Q02_22_00 + 3 Q02_22_01 + 3 Q02_22_02 + 3 Q02_23_00 + 3 Q02_23_01 + 3 Q02_23_02 + 3
↪ Q02_24_00 + 3 Q02_24_01 + 3 Q02_24_02 + 3 Q02_25_00 + 3 Q02_25_01 + 3 Q02_25_02 + 3 Q02_26_00 + 3 Q02_26_01 + 3 Q02_26_02 + 3
↪ Q02_27_00 + 3 Q02_27_01 + 3 Q02_27_02 + 3 Q02_28_00 + 3 Q02_28_01 + 3 Q02_28_02 + 3 Q02_29_00 + 3 Q02_29_01 + 3 Q02_29_02 + 3
↪ Q02_30_00 + 3 Q02_30_01 + 3 Q02_30_02 + 3 Q02_31_00 + 3 Q02_31_01 + 3 Q02_31_02 + 2 Q04_00_00 + 2 Q04_00_01 + 2 Q04_00_02 + 2
↪ Q04_01_00 + 2 Q04_01_01 + 2 Q04_01_02 + 2 Q04_02_00 + 2 Q04_02_01 + 2 Q04_02_02 + 2 Q04_03_00 + 2 Q04_03_01 + 2 Q04_03_02 + 2
↪ Q04_04_00 + 2 Q04_04_01 + 2 Q04_04_02 + 2 Q04_05_00 + 2 Q04_05_01 + 2 Q04_05_02 + 2 Q04_06_00 + 2 Q04_06_01 + 2 Q04_06_02 + 2
↪ Q04_07_00 + 2 Q04_07_01 + 2 Q04_07_02 + 2 Q04_08_00 + 2 Q04_08_01 + 2 Q04_08_02 + 2 Q04_09_00 + 2 Q04_09_01 + 2 Q04_09_02 + 2
↪ Q04_10_00 + 2 Q04_10_01 + 2 Q04_10_02 + 2 Q04_11_00 + 2 Q04_11_01 + 2 Q04_11_02 + 2 Q04_12_00 + 2 Q04_12_01 + 2 Q04_12_02 + 2
↪ Q04_13_00 + 2 Q04_13_01 + 2 Q04_13_02 + 2 Q04_14_00 + 2 Q04_14_01 + 2 Q04_14_02 + 2 Q04_15_00 + 2 Q04_15_01 + 2 Q04_15_02 + 2
↪ Q04_16_00 + 2 Q04_16_01 + 2 Q04_16_02 + 2 Q04_17_00 + 2 Q04_17_01 + 2 Q04_17_02 + 2 Q04_18_00 + 2 Q04_18_01 + 2 Q04_18_02 + 2
↪ Q04_19_00 + 2 Q04_19_01 + 2 Q04_19_02 + 2 Q04_20_00 + 2 Q04_20_01 + 2 Q04_20_02 + 2 Q04_21_00 + 2 Q04_21_01 + 2 Q04_21_02 + 2
↪ Q04_22_00 + 2 Q04_22_01 + 2 Q04_22_02 + 2 Q04_23_00 + 2 Q04_23_01 + 2 Q04_23_02 + 2 Q04_24_00 + 2 Q04_24_01 + 2 Q04_24_02 + 2
↪ Q04_25_00 + 2 Q04_25_01 + 2 Q04_25_02 + 2 Q04_26_00 + 2 Q04_26_01 + 2 Q04_26_02 + 2 Q04_27_00 + 2 Q04_27_01 + 2 Q04_27_02 + 2
↪ Q04_28_00 + 2 Q04_28_01 + 2 Q04_28_02 + 2 Q04_29_00 + 2 Q04_29_01 + 2 Q04_29_02 + 2 Q04_30_00 + 2 Q04_30_01 + 2 Q04_30_02 + 2
↪ Q04_31_00 + 2 Q04_31_01 + 2 Q04_31_02 + 1.415037 Q06_00_00 + 1.415037 Q06_00_01 + 1.415037 Q06_00_02 + 1.415037 Q06_01_00 +
↪ 1.415037 Q06_01_01 + 1.415037 Q06_01_02 + 1.415037 Q06_02_00 + 1.415037 Q06_02_01 + 1.415037 Q06_02_02 + 1.415037 Q06_03_00 +
↪ 1.415037 Q06_03_01 + 1.415037 Q06_03_02 + 1.415037 Q06_04_00 + 1.415037 Q06_04_01 + 1.415037 Q06_04_02 + 1.415037 Q06_05_00 +
↪ 1.415037 Q06_05_01 + 1.415037 Q06_05_02 + 1.415037 Q06_06_00 + 1.415037 Q06_06_01 + 1.415037 Q06_06_02 + 1.415037 Q06_07_00 +
↪ 1.415037 Q06_07_01 + 1.415037 Q06_07_02 + 1.415037 Q06_08_00 + 1.415037 Q06_08_01 + 1.415037 Q06_08_02 + 1.415037 Q06_09_00 +
↪ 1.415037 Q06_09_01 + 1.415037 Q06_09_02 + 1.415037 Q06_10_00 + 1.415037 Q06_10_01 + 1.415037 Q06_10_02 + 1.415037 Q06_11_00 +
↪ 1.415037 Q06_11_01 + 1.415037 Q06_11_02 + 1.415037 Q06_12_00 + 1.415037 Q06_12_01 + 1.415037 Q06_12_02 + 1.415037 Q06_13_00 +
↪ 1.415037 Q06_13_01 + 1.415037 Q06_13_02 + 1.415037 Q06_14_00 + 1.415037 Q06_14_01 + 1.415037 Q06_14_02 + 1.415037 Q06_15_00 +
↪ 1.415037 Q06_15_01 + 1.415037 Q06_15_02 + 1.415037 Q06_16_00 + 1.415037 Q06_16_01 + 1.415037 Q06_16_02 + 1.415037 Q06_17_00 +
↪ 1.415037 Q06_17_01 + 1.415037 Q06_17_02 + 1.415037 Q06_18_00 + 1.415037 Q06_18_01 + 1.415037 Q06_18_02 + 1.415037 Q06_19_00 +
↪ 1.415037 Q06_19_01 + 1.415037 Q06_19_02 + 1.415037 Q06_20_00 + 1.415037 Q06_20_01 + 1.415037 Q06_20_02 + 1.415037 Q06_21_00 +
↪ 1.415037 Q06_21_01 + 1.415037 Q06_21_02 + 1.415037 Q06_22_00 + 1.415037 Q06_22_01 + 1.415037 Q06_22_02 + 1.415037 Q06_23_00 +
↪ 1.415037 Q06_23_01 + 1.415037 Q06_23_02 + 1.415037 Q06_24_00 + 1.415037 Q06_24_01 + 1.415037 Q06_24_02 + 1.415037 Q06_25_00 +
↪ 1.415037 Q06_25_01 + 1.415037 Q06_25_02 + 1.415037 Q06_26_00 + 1.415037 Q06_26_01 + 1.415037 Q06_26_02 + 1.415037 Q06_27_00 +
↪ 1.415037 Q06_27_01 + 1.415037 Q06_27_02 + 1.415037 Q06_28_00 + 1.415037 Q06_28_01 + 1.415037 Q06_28_02 + 1.415037 Q06_29_00 +
↪ 1.415037 Q06_29_01 + 1.415037 Q06_29_02 + 1.415037 Q06_30_00 + 1.415037 Q06_30_01 + 1.415037 Q06_30_02 + 1.415037 Q06_31_00 +
↪ 1.415037 Q06_31_01 + 1.415037 Q06_31_02 >= 7.000000
3 Q02_00_01 + 3 Q02_00_02 + 3 Q02_00_03 + 3 Q02_01_01 + 3 Q02_01_02 + 3 Q02_01_03 + 3 Q02_02_01 + 3 Q02_02_02 + 3 Q02_02_03 + 3
↪ Q02_03_01 + 3 Q02_03_02 + 3 Q02_03_03 + 3 Q02_04_01 + 3 Q02_04_02 + 3 Q02_04_03 + 3 Q02_05_01 + 3 Q02_05_02 + 3 Q02_05_03 + 3
↪ Q02_06_01 + 3 Q02_06_02 + 3 Q02_06_03 + 3 Q02_07_01 + 3 Q02_07_02 + 3 Q02_07_03 + 3 Q02_08_01 + 3 Q02_08_02 + 3 Q02_08_03 + 3
↪ Q02_09_01 + 3 Q02_09_02 + 3 Q02_09_03 + 3 Q02_10_01 + 3 Q02_10_02 + 3 Q02_10_03 + 3 Q02_11_01 + 3 Q02_11_02 + 3 Q02_11_03 + 3
↪ Q02_12_01 + 3 Q02_12_02 + 3 Q02_12_03 + 3 Q02_13_01 + 3 Q02_13_02 + 3 Q02_13_03 + 3 Q02_14_01 + 3 Q02_14_02 + 3 Q02_14_03 + 3
↪ Q02_15_01 + 3 Q02_15_02 + 3 Q02_15_03 + 3 Q02_16_01 + 3 Q02_16_02 + 3 Q02_16_03 + 3 Q02_17_01 + 3 Q02_17_02 + 3 Q02_17_03 + 3
↪ Q02_18_01 + 3 Q02_18_02 + 3 Q02_18_03 + 3 Q02_19_01 + 3 Q02_19_02 + 3 Q02_19_03 + 3 Q02_20_01 + 3 Q02_20_02 + 3 Q02_20_03 + 3
↪ Q02_21_01 + 3 Q02_21_02 + 3 Q02_21_03 + 3 Q02_22_01 + 3 Q02_22_02 + 3 Q02_22_03 + 3 Q02_23_01 + 3 Q02_23_02 + 3 Q02_23_03 + 3
↪ Q02_24_01 + 3 Q02_24_02 + 3 Q02_24_03 + 3 Q02_25_01 + 3 Q02_25_02 + 3 Q02_25_03 + 3 Q02_26_01 + 3 Q02_26_02 + 3 Q02_26_03 + 3
↪ Q02_27_01 + 3 Q02_27_02 + 3 Q02_27_03 + 3 Q02_28_01 + 3 Q02_28_02 + 3 Q02_28_03 + 3 Q02_29_01 + 3 Q02_29_02 + 3 Q02_29_03 + 3
↪ Q02_30_01 + 3 Q02_30_02 + 3 Q02_30_03 + 3 Q02_31_01 + 3 Q02_31_02 + 3 Q02_31_03 + 2 Q04_00_01 + 2 Q04_00_02 + 2 Q04_00_03 + 2
↪ Q04_01_01 + 2 Q04_01_02 + 2 Q04_01_03 + 2 Q04_02_01 + 2 Q04_02_02 + 2 Q04_02_03 + 2 Q04_03_01 + 2 Q04_03_02 + 2 Q04_03_03 + 2
↪ Q04_04_01 + 2 Q04_04_02 + 2 Q04_04_03 + 2 Q04_05_01 + 2 Q04_05_02 + 2 Q04_05_03 + 2 Q04_06_01 + 2 Q04_06_02 + 2 Q04_06_03 + 2
↪ Q04_07_01 + 2 Q04_07_02 + 2 Q04_07_03 + 2 Q04_08_01 + 2 Q04_08_02 + 2 Q04_08_03 + 2 Q04_09_01 + 2 Q04_09_02 + 2 Q04_09_03 + 2
↪ Q04_10_01 + 2 Q04_10_02 + 2 Q04_10_03 + 2 Q04_11_01 + 2 Q04_11_02 + 2 Q04_11_03 + 2 Q04_12_01 + 2 Q04_12_02 + 2 Q04_12_03 + 2
↪ Q04_13_01 + 2 Q04_13_02 + 2 Q04_13_03 + 2 Q04_14_01 + 2 Q04_14_02 + 2 Q04_14_03 + 2 Q04_15_01 + 2 Q04_15_02 + 2 Q04_15_03 + 2
↪ Q04_16_01 + 2 Q04_16_02 + 2 Q04_16_03 + 2 Q04_17_01 + 2 Q04_17_02 + 2 Q04_17_03 + 2 Q04_18_01 + 2 Q04_18_02 + 2 Q04_18_03 + 2
↪ Q04_19_01 + 2 Q04_19_02 + 2 Q04_19_03 + 2 Q04_20_01 + 2 Q04_20_02 + 2 Q04_20_03 + 2 Q04_21_01 + 2 Q04_21_02 + 2 Q04_21_03 + 2
↪ Q04_22_01 + 2 Q04_22_02 + 2 Q04_22_03 + 2 Q04_23_01 + 2 Q04_23_02 + 2 Q04_23_03 + 2 Q04_24_01 + 2 Q04_24_02 + 2 Q04_24_03 + 2
↪ Q04_25_01 + 2 Q04_25_02 + 2 Q04_25_03 + 2 Q04_26_01 + 2 Q04_26_02 + 2 Q04_26_03 + 2 Q04_27_01 + 2 Q04_27_02 + 2 Q04_27_03 + 2
↪ Q04_28_01 + 2 Q04_28_02 + 2 Q04_28_03 + 2 Q04_29_01 + 2 Q04_29_02 + 2 Q04_29_03 + 2 Q04_30_01 + 2 Q04_30_02 + 2 Q04_30_03 + 2
↪ Q04_31_01 + 2 Q04_31_02 + 2 Q04_31_03 + 1.415037 Q06_00_01 + 1.415037 Q06_00_02 + 1.415037 Q06_00_03 + 1.415037 Q06_01_01 +
↪ 1.415037 Q06_01_02 + 1.415037 Q06_01_03 + 1.415037 Q06_02_01 + 1.415037 Q06_02_02 + 1.415037 Q06_02_03 + 1.415037 Q06_03_01 +
↪ 1.415037 Q06_03_02 + 1.415037 Q06_03_03 + 1.415037 Q06_04_01 + 1.415037 Q06_04_02 + 1.415037 Q06_04_03 + 1.415037 Q06_05_01 +
↪ 1.415037 Q06_05_02 + 1.415037 Q06_05_03 + 1.415037 Q06_06_01 + 1.415037 Q06_06_02 + 1.415037 Q06_06_03 + 1.415037 Q06_07_01 +
↪ 1.415037 Q06_07_02 + 1.415037 Q06_07_03 + 1.415037 Q06_08_01 + 1.415037 Q06_08_02 + 1.415037 Q06_08_03 + 1.415037 Q06_09_01 +
↪ 1.415037 Q06_09_02 + 1.415037 Q06_09_03 + 1.415037 Q06_10_01 + 1.415037 Q06_10_02 + 1.415037 Q06_10_03 + 1.415037 Q06_11_01 +
↪ 1.415037 Q06_11_02 + 1.415037 Q06_11_03 + 1.415037 Q06_12_01 + 1.415037 Q06_12_02 + 1.415037 Q06_12_03 + 1.415037 Q06_13_01 +
↪ 1.415037 Q06_13_02 + 1.415037 Q06_13_03 + 1.415037 Q06_14_01 + 1.415037 Q06_14_02 + 1.415037 Q06_14_03 + 1.415037 Q06_15_01 +
↪ 1.415037 Q06_15_02 + 1.415037 Q06_15_03 + 1.415037 Q06_16_01 + 1.415037 Q06_16_02 + 1.415037 Q06_16_03 + 1.415037 Q06_17_01 +
↪ 1.415037 Q06_17_02 + 1.415037 Q06_17_03 + 1.415037 Q06_18_01 + 1.415037 Q06_18_02 + 1.415037 Q06_18_03 + 1.415037 Q06_19_01 +
↪ 1.415037 Q06_19_02 + 1.415037 Q06_19_03 + 1.415037 Q06_20_01 + 1.415037 Q06_20_02 + 1.415037 Q06_20_03 + 1.415037 Q06_21_01 +
↪ 1.415037 Q06_21_02 + 1.415037 Q06_21_03 + 1.415037 Q06_22_01 + 1.415037 Q06_22_02 + 1.415037 Q06_22_03 + 1.415037 Q06_23_01 +
↪ 1.415037 Q06_23_02 + 1.415037 Q06_23_03 + 1.415037 Q06_24_01 + 1.415037 Q06_24_02 + 1.415037 Q06_24_03 + 1.415037 Q06_25_01 +
↪ 1.415037 Q06_25_02 + 1.415037 Q06_25_03 + 1.415037 Q06_26_01 + 1.415037 Q06_26_02 + 1.415037 Q06_26_03 + 1.415037 Q06_27_01 +
↪ 1.415037 Q06_27_02 + 1.415037 Q06_27_03 + 1.415037 Q06_28_01 + 1.415037 Q06_28_02 + 1.415037 Q06_28_03 + 1.415037 Q06_29_01 +
↪ 1.415037 Q06_29_02 + 1.415037 Q06_29_03 + 1.415037 Q06_30_01 + 1.415037 Q06_30_02 + 1.415037 Q06_30_03 + 1.415037 Q06_31_01 +
↪ 1.415037 Q06_31_02 + 1.415037 Q06_31_03 >= 7.000000
```

Fig. 5.8 Previous constraints for `GIFT-128` 4-Round differential bound in LP format (part III)

References

[MWG+11] N. Mouha, Q. Wang, D. Gu, B. Preneel. Differential and linear cryptanalysis using mixed-integer linear programming, in *Information Security and Cryptology - 7th International Conference, Inscrypt 2011, Beijing, China, November 30 - December 3, 2011*. Revised Selected Papers (2011), pp. 57–76. https://doi.org/10.1007/978-3-642-34704-7%5C_5

[SHW+13] S. Sun, L. Hu, P.Wang, K. Qiao, X. Ma, L. Song, Automatic security evaluation and (related-key) differential characteristic search: application to simon, present, lblock, des(l) and other bit-oriented block ciphers. Cryptology ePrint Archive, Report 2013/676 (2013), https://eprint.iacr.org/2013/676

[SHW+14] S. Sun, L. Hu, M. Wang, P. Wang, K. Qiao, X. Ma, D. Shi, L. Song, and K. Fu. Towards finding the best characteristics of some bit-oriented block ciphers and automatic enumeration of (related-key) differential and linear characteristics with predefined properties. IACR Cryptol. ePrint Archive, p. 747 (2014). http://eprint.iacr.org/2014/747

[ZDY18] B. Zhu, X. Dong, H. Yu, MILP-based differential attack on round-reduced gift. Cryptology ePrint Archive, Report 2018/390 (2018), https://eprint.iacr.org/2018/390

[LWZ+19] L. Li, W. Wu, Y. Zheng, L. Zhang.,The relationship between the construction and solution of the milp models and applications. Cryptology ePrint Archive, Report 2019/049 (2019), https://eprint.iacr.org/2019/049

[JZD19] F. Ji, W. Zhang, T. Ding, Improving matsui's search algorithm for the best differential/linear trails and its applications for des, desl and gift. Cryptology ePrint Archive, Report 2019/1190 (2019), https://eprint.iacr.org/2019/1190

[AST+17] A. Abdelkhalek, Y. Sasaki, Y. Todo, M. Tolba, A.M. Youssef, MILP modelling for (large) s-boxes to optimize probability of differential characteristics. IACR Trans. Symmetric Cryptol. **2017**(4), 99–129 (2017). https://doi.org/10.13154/tosc.v2017.i4.99-129

[ZZD+19] C. Zhou, W. Zhang, T. Ding, Z. Xiang, Improving the milp-based security evaluation algorithm against differential/linear cryptanalysis using a divide-and conquer approach. Cryptology ePrint Archive, Report 2019/019 (2019). https://eprint.iacr.org/2019/019

[BS90] E. Biham, A. Shamir, Differential cryptanalysis of des-like cryptosystems, in *Advances in Cryptology - CRYPTO '90, 10th Annual International Cryptology Conference, Santa Barbara, California, USA, August 11-15, 1990, Proceedings* (1990), pp. 2–21. https://doi.org/10.1007/3-540-38424-3_1

[Mat93] M. Matsui, Linear cryptanalysis method for DES cipher, in *Advances in Cryptology - EUROCRYPT '93, Workshop on the Theory and Application of of Cryptographic Techniques, Lofthus, Norway, May 23-27, 1993, Proceedings*, ed. by T. Helleseth. Lecture Notes in Computer Science, vol. 765 (Springer, 1993), pp. 386–397. https://doi.org/10.1007/3-540-48285-7%5C_33

[BPP+17] S. Banik, S.K. Pandey, T. Peyrin, Y. Sasaki, S.M. Sim, Y. Todo, Gift: a small present. Cryptology ePrint Archive, Report 2017/622 (2017), https://eprint.iacr.org/2017/622

[BGN05] E. Biham, L. Granboulan, P.Q. Nguyen, Impossible fault analysis of RC4 and differential fault analysis of RC4, in *Fast Software Encryption: 12th International Workshop, FSE 2005, Paris, France, February 21-23, 2005*, Revised Selected Papers (2005), pp. 359–367. https://doi.org/10.1007/11502760_24

[CJF+16] T. Cui, K. Jia, K. Fu, S. Chen, M. Wang, New automatic search tool for impossible differentials and zero-correlation linear approximations. IACR Cryptology ePrint Archive, 2016:689 (2016), http://eprint.iacr.org/2016/689

[ST17'2] Y. Sasaki, Y. Todo, New impossible differential search tool from design and cryptanalysis aspects - revealing structural properties of several ciphers, in *Advances in*

Cryptology - EUROCRYPT 2017 - 36th Annual International Conference on the Theory and Applications of Cryptographic Techniques, Paris, France, April 30 - May 4, 2017, Proceedings, Part III (2017), pp. 185–215. https://doi.org/10.1007/978-3-319-56617-7

[ZZ18] P. Zhang, W. Zhang, Differential cryptanalysis on block cipher skinny with milp program. Sec. Commun. Netw. **2018**, 1–11 (2018)-10

[ST17'1] Y. Sasaki, Y. Todo, New algorithm for modeling s-box in milp based differential and division trail search, in Innovative Security Solutions for Information Technology and Communications, ed. by P. Farshim, E. Simion (Springer International Publishing, Cham, 2017), pp. 150–165. ISBN: 978-3-319-69284-5

[BKL+07] A. Bogdanov, L.R. Knudsen, G. Leander, C. Paar, A. Poschmann, M.J. Robshaw, Y. Seurin, C. Vikkelsoe. PRESENT: An ultra-lightweight block cipher, in *CHES*, vol. 4727 (Springer, 2007), pp. 450–466

[LLL+19] Y. Liu, H. Liang, M. Li, L. Huang, K. Hu, C. Yang, M. Wang, STP models of optimal differential and linear trail for s-box based ciphers. Cryptology ePrint Archive, Report 2019/025 (2019). https://eprint.iacr.org/2019/025

[dBCvK+08] M. de Berg, O. Cheong, M. van Kreveld, M. Overmars, *Computational Geometry. Algorithms and Applications* (Springer, Berlin, Heidelberg, 2008) https://www.springer.com/gp/book/9783540779735

[Mou16] D.M. Mount, CMSC 754 - Computational Geometry (lecture notes) (2016), https://www.cs.umd.edu/class/fall2016/cmsc754/Lects/cmsc754-fall16-lects.pdf

[Sti06] D.R. Stinson, *Cryptography - Theory and Practice*. Discrete Mathematics and Its Applications Series (CRC Press, 2006). ISBN: 978-1-58488-508-5

[DeC07] C. De Cannière, Analysis and Design of Symmetric Encryption Algorithms. Katholieke Universiteit Leuven, Belgium (2007), https://www.esat.kuleuven.be/cosic/publications/thesis-139.pdf. Ph.D. Thesis

[BKD+21] A. Baksi, B. Karmakar, V.A. Dasu, D. Saha, A. Chattopadhyay, Further insights on implementation of the linear layer. SILC Workshop - Security and Implementation of Lightweight Cryptography (2021). https://www.esat.kuleuven.be/cosic/events/silc2020/wp-content/uploads/sites/4/2020/10/Submission1.pdf

Chapter 6
Machine Learning-Assisted Differential Distinguishers for Lightweight Ciphers

At CRYPTO'19, Gohr first introduces the deep learning-based cryptanalysis on round-reduced SPECK. Using a deep residual network, Gohr trains several neural network-based distinguishers on 8-round SPECK-32/64. The analysis follows an "all-in-one" differential cryptanalysis approach, which considers all the output differences effect under the same input difference.

Usually, the all-in-one differential cryptanalysis is more effective compared to the one using only one single differential trail. However, when the cipher is non-Markov or its block size is large, it is usually very hard to fully compute. Inspired by Gohr's work, we try to simulate the all-in-one differentials for non-Markov ciphers through machine learning.

Our idea here is to reduce a distinguishing problem to a classification problem so that it can be efficiently managed by machine learning. As a proof of concept, we show several distinguishers for four high-profile ciphers, each of which works with trivial complexity. In particular, we show differential distinguishers for 8-round GIMLI-HASH, GIMLI-CIPHER and GIMLI-PERMUTATION; 3-round ASCON-PERMUTATION; 10-round KNOT-256 permutation and 12-round KNOT-512 permutation; and 4-round CHASKEY- PERMUTATION. Finally, we explore more on choosing an efficient machine learning model and observe that only a three-layer neural network can be used. Our analysis shows the attacker is able to reduce the complexity of finding distinguishers by using machine learning techniques.

A. Baksi, *Classical and Physical Security of Symmetric Key Cryptographic Algorithms*, Computer Architecture and Design Methodologies,
https://doi.org/10.1007/978-981-16-6522-6_6

6.1 Introduction

Machine Learning (ML) tools have made a great progress in the last decade. At present, such techniques are widely used in various fields, such as computer vision [SHB14], machine translation [BCB14, WSC+16], and autonomous driving [CSK+15], to name a few. In the aspect of cryptography however, usage of ML is mainly confined in the context of side channel analysis, such as [MPP16, CDP17], to the best of our knowledge.

Therefore, the applicability of ML techniques in classical cryptanalysis is not much explored. It seems the community is rather skeptical regarding such an approach. For example, the authors in [AA16] comment, "Neural networks are generally not meant to be great at cryptography. Famously, the simplest neural networks cannot even compute XOR, which is basic to many cryptographic algorithms".

Although ML techniques have been used against legacy ciphers, like the Enigma (the German cipher used during Second World War) [Gre17], it was not until the work by Gohr [Goh19] at CRYPTO'19 that this line of research finally got its exposure. The idea here is applied to key recovery attacks on round-reduced `SPECK` using ML.

This work focuses on extending the commonly used model of differential distinguisher by using ML techniques. In the case of differential distinguisher, the attacker XORs a chosen input difference δ to the input of the state of the (reduced-round) cipher and watches for a particular output difference Δ, with randomly chosen inputs. If the (δ, Δ) pair occurs with a probability significantly higher for the (reduced-round) cipher than what it should be for a random case, the (reduced-round) cipher can be distinguished from the random case. This probability distribution of $\delta \to \Delta$ is modelled by differential branch number [DR20] or by automated tools like Mixed Integer Linear Programming (MILP) [MWG+11] (see also Chap. 5). We extend the modelling for differential distinguisher by incorporating machine learning algorithms.

We argue that the usual modelling with branch number or MILP underestimates attacker's power. Having been collected the output differences for the chosen input differences, the attacker can use any technique to distinguish the cipher from the random case. In such a situation, machine learning-based techniques can reduce the search complexity estimated by existing methods. In fact, we observe that ML can reduce the search complexity to the cube root of the previously estimated bound.

The machine learning models we propose can work with any number (say, t) of suitably chosen input differences. Our first model comes into play when $t \geq 2$ (see Sect. 6.4.1). In order to work with only one input difference, we propose our second model (see Sect. 6.4.2).

A basic introduction to ML is given in Sect. 2.8.3; for more information we refer to textbooks such as [Hay08]. We use TensorFlow[1] back-end with Keras[2] API. The optimizer function is based on Adam algorithm [KB14].

---

[1] https://www.tensorflow.org/.

[2] https://keras.io/.

Our Contributions

In this work, we construct machine learning models to simulate the *all-in-one differentials* from [AL12]. We consider the classical distinguisher game: Given ORACLE $\xleftarrow{\$}$ {CIPHER, RANDOM}, the attacker is to identify whether ORACLE = CIPHER with probability significantly $> \frac{1}{2}$ and with sufficiently small number of queries.

The core of our analysis lies in the actual problem of distinguishing the CIPHER from RANDOM into a classification problem. For this purpose, we propose two models (details are given in Sect. 6.4). Here we choose the most common ML tool, Multi-Layer Perceptron (MLP, Sect. 2.8.3) as a starting point (we also have tested with other ML tools, as described in Sect. 6.6).

As for transforming a differential distinguisher to a classification problem, we propose two models here. We apply the first model (described in Sect. 6.4.1) to round-reduced GIMLI [BKL+19], ASCON [DEM+19], and KNOT [ZDY+19]. All of the ciphers are the second-round candidates in the ongoing NIST Lightweight Cryptography (LWC) competition (see Sect. 2.8.4). Further, we also show the effectiveness of the second model (described in Sect. 6.4.2) over the lightweight MAC CHASKEY [MMH+14]. Brief description of the ciphers is given in Sect. 6.3 (except that of CHASKEY, which is given in Sect. 2.4.4).

Results are described in Sects. 6.5 and 6.6, which can be summarized as follows:

- For GIMLI, we obtain 8-round practical distinguishers on GIMLI-HASH and GIMLI-CIPHER in nonce respecting case in Sect. 6.5.1. The distinguishing complexity is $2^{17.6}$ offline data to train the model offline and $2^{14.3}$ online data to distinguish the cipher. Note that the authors of GIMLI [BKL+19] prove that the optimal 8-round differential trail is with probability of 2^{-52}. If we use this trail to distinguish 8-round GIMLI, we need at least 2^{52} online data. Thus, we are able to perform a differential distinguisher in around cube root complexity.
- For the ciphers, ASCON [DEM+19] and KNOT [ZDY+19], we show practical differential distinguishers for reduced-round versions of the underlying permutations in Sect. 6.5.2. For the 320-bit ASCON-PERMUTATION, we show two separate distinguishers that work till 3 rounds with 2^{19} training data. For the 256-bit permutation of KNOT (KNOT-256), we show distinguishers for up to 10 rounds; and for the 512-bit permutation (KNOT-512), we show the same for up to 12 rounds. All results on KNOT are obtained with 2^{19} training data.
- For CHASKEY [MMH+14, Mou15], we present our results in Sect. 6.5.3. With trivial training/testing complexity (2^{23} for training and $2^{14.3}$ for testing), we show the existence of a 4-round distinguisher. This contradicts the authors' claim that no 4-round differential distinguisher of complexity 2^{37} searches exists.

In Sect. 6.6, we discuss the effects of choosing different neural network architectures with respect to 8-round GIMLI-PERMUTATION as the target cipher. We show that even a shallow three-layer neural network works well for our purpose. We also report a few differential distinguishers for 8-round GIMLI-PERMUTATION in the process.

With regard to our analysis, we emphasize the following points:

- **Generality and practicality.** It is to be mentioned that our ML-assisted model is generic in nature and can be adopted to any symmetric key setting. All the results presented are practical and can be carried out in around an hour on a modern computer.
- **Compatibility with differential distinguisher.** We would like to point out that we use the same attack model as that of the pre-existing differential distinguisher. The only distinction is made during the analysis, as ours is done by machine learning (instead of the branch number or automated tools like MILP). We do not fix any output difference a priori, instead all the differentials are fed to the ML.
- **Effect of truncation.** The results (Sect. 6.5.1) for GIMLI-HASH and GIMLI-CIPHER are with truncated versions of the state, whereas the results for GIMLI-PERMUTATION (Sect. 6.6) are with the full state. Moreover, we argue that the truncation of the state does not fall outside the perceived model. Having been collected the differentials, the attacker can employ any method (including truncating a part) based on her preference for analysis. This is also noted (with an example) in [BBD+21].
- **Extensibility.** The results presented in our work do not constitute the theoretical upper bound for the ML-assisted distinguishers. By using a more sophisticated ML model and/or more training/testing data, it is likely that one can cover more rounds.[3]

6.2 Background

6.2.1 *Markov Ciphers*

Lai, Massey, and Murphy [LMM91] introduce the concept of *Markov ciphers* at Eurocrypt'91.

Definition 6.1 (*Markov Chain* [LMM91]) A sequence of discrete random variables $v_0, v_1, \ldots, v_r$ is a Markov chain, if for $0 \leq i < r$,

$$\Pr(v_{i+1} = \beta_{i+1} | v_i = \beta_i, v_{i-1} = \beta_{i-1}, \ldots, v_0 = \beta_0) = \Pr(v_{i+1} = \beta_{i+1} | v_i = \beta_i). \tag{6.1}$$

If $\Pr(v_{i+1} = \beta | v_i = \alpha)$ is independent of i for all α and β, the Markov chain is called homogeneous.

Given an r-round iterated cipher, the input of ith round is denoted as Y_i $(0 \leq i \leq r)$. Given a group operation $\otimes$, we define the corresponding input differences as

[3] New functionalities like higher order differential or key recovery can be incorporated too.

$\Delta Y_0, \Delta Y_1, \ldots, \Delta Y_r$, where $\Delta Y = Y \otimes Y'$. Then, Lai et al. introduce the following definition of Markov cipher as given in Definition 6.2.

Definition 6.2 (*Markov Cipher* [LMM91]) An iterated cipher with round function $Y = f(X, K)$ is a Markov cipher if there is a group operation $\otimes$ for defining differences such that, for all choices of α ($\alpha \neq 0$) and β ($\beta \neq 0$), $\Pr(\Delta Y = \beta | \Delta X = \alpha, X = \gamma)$ is independent of γ when the sub-key K is uniformly random, or equivalently, if $\Pr(\Delta Y = \beta | \Delta X = \alpha, X = \gamma) = \Pr(\Delta Y(1) = \beta_1 | \Delta X = \alpha)$ for all choices of γ when the sub-key K is uniformly random.

Theorem 6.1 *If an r-round iterated cipher is a Markov cipher and the r-round keys are independent and uniformly random, then the sequence of differences* $\Delta X = (\Delta Y_0, \Delta Y_1, \ldots, \Delta Y_r)$ *is a homogeneous Markov chain. Moreover, this Markov chain is stationary if* ΔX *is uniformly distributed over the non-neutral elements of the group.*

Theorem 6.1 is adopted from [LMM91]. According to Theorem 6.1, one can compute the probability of an r-round differential characteristic as

$$\Pr(\Delta Y_1 = \beta_1, \Delta Y_2 = \beta_2, \ldots, \Delta Y_r = \beta_r | \Delta X = \beta_0) = \prod_{i=1}^{r} \Pr(\Delta Y_1 = \beta_i | \Delta X = \beta_{i-1}). \quad (6.2)$$

However, the above theory cannot be applied to non-Markov ciphers, such as `GIMLI` [BKL+19] or `ASCON` [DEM+19]. One of the most important feature of those primitives is that there are no sub-keys in each iterated round. In this situation, the iterated cipher usually cannot be regarded as a Markov cipher or a homogeneous Markov chain. This is because the differences in the former rounds usually have a significant effect on the differences of the latter rounds, where Eq. (6.1) does not hold any more. Hence, we cannot use Eq. (6.2) to compute the probability of a characteristic any more.

As shown in Fig. 6.1, we introduce a toy cipher to explain the dependence of the differences between a 2-round cipher without sub-keys. We use the 4-bit SBox of `GIFT-64` [BPP+17] as an example. The Difference Distribution Table (DDT) SBox (`1A4C6F392DB7508E`) is omitted here for the interest of brevity and can be found in [BPP+17, Table 18]. Suppose $\Delta Y_1[0] = Y_1[0] \oplus Y_1'[0]$ and $\Delta Y_1[1] = Y_1[1] \oplus Y_1'[1]$ are the input difference of the upper and lower SBox in Fig. 6.1, respectively. We try to compute the probability of a differential characteristic, where $\Delta Y_1 = (2, 3)$, $\Delta W_1 = (5, 8)$, $\Delta Y_2 = (6, 2)$, $\Delta W_2 = (2, 5)$. From the DDT of the `GIFT` SBox, it is easy to know the probability of $\Delta Y_1 \rightarrow \Delta W_1$ is 2^{-5}. The valid tuples of $(Y_1[0], W_1[0], Y_1'[0], and\, W_1'[0])$ are $(0, 1, 2, 4)$, $(2, 4, 0, 1)$, $(4, 6, 6, 3)$, and $(6, 3, 4, 6)$. The valid tuples of $(Y_1[1], W_1[1], Y_1'[1], W_1'[1])$ are $(\mathtt{d}, 0, \mathtt{e}, 8)$ and $(\mathtt{e}, 8, \mathtt{d}, 0)$. Only input pairs with $(Y_1[0], Y_1[1]) = (0, \mathtt{d})$, $(0, \mathtt{e})$, $(2, \mathtt{d})$, and $(2, \mathtt{e})$ are valid for the differential characteristic. Hence the probability of the characteristic is 2^{-6}, instead of 2^{-9} computed by Eq. (6.2).

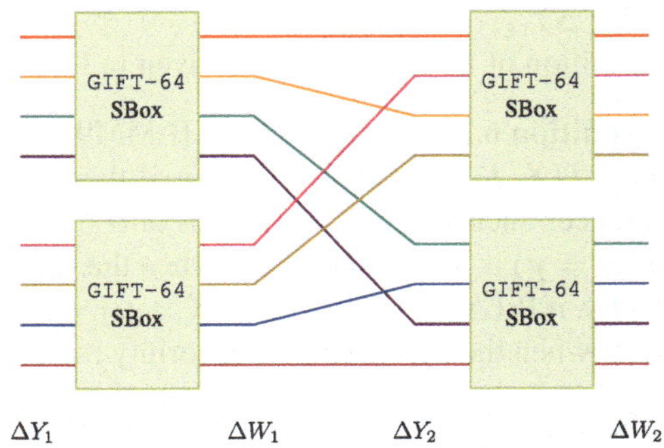

Fig. 6.1 One-round GIFT-128 (un-keyed) permutation

6.2.2 *Gohr's Work on* SPECK *(CRYPTO'19)*

Using deep residual neural networks, Gohr in [Goh19] achieves better results than the best classical cryptanalysis on 11-round SPECK [BSS+13]. In this work, first several machine-learned distinguishers for round-reduced SPECK based on an all-in-one differential cryptanalysis [AL12] are trained, where the attacker considers potentially all the output differences of a cipher for a given input difference and to combine the information derived from them. Under the Markov assumption, Gohr first computes the entire DDT of round-reduced SPECK with a fixed input difference, which can be achieved due to the small block size of SPECK-32/64. Hence, the all-in-one differentials for the 5-/6-/7-/8-round SPECK-32/64 following the Markov assumption are reported. After this, the same distinguishing task via the neural networks is performed.

Concretely, the following steps are used to train the model:

1. Collecting training data by generating uniformly distributed keys and plaintext pairs given a fixed input difference as well as the binary labels Y_i.
2. If the binary label $Y_i = 1$, then the plaintext pairs are encrypted by k-round SPECK to produce the ciphertext pairs. If otherwise, the random ciphertext pairs are generated.
3. Pre-process ciphertext pairs to fit the format required by the neural network and start to train them.

Finally, the author gains several machine-learned distinguishers on round-reduced SPECK which are a little more efficient than the distinguishers derived by computing the entire difference distribution table. After that, Gohr uses the neural network-based distinguishers to launch several key-recovery attacks on round-reduced SPECK.

Motivation of Our Work. Inspired from Gohr's work [Goh19], the following questions come to our mind, and subsequently we employ neural networks:

- As the all-in-one approach considers all the output differences with one input difference—it is more efficient to distinguish the ciphers than that using only one input difference and one output difference. However, computing the all-in-one distinguisher is infeasible when the state of the cipher is large, e.g. 128-bit state. However, as shown by Gohr's attack, the neural networks can match the all-in-one distinguisher well. Hence, we can take advantage of the neural networks to simulate the all-in-one distinguishers for larger state ciphers.

- As shown in Gohr's attack, under Markov assumption, one can compute the all-in-one distinguishers somehow. However, there are many ciphers which cannot be assumed as Markov, such as some permutation-based ciphers or stream ciphers, where there are no sub-key in the iterated rounds (the justification is given in Sect. 6.2.1). In these situations, it is hard to compute the all-in-one distinguishers.

6.3 Basic Description of the Ciphers

6.3.1 GIMLI

GIMLI [BKL+17] is a cross-platform permutation proposed by Bernstein, Kölbl, Lucks, Massolino, Mendel, Nawaz, Schneider, Schwabe, Standaert, Todo, and Viguier at CHES'17. Based on GIMLI permutation, the authors of [BKL+19] introduce the lightweight hashing and authenticated encryption algorithms, i.e., GIMLI-HASH and GIMLI-CIPHER, which were in the second round of NIST LWC (Sect. 2.8.4) competition, but not selected for the final round. GIMLI is also used in the open source cryptographic library LibHydrogen.[4]

GIMLI-PERMUTATION works on a 384-bit state which can be represented as a 3×4 matrix of 32-bit words, namely

$$\mathbf{s} = \begin{bmatrix} s_{0,0} & s_{0,1} & s_{0,2} & s_{0,3} \\ s_{1,0} & s_{1,1} & s_{1,2} & s_{1,3} \\ s_{2,0} & s_{2,1} & s_{2,2} & s_{2,3} \end{bmatrix}$$

where the jth column of $\mathbf{s}$ is $s_{*,j} = s_{0,j}, s_{1,j}, s_{2,j}$, the ith row of s is $s_{i,*} = s_{i,0}, s_{i,1}, s_{i,2}, s_{i,3}$. The details of GIMLI-PERMUTATION is given in Algorithm 6.1 (adopted from [BKL+17]).

GIMLI-HASH

Taking advantage of GIMLI-PERMUTATION and the sponge construction [BDP+07], GIMLI-HASH is defined as shown in Fig. 6.2. The message M is padded and divided into 128-bit blocks, i.e., $m_0, \ldots, m_t$. Those are XORed into the state in the *absorb* phase and the output 256-bit digest in the *squeeze* phase. Here we omit some padding rules since they do not affect our attacks. Our neural distinguisher happens in the processing of the last message block m_t.

[4] https://github.com/jedisct1/libhydrogen.

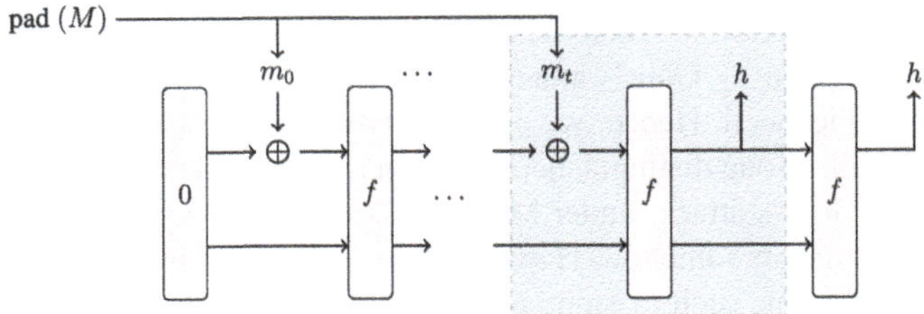

Fig. 6.2 Construction of GIMLI-HASH

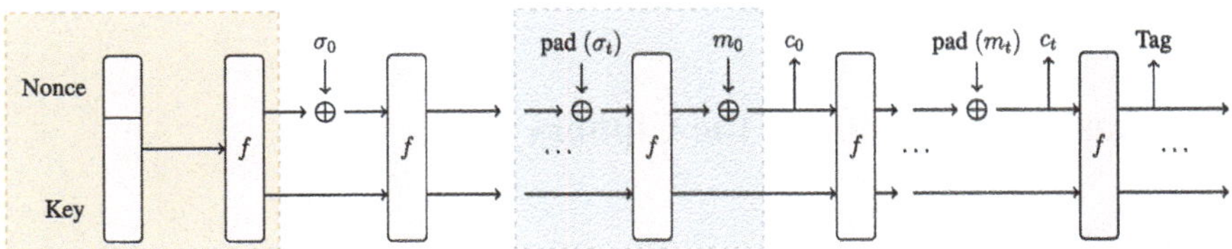

Fig. 6.3 Construction of GIMLI-CIPHER

GIMLI-CIPHER

Taking advantage of GIMLI-PERMUTATION and the Monkey Duplex construction [BDP+11] , GIMLI-CIPHER is defined as shown in Fig. 6.3. Here $\sigma_0, \ldots, \sigma_t$ are the associated data and padded to 128-bit blocks. Here $m_0, \ldots, m_t$ and $c_0, \ldots, c_t$ are the plaintext and ciphertext blocks, respectively. Note that at least one block of associated data is processed. So at least two GIMLI-PERMUTATION's (48 rounds) are used until the first ciphertext block c_0 generated. In this chapter, we target on the case where only one block of associated data is processed.

Algorithm 6.1 GIMLI-PERMUTATION

Input: $s = (\mathbf{s}_{i,j})$
Output: GIMLI-PERMUTATION(**s**)
for r from 24 down to 1 inclusive **do**
  for j from 0 to 3 inclusive **do**
    $x \leftarrow s_{0,j} \lll 24$ ▷ SP-box
    $y \leftarrow s_{1,j} \lll 9$
    $z \leftarrow s_{2,j}$
    $s_{2,j} \leftarrow x \oplus (z \ll 1) \oplus ((y \wedge z) \ll 2)$
    $s_{1,j} \leftarrow y \oplus x \oplus ((x \vee z) \ll 1)$
    $s_{0,j} \leftarrow z \oplus y \oplus ((x \wedge y) \ll 3)$
  if $r \mod 4 = 0$ **then** ▷ Linear layer
    $s_{0,0}, s_{0,1}, s_{0,2}, s_{0,3} \leftarrow s_{0,1}, s_{0,0}, s_{0,3}, s_{0,2}$ ▷ Small-Swap
  else if $r \mod 4 = 2$ **then**
    $s_{0,0}, s_{0,1}, s_{0,2}, s_{0,3} \leftarrow s_{0,2}, s_{0,3}, s_{0,0}, s_{0,1}$ ▷ Big-Swap
  if $r \mod 4 = 0$ **then**
    $s_{0,0} = s_{0,0} \oplus \texttt{9e377900} \oplus r$ ▷ Add constant **return** $s_{i,j}$

6.3.2 ASCON

ASCON is one of the winners of CAESAR (see Sect. 2.8.4 for more information on CAESAR, first choice for lightweight application), and one of the 10 finalists of the ongoing NIST LWC competition [DEM+19]. The mode of operation is Monkey Duplex [BDP+11]. It operates on a 320-bit state which are divided into five 64-bit words $x_0, \ldots, x_4$, i.e., $S = x_0 \parallel x_1 \parallel x_2 \parallel x_3 \parallel x_4$.

Since for the machine learning-assisted distinguisher we only consider the 320-bit ASCON-PERMUTATION, we give a concise description of it. For more information regarding the cipher, one may refer to [DEM+19]. ASCON-PERMUTATION iteratively applies a three-step round transformation for 12 rounds, which are (in order) denoted by p_C (addition of constants), p_S (substitution layer), and p_L (linear diffusion layer).

The constant addition, p_C, adds an 8-bit round constant to the least significant 8-bits of x_2 at round i. The round constants are as follows: `f0, e1, d2, c3, b4, a5, 96, 87, 78, 69, 5a, 4b`.

The substitution layer, p_S, updates the 320-bit state by applying a 5-bit SBox on it (the SBox is applied 64 times in parallel). The SBox in the look-up format is given as: `4, B, 1F, 14, 1A, 15, 9, 2, 1B, 5, 8, 12, 1D, 3, 6, 1C, 1E, 13, 7, E, 0, D, 11, 18, 10, C, 1, 19, 16, A, F, 17`.

The linear diffusion layer, p_L, is used to diffuse within each 64-bit register. It applies a linear function to each of the register, $x_i \leftarrow \Sigma_i(x_i)$ for $i = 0, \ldots, 4$. The linear functions, Σ_i's are as follows. Here $x \ggg i$ indicates right-rotation (circular shift) by i bits of 64-bit word x.

$$
\begin{aligned}
\Sigma_1(x_1) &= x_0 \oplus (x_0 \ggg 19) \oplus (x_0 \ggg 28),\\
\Sigma_1(x_1) &= x_1 \oplus (x_1 \ggg 61) \oplus (x_1 \ggg 39),\\
\Sigma_2(x_2) &= x_2 \oplus (x_2 \ggg 1) \oplus (x_2 \ggg 6),\\
\Sigma_3(x_3) &= x_3 \oplus (x_3 \ggg 10) \oplus (x_3 \ggg 17),\\
\Sigma_4(x_4) &= x_4 \oplus (x_4 \ggg 7) \oplus (x_4 \ggg 41).
\end{aligned}
$$

6.3.3 KNOT

KNOT [ZDY+19] was one of the 32 candidates in the second round (not selected for the final round) of the NIST LWC competition. We only consider the permutation of KNOT, therefore we only describe KNOT-PERMUTATION here.

KNOT-PERMUTATION is parameterized by width b, where $b \in \{256, 384, 512\}$, although we only use $b = 256$ and 512 for our experiment. KNOT-PERMUTATION has a state size of b-bits, which is organized in a two-dimensional matrix, as detailed in [ZDY+19, Chap. 2.1]. At each round, an SPN round transformation (denoted by

p_b) is iteratively applied. Each p_b consists of the following three steps (in order): `AddRoundConstant`$_b$, `SubColumn`$_b$, and `ShiftRow`$_b$.

The `AddRoundConstant`$_b$ adds round constants (which are generated by a LFSR) to the state; more description can be found at [ZDY+19, Chap. 2.2]. Next, the `SubColumn`$_b$ step updates the state by applying a 4-bit SBox in column-major fashion. The SBox chosen by the designers is: `40A7BE1D9F6852C3`. Finally, the `ShiftRow`$_b$ step left-rotates the state in a row-major fashion. The 0th bit is rotated c_i bits, $i = 0, 1, 2, 3$, where $c_0 = 0, c_1 = 1, c_3 = 25$ for both $b = 256, 512$ and $c_2 = 8$ for $b = 256$, $c_2 = 16$ for $b = 512$.

As for the number of rounds, the designers recommend to use 28 rounds for $b = 256$ and 52 rounds for $b = 512$ [ZDY+19, Table 2].

6.4 Machine Learning-Based Distinguishers

6.4.1 Model 1: Multiple Input Differences

Under this model, the attacker chooses t (≥ 2) input differences $\delta_0, \delta_1, \ldots, \delta_{t-1}$. In the training (offline) phase, the attacker tries to learn whether there is any pattern in the `CIPHER` outputs that the machine learning tool is capable of finding. To test that, she creates t differentials with those input differences and feeds all the data to the machine learning. If the accuracy (a) of ML training is higher than what it should be for random data (i.e., $\frac{1}{t}$), the attacker is able to extract pattern from the `CIPHER` outputs and proceed to the online phase. Otherwise (if the training accuracy is $\frac{1}{t}$), she aborts.

During the testing (online) phase, the attacker would check if the sequence of classes predicted by the already-trained machine learning model matches the expected sequence (in which she has queried the `ORACLE`) with the same probability as training. Since she queries in a specific sequence to the `ORACLE`, the classes predicted by the machine learning would follow the specific sequence with the same probability as training if `ORACLE` = `CIPHER`; or would be arbitrary if `ORACLE` = `RANDOM`. Since the `ORACLE` can only choose between `CIPHER` and `RANDOM`, the machine learning produced sequence (of predicted classes) would either match the attacker's pre-perceived sequence with the same probability as she has observed during training ($> \frac{1}{t}$); or that sequence would be arbitrary and hence would match the attacker's pre-perceived sequence with probability $\frac{1}{t}$. Therefore, if the accuracy of predicting the classes (a') is the same as a, we infer the chosen `ORACLE` = `CIPHER`. On the other hand, if $a' = \frac{1}{t}$, we conclude `ORACLE` = `RANDOM`.

A top-level description of the model is given in Algorithm 6.2. Although we describe the algorithm in terms of the entire state of the permutation for the sake of simplicity; in actual experimentation, we often consider part of the state (such as the *rate* part) part to make the permutation compatible with the primitives that use it.

Algorithm 6.2 Model 1 (multiple input differences) for differential distinguisher with machine learning

1: **procedure** OFFLINE PHASE (Training)
2: $TD \leftarrow (\cdot)$ ▷ Training data
3: Choose random P
4: $C \leftarrow$ `CIPHER` (P)
5: **for** $i = 0; i \leq t-1; i \leftarrow i+1$ **do**
6: $P_i \leftarrow P \oplus \delta_i$
7: $C_i \leftarrow$ `CIPHER`(P_i)
8: Append TD with $(i, C_i \oplus C)$ ▷ $C_i \oplus C$ is from class i
9: Repeat from Step 3 if required
10: Train ML model with TD
11: ML training reports accuracy a
12: **if** $a > \frac{1}{t}$ **then**
13: Proceed to Online phase
14: **else** ▷ $a = \frac{1}{t}$
15: Abort

1: **procedure** ONLINE PHASE (Testing)
2: $TD' \leftarrow (\cdot)$ ▷ Testing data
3: Choose random P
4: $C \leftarrow$ `ORACLE` (P)
5: **for** $i = 0; i \leq t-1; i \leftarrow i+1$ **do**
6: $P_i \leftarrow P \oplus \delta_i$
7: $C_i \leftarrow$ `ORACLE`(P_i)
8: Append TD' with $C_i \oplus C$
9: Test ML model with TD' to get $\mathcal{C}$ ▷ $\mathcal{C}$ is sequence of classes by ML
10: $a' =$ probability that $\mathcal{C}$ matches $(0, 1, \ldots, t-1)$
11: **if** $a' = a > \frac{1}{t}$ **then**
12: `ORACLE` = `CIPHER`
13: **else** ▷ $a' = \frac{1}{t}$
14: `ORACLE` = `RANDOM`
15: Repeat from Step 3 if required

Discussion on `ORACLE` = `RANDOM`

Given the `RANDOM` case, the machine learning tool will assign classes arbitrarily. Therefore, the accuracy for training will be (close to) $\frac{1}{t}$. For example, if $t = 2$, then the expected training accuracy is 0.5; if $t = 32$, the expected training accuracy is 0.03125. This is confirmed through our experiments. In fact, for all the results reported (in Sect. 6.5), we note that the accuracy for training drops to this bound for higher rounds of the ciphers.

Training and Testing the Model

Our distinguisher works on the (un-keyed) permutation with a suitably chosen number of rounds. For the sake of simplicity, we denote the (un-keyed) permutation with input P as `CIPHER`(P). The basic work-flow is given next.

Training (Offline).

1. Select t (≥ 2) non-zero input differences $\delta_0, \delta_1, \ldots, \delta_{t-1}$.
2. For each input difference δ_i, generate (an arbitrary number of) input pairs $(P, P_i = P \oplus \delta_i)$. Run the (un-keyed) permutation of the input pairs to get the output pairs: $C \leftarrow$ `CIPHER`(P), $C_i \leftarrow$ `CIPHER`(P_i) for all i. Then XOR the outputs within a pair to generate the output difference $(C_i \oplus C)$. The output difference together with its label i (i.e., this sample belongs to class i) form a training sample.

3. Check if the training accuracy is $> \frac{1}{t}$. Otherwise (i.e., if accuracy $= \frac{1}{t}$), the procedure is aborted.

Testing (Online).

1. Generate the input pairs in the same way as training. In other words, randomly generate an input P. With the same input differences chosen during training $\delta_0, \delta_1, \ldots, \delta_{t-1}$, generate new inputs $P_i = P \oplus \delta_i$, for all $i = 0(1)t - 1$.
2. Collect the outputs C and C_is by querying ORACLE with input P and P_is in order, for all $i = 0(1)t - 1$.
3. Generate the testing data as $C \oplus C_i$, for all i and in order.
4. Get the predicted classes from the trained model with the testing data.
5. Find the accuracy of class prediction. In other words, tally the classes returned by the trained ML with the sequence: $(0, 1, \ldots, t-1, 0, 1, \ldots, t-1, \ldots, 0, 1, \ldots, t-1)$, and find the probability that both match.
6. (a) If ORACLE = CIPHER, the ML would predict the class for $C \oplus C_i$ as i with the same probability as training. Therefore in this case, the accuracy for class prediction (in Step 5) would be the same (or close to) as the accuracy observed during training, i.e., $> \frac{1}{t}$.
   (b) If ORACLE = RANDOM, then ML would arbitrarily predict the classes. Therefore the accuracy for predicting classes by the trained ML (in Step 5) would be equal to (or close to) $\frac{1}{t}$.

Algorithm 6.3 Model 2 (one input difference) for differential distinguisher with machine learning

1: **procedure** OFFLINE PHASE (Training)
2: $TD \leftarrow (\cdot)$ ▷ Training data
3: Choose random $P_0, P_1 \ (\neq P_0 \oplus \delta)$
4: $P_2 = P_1 \oplus \delta$
5: $C_i \leftarrow$ CIPHER(P_i), for $i = 0, 1, 2$
6: Append TD with:
$(0, C_1 \parallel C_0)$, ▷ $C_1 \parallel C_0$ is from class 0
$(1, C_1 \parallel C_2)$ ▷ $C_1 \parallel C_2$ is from class 1
7: Repeat from Step 3 if required
8: Train ML model with TD
9: ML training reports accuracy a
10: **if** $a > \frac{1}{2}$ **then**
11: Proceed to Online phase
12: **else** ▷ $a = \frac{1}{2}$
13: Abort

1: **procedure** ONLINE PHASE (Testing)
2: $TD' \leftarrow (\cdot)$ ▷ Testing data
3: Choose random $P_0, P_1 \ (\neq P_0 \oplus \delta)$
4: $P_2 = P_1 \oplus \delta$
5: $C_i \leftarrow$ ORACLE(P_i), for $i = 0, 1, 2$
6: Append TD' with $C_1 \parallel C_0$ and $C_1 \parallel C_2$ in order
7: Test ML model with TD' to get $\mathcal{C}$ ▷ $\mathcal{C}$ is sequence of classes by ML
8: a' = probability that $\mathcal{C}$ matches $(0, 1)$
9: **if** $a' = a > \frac{1}{2}$ **then**
10: ORACLE = CIPHER
11: **else** ▷ $a' = \frac{1}{2}$
12: ORACLE = RANDOM
13: Repeat from Step 3 if required

6.4.2 *Model 2: One Input Difference*

While the first model (described earlier in Sect. 6.4.1) can work with an arbitrary number of input differences, $t \geq 2$, here we propose a different model that can work with only one input difference. A top level view can be found in Algorithm 6.3. As it can be seen, this model actually converts one input difference to a problem of classification with two classes. The case for ORACLE = RANDOM would result in training accuracy of $\frac{1}{2}$.

Training and Testing the Model

With the backdrop already presented in the previous model, here we present the basic work-flow.

Training (Offline).

1. Select the non-zero input difference, δ.
2. Generate (an arbitrary number of) pairs of inputs, P_0, P_1 ($\neq P_0 \oplus \delta$); and compute $P_2 = P_1 \oplus \delta$.
3. Run the (un-keyed) permutation of the inputs to get the corresponding outputs: $C_i \leftarrow \texttt{CIPHER}(P_i)$, for $i = 0, 1, 2$.
4. Label $C_1 \parallel C_0$ as class 0 and $C_1 \parallel C_2$ as class 1. Run the machine learning model with the data.
5. Check if the training accuracy is $> \frac{1}{2}$. If the accuracy equals (or close to) $\frac{1}{2}$, the procedure is aborted.

Testing (Online).

1. Generate the input triplets P_0, P_1, and P_2 in the same way as training.
2. Collect the outputs C_0, C_1, and C_2 by querying ORACLE with input P_0, P_1, and P_2 in order.
3. Feed the trained machine learning data $C_1 \parallel C_0$ and $C_1 \parallel C_2$ in order.
4. Get the predicted classes from the trained model with the testing data.
5. Find the accuracy of class prediction. More precisely, tally the classes returned by the trained ML with the sequence: $(0, 1, 0, 1, \ldots, 0, 1)$, and find the probability that both match.
6. (a) If ORACLE = CIPHER, the ML would predict the class for $C_1 \parallel C_0$ and $C_1 \parallel C_2$ as 0 and 1, respectively, wtih the same probability as training. Therefore, the accuracy for class prediction (in Step 5) would be the same (or close to) as the accuracy observed during training, i.e., $> \frac{1}{2}$.

(b) If ORACLE = RANDOM, the ML would arbitrarily predict the classes. So the accuracy for predicting classes by the trained ML (in Step 5) would be equal to (or close to) $\frac{1}{2}$.

As we can see for both the models, it is important to feed the machine learning (during the online phase) the queries returned by the ORACLE in a specific order. The ordering of the classes predicted by the trained machine learning is what determines the inference regarding the ORACLE. Also, this inference, in turn, is determined by the accuracy observed during the training phase.

6.4.3 Comparison with Existing Models

All-in-one Differential Cryptanalysis

All-in-one differential cryptanalysis [AL12] does not use the cases where the output difference is not the same as the pre-determined output difference. Such cases are simply discarded. We make use of all the cases and are able to obtain distinguishers with reduced search complexity in the process.

Gohr's Model

Our analysis focuses on different direction from that of Gohr's [Goh19]. Gohr tries to prove the ability of ML-based distinguishers, so he selects SPECK [BSS+13] with small state size. This way he is able to compute the full DDT of round-reduced SPECK to compare with his neural distinguishers. We focus on other directions. More precisely, we consider ciphers with bigger state and non-Markov ciphers. Both of them are hard to compute the all-in-one distinguishers and we successfully simulate them via machine learning. It can be stated that our algorithm is simpler to understand and implement.

6.5 Results on Round-Reduced Ciphers

6.5.1 GIMLI *(Model 1)*

Using a SAT/SMT-based approach, the authors of GIMLI present the optimal differential trails up to 8 rounds [BKL+19], which are given in Table 6.1. If we use the optimal 8-round differential trail to distinguish GIMLI-PERMUTATION, we need more than 2^{52} input pairs. Now, based on the ML assisted model presented in Sect. 6.4.1 (the first model), we present differential distinguishers till 8 rounds

Table 6.1 Optimal differential trails for the round-reduced GIMLI-PERMUTATION

Rounds	1	2	3	4	5	6	7	8
Weight	0	0	2	6	12	22	36	52

Table 6.2 Accuracy of ML training on round-reduced GIMLI-HASH and GIMLI-CIPHER

Rounds	Accuracy	
	GIMLI-HASH	GIMLI-CIPHER
6	0.9689	0.9528
7	0.7229	0.6340
8	0.5219	0.5099

of GIMLI-HASH and GIMLI-CIPHER, each with training data of $2^{17.6}$ samples (and testing data of size $2^{14.3}$ samples). In both the cases, we choose only two input differences, so $t = 2$. A summary of results can be found in Table 6.2.

We use the same MLP network for distinguishing 8-round GIMLI-HASH and GIMLI-CIPHER. The network has five middle layers with number of neurons as (296, 258, 207, 112, 160) . The activation function is ReLU, and the number of epochs is set to 20. We observe that our ML model is able to train the data collected with accuracy > 0.5 till 8 rounds (the accuracy drops to 0.5 from 9 rounds onwards) for both GIMLI-HASH and GIMLI-CIPHER.

GIMLI-HASH

For GIMLI-HASH, we focus on the processes of absorbing the last block of message M and squeezing the first 128-bit hash value h. Suppose the full message M is just 127 bytes (denote the ith byte as $M[i]$), then after padded with a zero byte, the 128-bit block is absorbed into the sponge function. We generate the training data by flipping the least significant bit of byte $M[4]$ and $M[12]$. In other words, we process the message pairs with difference 1 in the 4th byte (as δ_0) and 12th byte (as δ_1). Then we compute the first 128-bit hash values pairs accordingly and collect the difference of hash values as training samples.

GIMLI-CIPHER

For GIMLI-CIPHER in Fig. 6.3, we assume there is only one associated data block, so at least 2 GIMLI permutations (48 rounds) are involved until the first output of the ciphertext block. Of 48 rounds, here we take the reduced version, namely up to 8 rounds to show our distinguisher. For the data collection, we generate uniformly distributed 256-bit keys and 128-bit nonce pairs with difference 1 in the 4th byte (as

Table 6.3 Accuracy of ML training for reduced-round ASCON-PERMUTATION

(a) Mask: 1000

Rounds	Accuracy
1	0.7499
2	0.9847
3	0.9861

(b) Mask: 10001

Rounds	Accuracy
1	0.8749
2	0.9990
3	0.8314

Table 6.4 Accuracy of ML training for reduced-round KNOT-256 and KNOT-512

KNOT-256

Rounds	Accuracy
6	0.9508
7	0.8984
8	0.8293
9	0.7096
10	0.5912

KNOT-512

Rounds	Accuracy
8	0.9999
9	0.9989
10	0.9824
11	0.8472
12	0.6032

δ_0) or 12th byte (as δ_1) of the nonces (similar to GIMLI-HASH). We set the associated data and the first message block m_0 to be zero. Then compute the ciphertext c_0 and the differences of it.

6.5.2 ASCON ***and*** KNOT ***(Model 1)***

Here we present differential distinguishers on ASCON[5] [DEM+19] and KNOT [ZDY+19], using the first model (Sect. 6.4.1).

ASCON-PERMUTATION works with 320-bit state, and we only take 256 and 512-bit versions of KNOT-PERMUTATION. Our results show existence of differential distinguishers for up to 3 rounds of ASCON-PERMUTATION; up to 10 rounds of KNOT-256, and up to 12 rounds KNOT-512 with trivial complexity. Table 6.3 shows results for ASCON with 1–3 reduced rounds. Here, we XOR a mask value to the 64-bit register x_0 to get δ_0, and XOR the same mask value to the register x_1 to get δ_1. The results with the mask value 1000 are presented in Table 6.3a and the same for the mask value 10001 are presented in Table 6.3b. For both the variants of KNOT, the mask value 1 is XORed to the 0th and 1st bytes of the state to generate the input differences, respectively.

Similarly, Table 6.4 shows the results for KNOT-PERMUTATION. In particular, Table 6.4(a) shows the results for KNOT-256 with 6–10 reduced rounds, and Table 6.4(b) shows that of KNOT-512 with 8–12 reduced rounds.

[5] We denote the latest version, ASCONv1.2, as ASCON for simplicity.

The size of neurons for the middle layers of the MLP used for ASCON and KNOT-256 is (128, 1024, 1024, 1024), and for KNOT-512 it is (256, 1024, 1024, 1024). The activation function in the hidden layers is taken as ReLU. Training is run for 20 epochs with 2^{19} training data (validation is done with testing data of equal size) samples in all the cases.

6.5.3 CHASKEY *(Model 2)*

Using the Lipmaa–Moriai formula, the authors of CHASKEY present some differential trails up to 8 rounds [MMH+14], which are given in Table 6.5. Using the notion of pre-existing all-in-one differential, if we use the 4-round differential trail to distinguish CHASKEY, we need at least 2^{37} input pairs.

Here, we apply the second model of machine learning-based differential distinguisher (described in Sect. 6.4.2) on 4-round CHASKEY- PERMUTATION. The ML model from Gohr's [Goh19] is used here, except the number of epochs is set to 10.

First, we extend the 4-round differential given by the designers [MMH+14] to get a new 5-round trail presented in Table 6.6, which shows the complexity for distinguishing 4-round CHASKEY- PERMUTATION would be at minimum 2^{37} input pairs. Next for our analysis, first two inputs P_0, P_1 are randomly generated. Thereafter, we obtain two pairs of samples: $P_1 \parallel P_0$, and $P_1 \parallel P_0 \oplus$ (00008400 $\parallel$ 00000400 $\parallel$ 00000000 $\parallel$ 00000000). Then we follow the methodology as described in Sect. 6.4.2. With training data size of 2^{23}, (validation is done with $2^{14.3}$ data) we observe the accuracy of ML is 0.616 for 4-round CHASKEY-PERMUTATION (Tables 6.7 and 6.8).

6.6 Choice of Machine Learning Model

For finding a distinguisher, we opt for machine learning techniques that can reveal the hidden structures in the data without explicit feature selection. Our problem is treated as a classification problem where one tries to check whether a particular differential is inserted at the input to the (round-reduced) cipher or not. Results in this section are obtained by taking the 8-round distinguisher of GIMLI-PERMUTATION as the benchmark. For training, we use 2^{18} data samples. The number of epochs is set to 20 as for higher numbers the models tend to overfit.

Table 6.5 Differential trails for the round-reduced CHASKEY- PERMUTATION

Rounds	1	2	3	4	5	6	7	8
Weight	0	4	16	37	73	133	208	293

When using machine learning algorithms, one has to make a choice on hyper-parameters that would perform the best on the given problem. These parameters are normally chosen in an empirical way, by testing various network architectures and following best practices. There are automated techniques to tune the hyper-parameters [Ben00, BB12]; however, these require significant resources which can be hard to emulate. Here, we report the outcome from the manual architecture search next. The results in this section are obtained by using Intel Xeon Silver 4215 processor with 256 GB RAM.

We have tried several different neural network types, including the basic MLP, *Convolutional Neural Network* (CNN), and *Long Short-Term Memory Network* (LSTM). We have varied the width (number of neurons per layer) and the depth (number of layers) of these to find out the best accuracy and speed of learning. We have also tried several different types of activation functions.

Here, we choose our first model (Sect. 6.4.1), with two differences, $\delta_0 = 1$ and $\delta_1 = 2$ (i.e., the least significant and the second least significant bits are flipped, respectively). The input layer size of the model is 128 (first 128-bits are squeezed), and the output layer size, $t = 2$. Our findings, which contain several distinguishers for 8-round `GIMLI-PERMUTATION` (highlighted), are stated in Table 6.8. Architecture denotes the number of neurons per layer starting from the input layer. The activation function denotes the function that is used in the hidden layers, as the output layer always used softmax. A summary can be given as follows:

- **CNNs** are not suitable for the purpose of finding a distinguisher. We have tried several architectures, and the accuracy is always 0.5. This is an expected result, as CNNs are aimed at recognizing patterns in input data, which helps in image recognition or natural language processing but does not work for cipher input where the bits are not related in any way. In fact, we observe distinguisher for 8-round `GIMLI-PERMUTATION` only except these models. It is, however, possible to design a CNN that would successfully work.

Table 6.6 Differential trail for 5-round `CHASKEY- PERMUTATION`

Round(s)	Δv_0	Δv_1	Δv_2	Δv_3	Weight
0	c0240100	44202100	0c200008	0c200000	–
1	00008400	00000400	00000000	00000000	15
2	80000000	00000000	00000000	80000000	2
3	80008080	00000040	00000000	80109080	2
4	10409000	00542800	08400010	12408210	18
5	42828202	48540a0d	0a000090	10c0cb52	35

Table 6.7 Accuracy of ML training on reduced-round `CHASKEY- PERMUTATION`

Rounds	Differential probability	Accuracy
$0 \rightarrow 4$	2^{-37}	0.61618899

Table 6.8 Results for architecture search with 8-round GIMLI-PERMUTATION

Network	Architecture	Activation function	Number of parameters	Training time (s)	Accuracy
MLP I	128, 296, 258, 207, 112, 160, 2	ReLU	226,633	1267.39	0.5588
MLP II	128, 1024, 2	ReLU	150,658	1081.67	0.5569
MLP III	128, 1024, 1024, 2	ReLU	1,200,256	1162.21	0.5652
MLP IV	128, 256, 128, 64, 2	LeakyReLU	90,818	510.9	0.5478
MLP VI	128, 1024, 1024, 2	LeakyReLU	1,200,256	1778.5	0.5509
MLP VII	128, 1024, 1024, 1024, 2	ReLU	2,249,858	2410.1	0.5689
CNN I	128, 128, 128, 100, 2	ReLU	128,046	2951.7	0.5000
CNN II	128, 1024, 128, 128, 100, 2	ReLU	604,206	11503.0	0.5000
LSTM I	128, 256, 128, 2	tanh/sigmoid	444,162	50460.7	0.5316
LSTM II	128, 200, 100, 128, 2	tanh/sigmoid	313,170	39825.9	0.5325

- **LSTMs** perform better than CNNs, but worse than fine-tuned MLP. The main drawback of LSTMs is the training speed—due to the presence of recurrent layers, these have high memory requirements and more computations are required.
- **MLPs** provide the best accuracy and can be tuned to train in a very fast time. Our best result is achieved by MLP with three hidden layers and 1024 neurons per hidden layer (MLP VII in Table 6.8), respectively. One can notice that in some cases we used Leaky ReLU as activation function [MHN13], which allows a small, positive gradient when the unit is not active (e.g., input is negative), and therefore is considered more "balanced". This is an advantage for smaller networks compared to normal ReLU, but for the network with 1.2M parameters, its performance is slightly worse.

6.7 Conclusion and Follow-Up Problems

In this work, we propose two novel methods for finding distinguishers on symmetric key primitives by using machine learning. Our methods work with any number (non-zero) input differences. At the core, we use multiple differentials and convert the problem of distinguishing CIPHER from RANDOM into a classification problem, which is then tackled by a machine learning technique.

Overall, here we generalize the classical (all-in-one) differential distinguisher [BS90, AL12]. Unlike the usual differential distinguisher that relies on rigorous observation and specifics of the target cipher, both our approaches are much simpler and only rely on analysing a set of input difference–output difference by machine learning. Thus, we propose the first proof of concept on how machine learning can be used as a generic tool in symmetric key cryptanalysis.

Our methods are also efficient as it can be carried out in around an hour by a modern computer. On the non-Markov ciphers GIMLI [BKL+19], ASCON [DEM+19], KNOT [ZDY+19], and CHASKEY [MMH+14, Mou15]; we show drastic reduction of search complexity reported by the designers for the round-reduced versions (typically of the order of cube root).

A summary of advantages and limitations of our work can be given as:

- ✓ The attack model is simple. In terms of neural network, our method works with as simple as a three-layer neural network.
- ✓ For round-reduced ciphers, we show that the complexity for differential cryptanalysis can be reduced to around the cube root of the claimed complexity. For example, the designers of GIMLI [BKL+19] claim the complexity of mounting a differential distinguisher by existing modelling methods would be at least 2^{52} data. However, we are able to find the same with complexity of around $2^{17.6}$ data (details can be found in Sects. 6.5.1 and 6.6).
- ✓ Our work is generic and can be applied to any symmetric key primitive where differential cryptanalysis can be applied. Here we target non-Markov ciphers as these are typically more complicated to analyse.
- ✗ So far, we are not able to extend our model to cover further rounds.
- ✗ For the time being, our model does not have a key recovery functionality.

That said, we would like to emphasize that our work does not indicate the theoretical limit of the application of machine learning in symmetric key cryptography. Future works in this direction will likely cover more rounds with advanced ML modelling and/or more training/testing data. As noted in [BBD+21], new functionalities like key recovery, higher order differential can be achieved by switching to a *Support Vector Machine* (SVM).[6]

As this is the first research work of its kind, there are scopes to improve the coverage of our experiments in the future. For example, training with more data can be performed to see if a distinguisher for 9-round GIMLI-PERMUTATION is achieved. We only take two input differences with our experiments with the first model (Sect. 6.4.1), so one may also be interested in experimenting with more differences. Apart from that, one may look for optimal choices for the input differences so that the size of the training/testing data can be reduced. The neural network models

[6] It is possible to get the same functionalities with artificial neural network, but it is computationally expensive compared to a support vector machine.

used here are somewhat arbitrarily chosen, therefore one may be interested in finding a model specifically tailored for this kind of problem in the future.

We are optimistic that the same methodologies can be applied to Markov ciphers like GIFT [BPP+17] as well, which we leave as future work. Since we use a classification problem, other machine learning techniques that specialize in classification can also be used instead of neural networks.

References

[SHB14] M. Sonka, V. Hlavac, R. Boyle, Image processing, analysis, and machine vision. Cengage Learning (2014)

[BCB14] D. Bahdanau, K. Cho, Y. Bengio, Neural machine translation by jointly learning to align and translate (2014). arXiv:1409.0473

[WSC+16] Y. Wu, M. Schuster, Z. Chen, Q.V. Le, M. Norouzi,W. Macherey, M. Krikun,Y. Cao, Q. Gao, K. Macherey, et al. Google's neural machine translation system: bridging the gap between human and machine translation (2016). arXiv:1609.08144

[CSK+15] C. Chen, A. Seff, A. Kornhauser, J. Xiao, Deepdriving: learning affordance for direct perception in autonomous driving, in *Proceedings of the IEEE International Conference on Computer Vision* (2015), pp. 2722–2730

[Gre17] S. Greydanus, Learning the enigma with recurrent neural networks (2017). arXiv:1708.07576

[Hay08] S. Haykin, *Neural Networks and Learning Machines*, 3rd (Pearson, 2008)

[KB14] D.P. Kingma, J. Ba, Adam: a method for stochastic optimization (2014). arXiv:1412.6980

[DEM+19] C. Dobraunig, M. Eichlseder, F. Mendel, M. Schläffer, Ascon v1.2. Submission to NIST (2019). https://csrc.nist.gov/CSRC/media/Projects/lightweight-cryptography/documents/round-2/spec-doc-rnd2/ascon-spec-round2.pdf

[ZDY+19] W. Zhang, T. Ding, B. Yang, Z. Bao, Z. Xiang, F. Ji, X. Zhao, KNOT: algorithm specifications and supporting document. Submission to NIST (2019). https://csrc.nist.gov/CSRC/media/Projects/lightweight-cryptography/documents/round-2/spec-doc-rnd2/knot-spec-round.pdf

[MMH+14] N. Mouha, B. Mennink, A. V. Herrewege, D. Watanabe, B. Preneel, I. Verbauwhede, Chaskey: an efficient MAC algorithm for 32-bit microcontrollers, in *Selected Areas in Cryptography - SAC 2014 - 21st International Conference, Montreal, QC, Canada, August 14-15, 2014*, Revised Selected Papers (2014). pp. 306–323. https://doi.org/10.1007/978-3-319-13051-4

[Mou15] N. Mouha, Chaskey: a MAC algorithm for microcontrollers - status update and proposal of chaskey-12 -. IACR Cryptology **1182**, 201 (2015)

[BDP+07] G. Bertoni, J. Daemen, M. Peeters, G.V. Assche, Sponge function, in *Ecrypt Hash Workshop (May 2007)* (2007). https://keccak.team/files/CSF-0.1.pdf

[Ben00] Y. Bengio, Gradient-based optimization of hyperparameters. Neural Comput. **12**(8), 1889–1900 (2000)

[BB12] J. Bergstra Y. Bengio, Random search for hyper-parameter optimization. J. Mach. Learn. Res. **13**(Feb), 281–305 (2012)

[MHN13] A.L. Maas, A.Y. Hannun, A.Y. Ng, Rectifier nonlinearities improve neural network acoustic models, in *Proceedings of ICML*, vol. 30, Issue 1, p. 3 (2013)

[MPP16] H. Maghrebi, T. Portigliatti, E. Prouff, Breaking cryptographic implementations using deep learning techniques, in *Security, Privacy, and Applied Cryptography Engineering - 6th International Conference, SPACE 2016, Hyderabad, India, December 14-18, 2016, Proceedings* (2016), pp. 3–26. https://doi.org/10.1007/978-3-319-49445-6

[CDP17] E. Cagli, C. Dumas, E. Prouff, Convolutional neural networks with data augmentation against jitter-based countermeasures - profiling attacks without preprocessing, in *Cryptographic Hardware and Embedded Systems - CHES 2017 - 19th International Conference, Taipei, Taiwan, September 25-28, 2017, Proceedings* (2017), pp. 45–68 https://doi.org/10.1007/978-3-319-66787-4

[AA16] M. Abadi, D.G. Andersen, Learning to protect communications with adversarial neural cryptography (2016). CoRR, abs/1610.06918. arXiv: 1610.06918

[DR20] J. Daemen, V. Rijmen, *The Design of Rijndael: The Advanced Encryption Standard (AES)*, 2nd edn. (Springer, Berlin, Heidelberg, 2020)

[AL12] M.R. Albrecht, G. Leander, An all-in-one approach to differential cryptanalysis for small block ciphers, in *Selected Areas in Cryptography, 19th International Conference, SAC 2012, Windsor, ON, Canada, August 15-16, 2012*, Revised Selected Papers (2012), pp. 1–15. https://doi.org/10.1007/978-3-642-35999-6%5C_1

[BKL+19] D.J. Bernstein, S. Kölbl, S. Lucks, P.M.C. Massolino, F. Mendel, K. Nawaz, T. Schneider, P. Schwabe, F. Standaert, Y. Todo, B. Viguier, Gimli, 2019. https://csrc.nist.gov/CSRC/media/Projects/lightweight-cryptography/documents/round-2/spec-doc-rnd2/gimli-spec-round2.pdf

[BBD+21] A. Baksi, J. Breier, V.A. Dasu, X. Dong, C. Yi, Following-up on machine learning assisted differential distinguishers, in *SILC Workshop - Security and Implementation of Lightweight Cryptography* (2021). https://www.esat.kuleuven.be/cosic/events/silc2020/wp-content/uploads/sites/4/2020/10/Submission4.pdf

[LMM91] X. Lai, J.L. Massey, S. Murphy, Markov ciphers and differential cryptanalysis, in *Advances in Cryptology-EUROCRYPT '91*, ed. by D.W. Davies (Springer, Berlin, Heidelberg, 1991) pp. 17–38. ISBN: 978-3-540-46416-7

[BPP+17] S. Banik, S.K. Pandey, T. Peyrin, Y. Sasaki, S.M. Sim, Y. Todo, Gift: a small present. Cryptology ePrint Archive, Report 2017/622 (2017) https://eprint.iacr.org/2017/622

[BSS+13] R. Beaulieu, D. Shors, J. Smith, S. Treatman-Clark, B.Weeks, L.Wingers, The simon and speck families of lightweight block ciphers. Cryptology ePrint Archive, Report 2013/404 (2013). https://eprint.iacr.org/2013/404

[BKL+17] D.J. Bernstein, S. Kölbl, S. Lucks, P.M.C. Massolino, F. Mendel, K. Nawaz, T. Schneider, P. Schwabe, F. Standaert, Y. Todo, B. Viguier, Gimli : A crossplatform permutation, in *Cryptographic Hardware and Embedded Systems - CHES 2017 - 19th International Conference, Taipei, Taiwan, September 25-28, 2017, Proceedings* (2017), pp. 299–320. https://doi.org/10.1007/978-3-319-66787-4

[BDP+11] G. Bertoni, J. Daemen, M. Peeters, G.V. Assche, Duplexing the sponge: single-pass authenticated encryption and other applications, in *Selected Areas in Cryptography - 18th International Workshop, SAC 2011, Toronto, ON, Canada, August 11-12, 2011*, Revised Selected Papers (2011), pp. 320–337. http://dx.doi.org/10.1007/978-3-642-28496-0_19

[BS90] E. Biham, A. Shamir, Differential cryptanalysis of des-like cryptosystems, in *Advances in Cryptology - CRYPTO '90, 10th Annual International Cryptology Conference, Santa Barbara, California, USA, August 11-15, 1990, Proceedings* (1990), pp. 2–21. https://doi.org/10.1007/3-540-38424-3_1

[Goh19] A. Gohr, Improving attacks on round-reduced speck32/64 using deep learning, in *Advances in Cryptology - CRYPTO 2019*, ed. by A. Boldyreva, D. Micciancio (2019), pp. 150–179 (Springer International Publishing, Cham, 2019). ISBN: 978-3-030-26951-7

[MWG+11] N. Mouha, Q. Wang, D. Gu, B. Preneel, Differential and linear cryptanalysis using mixed-integer linear programming, in *Information Security and Cryptology - 7th International Conference, Inscrypt 2011, Beijing, China, November 30 - December 3, 2011*. Revised Selected Papers (2011), pp. 57–76. https://doi.org/10.1007/978-3-642-34704-7%5C_5

Chapter 7
Differential Paradox: How an SBox Plays Against Differential Fault Analysis

Differential Fault Analysis (DFA) is a well-known cryptanalytic method that has been successfully applied to many block ciphers based on Substitution Permutation Network (SPN). In this work we seek the answer: How exactly DFA works and how can we possibly build a cipher level protection against it. Our study shows that SBoxes play a crucial role for DFA to succeed. Interestingly, SBoxes that are better against DFA are proved to be worse against differential cryptanalysis, and vice versa.

7.1 Introduction

In modern cipher design, an SBox (see Sect. 2.1.2) plays an important role. In the popular SPN family of block ciphers, the SBox is the only component which introduces non-linearity. As a result, no matter where the fault is injected, it affects the SBox in some way (injecting linear difference in other parts of the design does not yield meaningful information).

When it comes to Differential Fault Analysis (DFA) of block ciphers (see Sect. 3.5.1), each work usually aims at attacking a particular cipher, thus making the result cipher-specific. In our work, we look at DFA from a general view and try to understand how an SBox plays a role against this attack. In the process we provide several results that are useful to determine the properties of an SBox with respect to fault attacks.

We establish one very interesting property: If an SBox has a non-zero linear structure (see Definition 2.5), it is not possible to mount a (completely) successful DFA (see Theorem 7.1). In such a situation, the attacker cannot get as much information as she gets from an SBox with no linear structure, i.e., these SBoxes provide the best resistance against DFA. Note that an SBox with linear structure easily falls for differential cryptanalysis. Our results show how it is theoretically possible to design

A. Baksi, *Classical and Physical Security of Symmetric Key Cryptographic Algorithms*, Computer Architecture and Design Methodologies,
https://doi.org/10.1007/978-981-16-6522-6_7

a block cipher, which, by virtue of its design has a somewhat resistance against DFA; see Sect. 7.4 for the discussion.

It is to be noted that, our objective here is not to build a cipher (which can have a security against DFA). Instead, we discuss the theoretical background that will help researchers to understand the problem of DFA. For the time being, the results presented here are of theoretical interest. Those results, in future, can be used to design a DFA-resistant cipher.

7.2 Difference Distribution Table-Related Properties

We adopt the following notations: S is the SBox with the input α (unknown to the attacker). She tries to recover the input α by the input difference (fault) δ and observes the output difference Δ; i.e., $S(\alpha) \oplus S(\alpha \oplus \delta) = \Delta$. We often denote Δ as a function of α and δ, i.e., as $\Delta(\alpha, \delta)$. We implicitly assume that neither δ or Δ is 0 and S is a bijection from now on (unless mentioned otherwise).

In order for an SBox to be better resistant against DA, the (non-zero) maximal values in the DDT have to be small, otherwise DA will be more effective. Thus, symmetric key cryptography designers almost exclusively look for SBoxes which have smaller values in the DDT. The DDT values are non-negative and even; hence an ideal (with respect to DA) DDT would contain only 0, 2 values. Whether or not an $n \times n$ SBox exists with only 0, 2 entries in its DDT is known as the *Big APN Problem* and it is still open when $n > 6$ and n is even [BDM+10]/Sect. 2.1.2. However, the situation for DFA is completely opposite. Here, if the (non-zero) DDT values are small, then the attacker has less number of solutions for the unknown input. Thus, she is able to narrow down the search space more efficiently. So, a DDT with smaller (non-zero) values will make the DFA easier. Hence we see that the requirement to thwart DA is exactly opposite to that of DFA. This paradoxical situation is among the challenges to build a cipher-level DFA protection.

Definition 7.1 (*Expanded DDT*) Expanded DDT of the SBox S is a matrix having the same dimension as its DDT, where entry corresponding to the input difference δ and output difference Δ is the set of solutions for the equation $S(x) \oplus S(x \oplus \delta) = \Delta(x, \delta)$.

An expanded DDT can be interpreted as: It gives the actual solutions, whereas DDT only shows the cardinality of each solutions instead. Table 7.1 shows the expanded DDT for the SBox `126CDE39F58BA047` (see Table 2.2 for its DDT), where Ω denotes the set of all inputs, i.e., $\Omega = \{\texttt{0}, \texttt{1}, \texttt{2}, \texttt{3}, \texttt{4}, \texttt{5}, \texttt{6}, \texttt{7}, \texttt{8}, \texttt{9}, \texttt{a}, \texttt{b}, \texttt{c}, \texttt{d}, \texttt{e}, \texttt{f}\}$.

Definition 7.2 ($S_\alpha\langle\delta\rangle$) For the SBox S and the fault δ, the set of solutions of the equation $S(x) \oplus S(x \oplus \delta) = S(\alpha) \oplus S(\alpha \oplus \delta)$ is the set $S_\alpha\langle\delta\rangle$.

Notice that both α and $\alpha \oplus \delta \in S_\alpha\langle\delta\rangle$. Basically, $S_\alpha\langle\delta\rangle$ gives the entry of the DDT at the δth row that contains α, which is at the column $\Delta(\alpha, \delta) = S(\alpha) \oplus S(\alpha \oplus \delta)$. By

Table 7.1 Expanded DDT for SBox 126CDE39F58BA047

δ \ Δ	1	2	3	4	5	6	7	8	9	a	b	c	d	e	f
1			0145abef							236789cd					
2							02578adf							13469bce	
3				12478bde									03569acf		
4					236789cd							0145abef			
5															Ω
6		03569acf									12478bde				
7	13469bce							02578adf							
8							13469bce							02578adf	
9				03569acf									12478bde		
a									Ω						
b			236789cd							0145abef					
c		12478bde									03569acf				
d	02578adf							13469bce							
e					0145abef							236789cd			
f						Ω									

applying fault δ, the attacker cannot identify α from other elements which belong to $S_\alpha\langle\delta\rangle$.

Example 7.1 Take the SBox 20135467A98BCDEF (# 288 in [DeC07]). Notice from the expanded DDT (Table 7.2) that, for fault $\delta = \texttt{1}$ (column 3), we have $S_{\texttt{8}}\langle\texttt{1}\rangle = S_{\texttt{9}}\langle\texttt{1}\rangle = S_{\texttt{a}}\langle\texttt{1}\rangle = S_{\texttt{b}}\langle\texttt{1}\rangle = \{\texttt{8}, \texttt{9}, \texttt{a}, \texttt{b}\}$.

Definition 7.3 ($\text{MinF}_S(\alpha)$) For an $n \times n$ SBox S with input α,

$$\text{MinF}_S(\alpha) = \begin{cases} -1 & \text{if } \bigcap_{\delta=1}^{2^n-1} S_\alpha\langle\delta\rangle \neq \{\alpha\}; \\ t & \text{where } t = \min_k \text{ such that } \bigcap_{i=1}^{k} S_\alpha\langle\delta_i\rangle = \{\alpha\}. \end{cases}$$

$\text{MinF}_S(\alpha) = -1$ means that no matter what fault values that Eve chooses, she will be left with more than one choice for α (refer to Theorem 7.1 and Example 7.3). Also, notice that if $\text{MinF}_S(\alpha) \neq -1$, then it must be ≥ 2.

Example 7.2 Consider the previous SBox 20135467A98BCDEF (expanded DDT is in Table 7.2). Assume the input, $\alpha = \texttt{8}$. If she inserts the fault values $\delta_1 = \texttt{d}$ and $\delta_2 = \texttt{1}$, then she is able to uniquely retrieve the input $\alpha = \texttt{8}$, since $S_{\texttt{8}}\langle\texttt{d}\rangle \cap S_{\texttt{8}}\langle\texttt{1}\rangle = \{\texttt{2}, \texttt{5}, \texttt{8}, \texttt{f}\} \cap \{\texttt{8}, \texttt{9}, \texttt{a}, \texttt{b}\} = \{\texttt{8}\}$. So, $\text{MinF}_{\texttt{20135467A98BCDEF}}(\texttt{8}) = 2$.

Definition 7.4 (MinF_S) Given an $n \times n$ SBox S, MinF_S is defined as

$$\text{MinF}_S = \begin{cases} \max_{0 \leq \alpha \leq 2^n-1} \text{MinF}_S(\alpha) & \text{if } \text{MinF}_S(\alpha) \neq -1, \ \forall \alpha \in \{0, 1 \ldots, 2^n - 1\}; \\ -1 & \text{otherwise.} \end{cases}$$

The subscript S is dropped if understood from context.

MinF_S can be interpreted as: Given an SBox S, it is the lowest number of faults required to uniquely solve any input.

Table 7.2 Expanded DDT for SBox `20135467A98BCDEF`

Δ δ	1	2	3	4	5	6	7	8	9	a	b	c	d	e	f
1	4 5 6 7 c d e f	0 1 2 3	8 9 a b												
2		8 9 a b c d e f	0 1 2 3 4 5 6 7												
3	0 1 2 3 8 9 a b	4 5 6 7	c d e f												
4				1 3 5 7 9 b d f		8 a c e	0 2 4 6								
5					1 3 4 6 9 b c e	0 2 5 7	8 a d f								
6				0 2 4 6 8 a c e		9 b d f	1 3 5 7								
7					0 2 5 7 8 a d f	1 3 4 6	9 b c e								
8								0 3 6 7 8 b e f	1 2 4 5 9 a c d						
9								4 5 c d	6 7 e f	1 2 8 b	0 3 9 a				
a										0 3 6 7 9 a c d	1 2 4 5 8 b e f				
b								1 2 9 a	0 3 8 b	4 5 e f	6 7 c d				
c												3 7 b f	1 5 9 d	0 6 a c	2 4 8 e
d												1 4 9 c	3 6 b e	2 5 8 f	0 7 a d
e												0 6 8 e	2 4 a c	3 7 9 d	1 5 b f
f												2 5 a d	0 7 8 f	1 4 b e	3 6 9 c

7.3 Characterizing SBoxes in View of DFA

Theorem 7.1 *Let* $S(x) \oplus S(x \oplus \delta) = \Delta(\alpha, \delta)$ *have a solution* $x = \alpha$. *Further, let* a *be a (non-zero) linear structure of* S. *Then,* $(\alpha \oplus a)$ *is also a solution of* $S(x) \oplus S(x \oplus \delta) = \Delta(\alpha, \delta)$, *i.e., the coset* $\alpha \oplus \mathcal{L}(S)$ *is a subset of solutions of* $S(x) \oplus S(x \oplus \delta) = \Delta(\alpha, \delta)$. *So,* $MinF_S = -1$.

Proof As a is a linear structure of S, we have $S(x) \oplus S(x \oplus a) =$ constant. Taking derivative with respect to δ, we get $S(x) \oplus S(x \oplus a) \oplus S(x \oplus \delta) \oplus S(x \oplus a \oplus \delta) = 0, \forall x$. Putting $x = \alpha$,

$$\begin{aligned} & S(\alpha) \oplus S(\alpha \oplus \delta) \oplus S(\alpha \oplus a) \oplus S(\alpha \oplus a \oplus \delta) = 0 \\ \Longrightarrow \ & S(\alpha \oplus a) \oplus S(\alpha \oplus a \oplus \delta) = S(\alpha) \oplus S(\alpha \oplus \delta) = \Delta(\alpha, \delta). \end{aligned}$$

Hence, $(\alpha \oplus a)$ is also a solution of $S(x) \oplus S(x \oplus \delta) = \Delta(\alpha, \delta)$.

Theorem 7.1 gives an interesting insight regarding DFA resistance in SBoxes. If an SBox has a linear structure, then it is not possible to find the input to the SBox just by analysing the effect of faults, no matter how many faults are injected. In such cases, the attacker has to search exhaustively among the set of solutions to find the proper input. This increases the search complexity associated with DFA.

Example 7.3 The SBox `80A23517496BCDEF` (the representative of class #293 in [DeC07, Chap. 4.2]) has a linear structure at $a = 2$, which can be seen from its DDT in Table 7.3(a). Zeros, and the rows–columns corresponding to $\delta = 0$ and $\Delta = 0$ are not shown here for the sake of better clarity. Table 7.3(b) gives the $S_0\langle\delta\rangle$ for varying δ on S.

Lemma 7.1 (Converse of Theorem 7.1) *For the input* α *to* S, *if* $\alpha \oplus a$ *is a solution of* $S(x) \oplus S(x \oplus \delta) = \Delta(\alpha, \delta)$ *for all input difference* δ, *then* a $(\neq 0)$ *is a linear structure of* S.

Proof Since both α and $\alpha \oplus a$ are solutions of $S(x) \oplus S(x \oplus \delta) = \Delta(\alpha, \delta)$, for any δ, so we have $S(\alpha) \oplus S(\alpha \oplus \delta) = \Delta(\alpha, \delta) = S(\alpha \oplus a) \oplus S(\alpha \oplus a \oplus \delta)$. From this, we have

$$\begin{aligned} & S(\alpha) \oplus S(\alpha \oplus \delta) = S(\alpha \oplus a) \oplus S(\alpha \oplus a \oplus \delta) \\ \Longrightarrow \ & S(\alpha \oplus \delta) \oplus S(\alpha \oplus \delta \oplus a) = S(\alpha) \oplus S(\alpha \oplus a). \end{aligned}$$

As the above relation holds for any δ, so we can write $S(x) \oplus S(x \oplus a) = S(\alpha) \oplus S(\alpha \oplus a)$ holds for all x. This means that a is a linear structure of S.

Remark 7.1 Theorem 7.1 and Lemma 7.1 are valid for all (non-trivial) linear structure(s) of S. In other words, the more the number of (non-trivial) linear structures, the higher the number of candidates will be in the intersection of solution sets of all faults.

Table 7.3 Effect of DFA on SBox `80A23517496BCDEF`

(**a**) DDT

δ \ d	1	2	3	4	5	6	7	8	9	a	b	c	d	e	f
1	4					4		4					4		
2		16													
3			4	4						4					4
4				4	4			4			4				
5			4		4				4				4		
6						4	4		4	4					
7	4						4				4				4
8								4	4			4			4
9	4			4					4					4	
a										4	4		4	4	
b			4			4					4	4			
c				4			4					4	4		
d	4				4					4		4			
e					4	4								4	4
f			4				4	4						4	

(**b**) Analysis for input $\alpha = 0$

δ	$S_0\langle\delta\rangle$	$\lvert S_0\langle\delta\rangle\rvert$
1	0, 1, 2, 3	4
2	0, 1, 2, 3, 4, 5, 6, 7, 8, 9, a, b, c, d, e, f	16
3	0, 1, 2, 3	4
4	0, 2, 4, 6	4
5	0, 2, 5, 7	4
6	0, 2, 4, 6	4
7	0, 2, 5, 7	4
8	0, 2, 8, a	4
9	0, 2, 9, b	4
a	0, 2, 8, a	4
b	0, 2, 9, b	4
c	0, 2, c, e	4
d	0, 2, d, f	4
e	0, 2, c, e	4
f	0, 2, d, f	4

Theorem 7.2 *If $S(x) \oplus S(x \oplus \delta) = \Delta(\alpha, \delta)$ has only two solutions, then there exists a δ', such that α is the unique common solution for $S(x) \oplus S(x \oplus \delta) = \Delta(\alpha, \delta)$ and $S(x) \oplus S(x \oplus \delta') = \Delta(\alpha, \delta')$.*

Proof Consider $S(x) \oplus S(x \oplus \delta) = \Delta(\alpha, \delta)$ which has only two solutions, α and $\alpha \oplus \delta$. On the contrary, assume that for all input differences d, if α is a solution of $S(x) \oplus S(x \oplus d) = \Delta(\alpha, d)$, so is $\alpha \oplus \delta$. Then by Lemma 7.1, we have that δ is a linear structure of S, which is a contradiction. Thus there will be a δ' such that α is a solution of $S(x) \oplus S(x \oplus \delta') = \Delta(\alpha, \delta')$, but $\alpha \oplus \delta$ is not. □

Note that Theorem 7.2 is a generalization of [LCF+17, Proposition 1]. In the proof [LCF+17, Appendix A.1], we see that the common solution depends on the span of two faults. So, for the applicability of [LCF+17, Proposition 1], those two faults must exist. In contrast, our proof shows that, if there is any fault δ that has two solutions, there will be another fault such that the solution is unique. Also, we propose an extension to it in Theorem 7.5.

Lemma 7.2 *Suppose S_1 and S_2 are two $n \times n$ SBoxes having ℓ_1 and ℓ_2 linear structures (including the trivial linear structure 0), respectively, then the $2n \times 2n$ SBox (S_1, S_2) will have $\ell_1\ell_2$ linear structures (including the trivial linear structure $(0, 0)$).*

Proof We show that (a_1, a_2) is a linear structure of (S_1, S_2) if and only if a_1, a_2 are linear structures of S_1 and S_2, respectively. If (a_1, a_2) is a linear structure of (S_1, S_2), then for some constant $(c_1, c_2) \in \mathbb{F}_2^n \times \mathbb{F}_2^n$, $(S_1(x), S_2(y)) \oplus (S_1(x \oplus a_1), S_2(y \oplus a_2)) = (c_1, c_2)$, for all $x, y \in \mathbb{F}_2^n$.

Therefore, $(S_1(x), S_2(0)) \oplus (S_1(x \oplus a_1), S_2(a_2)) = (c_1, c_2)$ for all $x \in \mathbb{F}_2^n$, from which we have $(S_1(x) \oplus S_1(x \oplus a_1), S_2(0)) \oplus S_2(a_2)) = (c_1, c_2)$, that is $S_1(x) \oplus S_1(x \oplus a_1) = c_1$ for all $x \in \mathbb{F}_2^n$. Therefore, a_1 is a linear structure of S_1. Similarly, it can be proved that a_2 is a linear structure of S_2.

Conversely if a_1, a_2 are linear structures of S_1 and S_2, respectively, then it is easy to prove that (a_1, a_2) of (S_1, S_2). Thus the total number of linear structures of (S_1, S_2) is $\ell_1\ell_2$ (including the trivial linear structure $(0, 0)$).

Lemma 7.3 *Suppose $F : \mathbb{F}_2^n \rightarrow \mathbb{F}_2^n$ is any function and $L : \mathbb{F}_2^n \rightarrow \mathbb{F}_2^n$ is affine. Then $L \circ F$ and F have the same number of linear structures.*

Proof Suppose $a \in \mathbb{F}_2^n$ is a linear structure of $L \circ F$. Then we have $L(F(x)) \oplus L(F(x \oplus a)) = c$, for some constant $c \in \mathbb{F}_2^n$. This implies that $L((F(x) \oplus F(x \oplus a)) = c$, that is $F(x) \oplus F(x \oplus a) = L^{-1}(c)$. Therefore, a is also a linear structure of F. To prove the converse, note that if a is a linear structure of F, then a is also a linear structure of $L \circ F$. Therefore, a is a linear structure of $L \circ F$ if and only if a is a linear structure of F.

Remark 7.2 Lemma 7.2 and Lemma 7.3 give another interesting view-point: If an un-keyed SPN permutation is constructed by repeating an SBox with l LS m times (in each round), then the total number of linear structures for the super SBox (which is the round function) is l^m.

Table 7.4 Number of linear structures versus DFA security (in terms of search complexity) for SPN ciphers

(a) 4×4

♯ LS	State size	DFA security
2	128	2^{32}
	256	2^{64}
4	128	2^{64}
	256	2^{128}

(b) 8×8

♯ LS	State size	DFA security
8	128	2^{48}
	256	2^{96}
64	128	2^{96}
	256	2^{192}
128	128	2^{112}
	256	2^{224}

Remark 7.3 For simplicity, denote $\Delta(\alpha, \delta)$ by Δ and $\Delta(\alpha, \delta')$ by Δ'. If $|S_\alpha\langle\delta\rangle| = 2$, $S(\alpha) \neq 0$, and $S(\alpha \oplus \delta) \neq 0$, then δ' can be chosen as $\delta' = \alpha \oplus S^{-1}(\Delta)$, such that $S_\alpha\langle\delta\rangle \cap S_\alpha\langle\delta'\rangle = \alpha$. Notice that, in this case, $\Delta' = S(\alpha) \oplus S(\alpha \oplus \delta') = S(\alpha) \oplus \Delta = S(\alpha \oplus \delta)$.

In order to better visualize the DFA security with respect to the number of linear structures for SPN ciphers (for a given size of the SBox), we present detailed information in Table 7.4 for varying state sizes. Note that the last cases (i.e., a 4×4 SBox with 4 and an 8×8 SBox with 128 linear structures) is the theoretical limit for DFA protection (as any more number of LS would imply the SBox is linear). Hence in theory, we can achieve DFA security up to 2^{64} (for 128-bit state) or 2^{128} (for 256-bit state) using 4-bit SBoxes; and 2^{112} (for 128-bit state) or 2^{224} (for 256-bit state) using 8-bit SBoxes. It may be noted that DFA security is always lower than that of the margin of the differential attack.

Theorem 7.3 *If an $n \times n$ SBox has ℓ non-zero linear structures, then the minimum non-zero value in its DDT* $= \min(2^n, 2\ell + 2)$.

Proof Let the ℓ non-zero linear structures are $a_1, \dots, a_\ell$. Hence, for $\delta \in \{a_1, \dots, a_\ell\}$, the minimum non-zero value in the corresponding rows are 2^n (in fact, this is the only non-zero value in its DDT if the SBox is linear). For $\delta \notin \{a_1, \dots, a_\ell\}$, the elements $\alpha, \alpha \oplus \delta, \alpha \oplus \alpha_1, \alpha \oplus \alpha_1 \oplus \delta, \dots, \alpha \oplus \alpha_\ell, \alpha \oplus \alpha_\ell \oplus \delta \in S_\alpha\langle\delta\rangle$. Hence the result follows.

Theorem 7.3 indicates that if we want better protection against DFA (by increasing linear structures in the SBox), then at the same time the minimum entry in its DDT will also grow, making it harder to resist against differential attack.

Remark 7.4 By choosing an SBox for which large number of coordinate functions are affine, it is possible to get more LS. For an $n \times n$ SBox S, if it has c ($\leq n - 2$) affine coordinate functions, then it can have 2^c linear structures. For example, the 8×8 SBox given by the coordinate functions, $S(x_0, x_1, x_2, x_3, x_4, x_5, x_6, x_7) = (x_0, x_1, x_2, x_3, x_4, x_5, x_0x_1 \oplus x_6, x_0x_1 \oplus x_7)$ has $2^6 = 64$ LS. As a side note, it can be mentioned that the SBox $S(x_0, x_1, x_2, x_3, x_4, x_5, x_6, x_7) = (x_0, x_1, x_2, x_3, x_4, x_5, x_6, x_0x_1 \oplus x_7)$ has 64 LS too.

Remark 7.5 Although (in terms of maximum DFA security) the maximum number of LS an $n \times n$ SBox can have is 2^{n-1} (as 2^n LS would make the SBox linear), so far we are able to find SBoxes with at most 2^{n-2} LS (see Remark 7.4). Whether or not such SBox exists is left as an open problem. This problem can be considered orthogonal to the big APN problem (Sect. 2.1.2).

Remark 7.6 Having a non-zero linear structure does not imply zero non-linearity. For example, the SBox `0123458967CDEFBA` has a linear structure at `1` but has non-linearity of 2.

Definition 7.5 ($D_\alpha\langle z\rangle$) For an SBox S with α as the input, $D_\alpha\langle z\rangle$ is defined as $D_\alpha\langle z\rangle = \{\text{fault value } d : z \text{ is a solution to } S(x) \oplus S(x \oplus d) = \Delta(\alpha, d)\}$, where z is an arbitrary input value.

So, for $d \in D_\alpha\langle z\rangle$, we have, $\Delta(\alpha, d) = S(\alpha) \oplus S(\alpha \oplus d) = S(z) \oplus S(z \oplus d) = \Delta(z, d)$. In other words, the attacker will not be able to distinguish α from z under fault $d \in D_\alpha\langle z\rangle$.

Lemma 7.4 *Suppose, for the SBox S, $S(x) \oplus S(x \oplus \delta) = \Delta(\alpha, \delta)$ has exactly $2m + 2$ solutions, $\{\alpha, \alpha \oplus \delta, \beta_1, \beta_1 \oplus \delta, \ldots, \beta_m, \beta_m \oplus \delta\}$ where $2m + 2 \geq 4$. Then for any fault d ($\neq \delta$), define d^* as $d^* = d \oplus \delta$. We show the following results:*

(i) $d \in D_\alpha\langle\alpha \oplus \delta\rangle \implies d^* \in D_\alpha\langle\alpha \oplus \delta\rangle$;
(ii) $d \in D_\alpha\langle\beta_i\rangle \implies d^* \in D_\alpha\langle\beta_i\rangle$, *for* $i = 1, \ldots, m$;
(iii) $d \in D_\alpha\langle\alpha \oplus \delta\rangle \cap D_\alpha\langle\beta_i\rangle \implies d^* \in D_\alpha\langle\alpha \oplus \delta\rangle \cap D_\alpha\langle\beta_i\rangle$, *for* $i = 1, \ldots, m$.

Proof Under fault d, we have:

$$\begin{aligned}
\Delta(\alpha, d) &= S(\alpha) \oplus S(\alpha \oplus d)\\
\Delta(\alpha \oplus \delta, d) &= S(\alpha \oplus \delta) \oplus S(\alpha \oplus \delta \oplus d)\\
\Delta(\beta_i, d) &= S(\beta_i) \oplus S(\beta_i \oplus d)\\
\Delta(\beta_i \oplus \delta, d) &= S(\beta_i \oplus \delta) \oplus S(\beta_i \oplus \delta \oplus d)
\end{aligned}$$

If i holds, then $\Delta(\alpha, d) = \Delta(\alpha \oplus \delta, d)$; if ii holds, then $\Delta(\beta_i, d) = \Delta(\beta_i \oplus \delta, d)$; if iii holds, then $\Delta(\alpha, d) = \Delta(\alpha \oplus \delta, d) = \Delta(\beta_i, d) = \Delta(\beta_i \oplus \delta, d)$.

Similarly, under fault $d^* = d \oplus \delta$, we have

$$\begin{aligned}
\Delta(\alpha, d^*) &= S(\alpha) \oplus S(\alpha \oplus d^*) = S(\alpha) \oplus S(\alpha \oplus \delta \oplus d)\\
\Delta(\alpha \oplus \delta, d^*) &= S(\alpha \oplus \delta) \oplus S(\alpha \oplus \delta \oplus d^*) = S(\alpha \oplus \delta) \oplus S(\alpha \oplus d)\\
\Delta(\beta_i, d^*) &= S(\beta_i) \oplus S(\beta_i \oplus d^*) = S(\beta_i) \oplus S(\beta_i \oplus \delta \oplus d)\\
\Delta(\beta_i \oplus \delta, d^*) &= S(\beta_i \oplus \delta) \oplus S(\beta_i \oplus \delta \oplus d^*) = S(\beta_i \oplus \delta) \oplus S(\beta_i \oplus d)
\end{aligned}$$

So, using previous relations, if i holds, then $\Delta(\alpha, d^*) = \Delta(\alpha \oplus \delta, d^*)$; if ii holds, then $\Delta(\beta_i, d^*) = \Delta(\beta_i \oplus \delta, d^*)$; if iii holds, then $\Delta(\alpha, d^*) = \Delta(\alpha \oplus \delta, d^*) = \Delta(\beta_i, d^*) = \Delta(\beta_i \oplus \delta, d^*)$.

Theorem 7.4 *Let the SBox S does not have any (non-zero) linear structure and $S(x) \oplus S(x \oplus \delta) = \Delta(\alpha, \delta)$ has exactly $2m + 2$ solutions, where $2m + 2 \geq 4$. Then there exist $m + 1$ faults $\{\delta', \delta_1, \ldots, \delta_m\}$, such that the system of equations*

$$\begin{aligned} S(x) \oplus S(x \oplus \delta) &= \Delta(\alpha, \delta) \\ S(x) \oplus S(x \oplus \delta') &= \Delta(\alpha, \delta') \\ S(x) \oplus S(x \oplus \delta_1) &= \Delta(\alpha, \delta_1) \\ &\vdots \\ S(x) \oplus S(x \oplus \delta_m) &= \Delta(\alpha, \delta_m) \end{aligned}$$

has a unique solution. Hence, $MinF_S(\alpha) \leq m + 2$.

Proof For the input α and for the fault δ, the output difference is $S(\alpha) \oplus S(\alpha \oplus \delta) = \Delta(\alpha, \delta)$. Suppose $S(x) \oplus S(x \oplus \delta) = \Delta(\alpha, \delta)$, has exactly $2m + 2$ solutions: $\{\alpha, \alpha \oplus \delta, \beta_1, \beta_1 \oplus \delta, \ldots, \beta_m, \beta_m \oplus \delta\}$. We will show that there exists a set of $m + 1$ faults $\{\delta', \delta_1, \ldots, \delta_m\}$ such that $S(x) \oplus S(x \oplus \delta) = \Delta(\alpha, \delta)$; $S(x) \oplus S(x \oplus \delta') = \Delta(\alpha, \delta')$; $S(x) \oplus S(x \oplus \delta_1) = \Delta(\alpha, \delta_1)$; ...; and $S(x) \oplus S(x \oplus \delta_m)=\Delta(\alpha, \delta_m)$ have only one common solution α; i.e., $S_\alpha\langle\delta\rangle \cap S_\alpha\langle\delta'\rangle \cap S_\alpha\langle\delta_1\rangle \cap \cdots \cap S_\alpha\langle\delta_m\rangle = \{\alpha\}$. So, to uniquely identify α, one needs not more than $m + 2$ faults: $\{\delta, \delta', \delta_1, \ldots, \delta_m\}$.

Since all of $\{\alpha, \alpha \oplus \delta, \beta_1, \beta_1 \oplus \delta, \ldots, \beta_m, \beta_m \oplus \delta\}$ are solutions of $\Delta(\alpha, \delta)$, we have

$$\begin{aligned} \Delta(\alpha, \delta) &= S(\alpha) \oplus S(\alpha \oplus \delta) = \Delta(\alpha \oplus \delta, \delta) \\ &= \Delta(\beta_i, \delta) = S(\beta_i) \oplus S(\beta_i \oplus \delta) = \Delta(\beta_i \oplus \delta, \delta); \text{ for } i = 1, \ldots, m. \end{aligned} \tag{7.1}$$

Consider the two sets of faults $D_\alpha\langle\alpha \oplus \delta\rangle$ and $D_{\alpha\beta} = \bigcup_{i=1}^m D_\alpha\langle\beta_i\rangle$. Note that $\delta \in D_\alpha\langle\alpha \oplus \delta\rangle \cap (\bigcap_{i=1}^m D_\alpha\langle\beta_i\rangle)$. Hence, none of the sets is empty. Define Ω_S to be the set of all $2^n - 1$ faults. We now treat rest of the proof in two cases.

Case 7.8.1. $D_\alpha\langle\alpha \oplus \delta\rangle \cup D_{\alpha\beta} \neq \Omega_S$.

Consider a $\delta' \in \Omega_S \setminus (D_\alpha\langle\alpha \oplus \delta\rangle \cup D_{\alpha\beta})$, then the equations $S(x) \oplus S(x \oplus \delta) = \Delta(\alpha, \delta)$ and $S(x) \oplus S(x \oplus \delta') = \Delta(\alpha, \delta')$ will have α as a solution and $\beta_i \oplus \delta$'s could be other solution(s). If no $\beta_i \oplus \delta$ appears as a solution, then the two faults δ and δ' would be enough to determine α uniquely.

Now suppose that, for a set of is, each of $\beta_i \oplus \delta$ is also included in the solution. Then, for each i, there must be one fault $\delta_i \in \Omega_S$ such that $\beta_i \oplus \delta$ is not a solution of $S(x) \oplus S(x \oplus \delta_i) = \Delta(\alpha, \delta_i)$; i.e., $\beta_i \oplus \delta \notin S_\alpha\langle\delta_i\rangle$. Otherwise if both α and $\beta_i \oplus \delta$ were common solutions of $S(x) \oplus S(x \oplus d) = \Delta(\alpha, d)$ for all input difference d, then by Lemma 7.1, $\alpha \oplus \beta_i \oplus \delta$ would be a linear structure of S, which is a contradiction. Therefore, these (at most $m + 2$) faults δ, δ', and δ_is will uniquely determine α.

Case 7.8.2. $D_\alpha\langle\alpha \oplus \delta\rangle \cup D_{\alpha\beta} = \Omega_S$.

First, we prove $D_{\alpha\beta}$ is not a subset of $D_\alpha\langle\alpha \oplus \delta\rangle$. If possible, assume $D_{\alpha\beta} \subseteq D_\alpha\langle\alpha \oplus \delta\rangle$. Since $D_\alpha\langle\alpha \oplus \delta\rangle \cup D_{\alpha\beta} = \Omega_S$ by assumption, we have $D_\alpha\langle\alpha \oplus \delta\rangle =$

Ω_S, implying δ is a linear structure, which is a contradiction. Hence, $D_{\alpha\beta} \setminus D_\alpha\langle\alpha \oplus \delta\rangle$ is non-empty.

Consider the fault: $\delta' \in D_{\alpha\beta} \setminus D_\alpha\langle\alpha \oplus \delta\rangle$. So, $S_\alpha\langle\delta\rangle \cap S_\alpha\langle\delta'\rangle$ can contain β_is or $\beta_i \oplus \delta$s along with α (but not $\alpha \oplus \delta$). Next, we prove $\beta_i \oplus \delta \notin S_\alpha\langle\delta\rangle \cap S_\alpha\langle\delta'\rangle$ by showing $\beta_i \oplus \delta \notin S_\alpha\langle\delta'\rangle$ for any i. Otherwise, if $\beta_i \oplus \delta \in S_\alpha\langle\delta'\rangle$ for some i, then $\Delta(\alpha, \delta') = \Delta(\beta_i \oplus \delta, \delta')$; i.e.,

$$\begin{aligned}
S(\alpha) \oplus S(\alpha \oplus \delta') &= S(\beta_i \oplus \delta) \oplus S(\beta_i \oplus \delta \oplus \delta') \\
\implies S(\alpha \oplus \delta') \oplus S(\beta_i \oplus \delta \oplus \delta') &= S(\alpha) \oplus S(\beta_i \oplus \delta) \\
&= S(\alpha \oplus \delta) \oplus S(\beta_i), \text{ using Equation (7.1)} \\
\implies S(\alpha \oplus \delta) \oplus S(\alpha \oplus \delta') &= S(\beta_i) \oplus S(\beta_i \oplus \delta \oplus \delta') \\
&\implies \delta \oplus \delta' \in D_\alpha\langle\alpha \oplus \delta\rangle \cap D_\alpha\langle\beta_i\rangle
\end{aligned}$$

Notice from Lemma 7.4 that if $\delta \oplus \delta' \in D_\alpha\langle\alpha \oplus \delta\rangle \cap D_\alpha\langle\beta_i\rangle$, then so will be $(\delta \oplus \delta') \oplus \delta = \delta'$, which is a contradiction, as $\delta' \in D_{\alpha\beta} \setminus D_\alpha\langle\alpha \oplus \delta\rangle$.

So, with faults δ and δ', assume we have α and β_i's in the solution, for a set of i's. Now, consider fault(s) $\delta_i \in \Omega_S \setminus D_\alpha\langle\beta_i\rangle$. Consequently, the intersection, $S_\alpha\langle\delta\rangle \cap S_\alpha\langle\delta'\rangle \cap S_\alpha\langle\delta_i\rangle$, cannot contain any of $\{\alpha \oplus \delta, \beta_i, \beta_i \oplus \delta\}$. Besides, such fault δ_i exists; otherwise, we will have, $D_\alpha\langle\beta_i\rangle = \Omega_S$, implying $\alpha \oplus \beta_i$ is a linear structure; which is a contradiction.

Hence the result follows.

From Theorem 7.4, we see that it is possible to uniquely recover the input/output value of each SBox with not more than $\frac{\mathrm{DU}_S}{2} + 1$ faults (unless there is a linear structure), when attacking the last round. This gives a provable upper bound for the attacker that she needs at most (if faults values are judiciously chosen) $\frac{\mathrm{DU}_S}{2} + 1$ faults per SBox in order the find out its input uniquely, given the SBox does not have a linear structure.

Corollary 7.1 (From Theorems 7.2, 7.4) $MinF_S \leq \dfrac{DU_S}{2} + 1.$

Remark 7.7 Although it is theoretically possible (refer to Corollary 7.1), we cannot find an SBox with $\mathrm{MinF}_S \geq 3$, even for bigger (such as 8×8) SBoxes. Whether or not this is the tight bound is left open for future research.

Differential cryptanalysis is one of the core attacks on block ciphers, and to protect from this attack the SBox needs to have lower values in the DDT. However, we see from Theorem 7.4 that if DDT has smaller values, then attacker can derive the SBox just by applying a small number of faults. In fact, so far we have seen that only two faults are enough, as stated in Remark 7.7 (if there is no linear structure). On the other hand, having more number of linear structures will increase the solution space of difference equations. However, such SBoxes will have poor protection against differential cryptanalysis. Therefore, the requirement for the resistance against DFA and differential cryptanalysis are conflicting with each other. So, we conclude that it may not be possible to achieve protection against DFA and DA by using a single SBox, rather a two different SBoxes may be necessary.

Theorem 7.5 (Extension of [LCF+17, Appendix A.1]) *For an $n \times n$ SBox S with input α, suppose, upon applying two distinct faults δ_1 and δ_2, α is not uniquely retrieved, since $\alpha \oplus \delta \in S_\alpha\langle\delta_1\rangle \cap S_\alpha\langle\delta_2\rangle$ $(\delta \neq \delta_1, \delta_2)$.*

(i) If $\delta = \delta_1 \oplus \delta_2$:
$\alpha, \alpha \oplus \delta_1, \alpha \oplus \delta_2, \alpha \oplus \delta_1 \oplus \delta_2 \in S_\alpha\langle\delta_1\rangle \cap S_\alpha\langle\delta_2\rangle \cap S_\alpha\langle\delta_1 \oplus \delta_2\rangle$.
Hence, $|S_\alpha\langle\delta_1\rangle|$, $|S_\alpha\langle\delta_2\rangle|$ and $|S_\alpha\langle\delta_1 \oplus \delta_2\rangle| \geq 4$.

(ii) Else:
$\alpha, \alpha \oplus \delta_i, \alpha \oplus \delta, \alpha \oplus \delta \oplus \delta_i \in S_\alpha\langle\delta_i\rangle$ *for* $i = 1, 2$.
$\alpha, \alpha \oplus \delta, \alpha \oplus \delta_1, \alpha \oplus \delta_2, \alpha \oplus \delta \oplus \delta_1, \alpha \oplus \delta \oplus \delta_2 \in S_\alpha\langle\delta\rangle$.
Hence, $|S_\alpha\langle\delta_1\rangle|$, $|S_\alpha\langle\delta_2\rangle| \geq 4$ and $|S_\alpha\langle\delta\rangle| \geq 6$.

Proof We have

$$S(\alpha) \oplus S(\alpha \oplus \delta_1) = \Delta(\alpha, \delta_1) = \Delta(\alpha \oplus \delta, \delta_1) = S(\alpha \oplus \delta) \oplus S(\alpha \oplus \delta \oplus \delta_1),$$
$$S(\alpha) \oplus S(\alpha \oplus \delta_2) = \Delta(\alpha, \delta_2) = \Delta(\alpha \oplus \delta, \delta_2) = S(\alpha \oplus \delta) \oplus S(\alpha \oplus \delta \oplus \delta_2),$$
$$\implies S(\alpha \oplus \delta_1) \oplus S(\alpha \oplus \delta_2) = S(\alpha \oplus \delta \oplus \delta_1) \oplus S(\alpha \oplus \delta \oplus \delta_2).$$

(i) Given, $\delta = \delta_1 \oplus \delta_2$. Now we have, $S(\alpha) \oplus S(\alpha \oplus \delta_1) = S(\alpha \oplus \delta_2) \oplus S(\alpha \oplus \delta_1 \oplus \delta_2) \implies \alpha \oplus \delta_2, \alpha \oplus \delta_1 \oplus \delta_2 \in S_\alpha\langle\delta_1\rangle$.

Again, $S(\alpha) \oplus S(\alpha \oplus \delta_2) = S(\alpha \oplus \delta_1) \oplus S(\alpha \oplus \delta_1 \oplus \delta_2) \implies \alpha \oplus \delta_1, \alpha \oplus \delta_1 \oplus \delta_2 \in S_\alpha\langle\delta_2\rangle$.

Also, $S(\alpha) \oplus S(\alpha \oplus \delta_1) = S(\alpha \oplus \delta_2) \oplus S(\alpha \oplus \delta_1 \oplus \delta_2) \implies S(\alpha) \oplus S(\alpha \oplus \delta_1 \oplus \delta_2) = S(\alpha \oplus \delta_1) \oplus S(\alpha \oplus \delta_2) \implies \alpha \oplus \delta_1, \alpha \oplus \delta_2 \in S_\alpha\langle\delta_1 \oplus \delta_2\rangle$.

(ii) $\delta \neq \delta_1 \oplus \delta_2$.
We get, $S(\alpha) \oplus S(\alpha \oplus \delta_i) = S(\alpha \oplus \delta) \oplus S(\alpha \oplus \delta \oplus \delta_i) \implies \alpha \oplus \delta, \alpha \oplus \delta \oplus \delta_i \in S_\alpha\langle\delta_i\rangle$; for $i = 1, 2$. Next, $S(\alpha) \oplus S(\alpha \oplus \delta) = S(\alpha \oplus \delta_1) \oplus S(\alpha \oplus \delta \oplus \delta_1) \implies \alpha \oplus \delta_1, \alpha \oplus \delta \oplus \delta_1 \in S_\alpha\langle\delta\rangle$. Similarly, $\alpha \oplus \delta_2, \alpha \oplus \delta \oplus \delta_2 \in S_\alpha\langle\delta\rangle$.

Hence the results follow.

Example 7.4 (*Theorem* 7.5ii) Take the SBox `20135467A98BCDEF` (its expanded DDT is in Table 7.2). Let $\alpha = 8, \delta_1 = 1, \delta_2 = 4$ with $\delta = 2 \neq \delta_1 \oplus \delta_2$. We have $|S_{\delta_1}\langle\alpha\rangle| = |\{8, 9, a, b\}| = 4$ and $|S_{\delta_2}\langle\alpha\rangle| = |\{8, a, c, e\}| = 4$ and $|S_\delta\langle\alpha\rangle| = |\{8, 9, a, b, c, d, e, f\}| \geq 6$.

7.4 Implication of Our Analysis and Future Work

Our analysis indicates that an SBox cannot provide the resistance to both the differential cryptanalysis and DFA simultaneously. In particular, the only resistance an SBox can have is (non-zero) linear structure(s). However, this makes the cipher weak against differential cryptanalysis as there is at least one differential transition with a probability 1. Also, the minimum non-zero value in the DDT increases with the number of linear structures (see Theorem 7.3), making it worse with respect to differential attack. On top of that, having a linear structure will likely increase the number of impossible transitions, thus arguably making impossible differential attack easier.

Thus, the requirement to thwart DFA is exactly opposite to that of DA. Therefore, one may think of a two-phase direction to design a cipher with in-built DFA security. More precisely, one can use a heterogeneous design with two different SBoxes, one with low DU (to protect against DA), while the other will have as many linear structures as possible. This way, the full security against DA as well as non-trivial security against DFA[1] can be achieved.

At a deeper level, one may observe that the DFA only works the last few rounds of the cipher. In contrast to that, DA work at the initial rounds. Hence the concept of DFA protecting layer can be loosely thought of as follows. First, a cipher can be designed in the usual way, i.e., with an SBox with lower DU, so that protection against classical attacks be ensured. Then, it can be augmented with a different design which only aims at protecting DFA by means of an SBox with many linear structures. If security against decryption is also needed, then this new design can also be prepended to the actual cipher and appended to the inverse of the actual cipher.

Therefore, to summarize, one can visualize the cipher-level protection of DFA as a two-level approach. The outer level with SBoxes, which are considered bad for a cipher design, protects against DFA. The inner level contains SBoxes with good cryptographic properties, hence can protect against classical attacks but not against DFA. The outer level thus protects the inner level against DFA. This is considered a future problem for this work and explored in adequate detail in Chap. 8.

References

[BDM+10] K. Browning, J. Dillon, M. McQuistan, A. Wolfe, An apn permutation in dimension six. Finite Fields: Theory Appl. **518**, 33–42 (2010)

[DeC07] C. De Cannière. Analysis and Design of Symmetric Encryption Algorithms, Ph.D. Thesis (Katholieke Universiteit Leuven, Belgium, 2007). https://www.esat.kuleuven.be/cosic/publications/thesis-139.pdf

[LCF+17] B. Lac, A. Canteaut, J. Fournier, and R. Sirdey. DFA on ls-designs with a practical implementation on SCREAM (extended version). IACR Cryptology ePrint Archive, 2017:76 (2017), http://eprint.iacr.org/2017/076

[1] Using SPN paradigm, the theoretically maximum DFA security that can be achieved is given in Table spsreftab:lsspsdfa. However, using more complicated designs, it may be possible to achieve full DFA security.

Chapter 8
DEFAULT: Cipher-Level Resistance Against Differential Fault Attack

Differential Fault Analysis (DFA) is a well-known cryptanalytic technique that exploits faulty outputs of an encryption device. Despite its popularity and similarity with the classical Differential Attack (DA), a thorough analysis explaining DFA from a designer's point of view is missing in the literature. To the best of our knowledge, no DFA immune cipher at an algorithmic level is proposed. Further, all known DFA countermeasures somehow depend on the device/protocol or on the implementation such as duplicate and compare. As all of these are outside the scope of the cipher designer, we focus on designing a cipher which can protect from DFA on its own. We present the first concept of cipher-level DFA resistance which does not rely on any device/protocol-related assumption, nor does it depend on any form of duplication. Our construction is simple, software/hardware-friendly, and DFA security scales up with the state size. It can be plugged before and/or after (almost) any symmetric key cipher and will ensure a non-trivial search complexity against DFA. One key component in our DFA protection layer is SBoxes with linear structures. Such SBoxes have never been used in cipher design as they generally perform poorly against differential attacks. We argue that they in fact represent an interesting trade-off between good cryptographic properties and DFA resistance. As a proof of concept, we construct a DFA protecting layer, named DEFAULT-LAYER, as well as a full-fledged block cipher, DEFAULT. Our solutions compare favourably to the state of the art, offering advantages over the sophisticated duplication-based solutions like CRAFT or infective countermeasures.

8.1 Introduction

Fault Attack (FA, see Chap. 3 for more details) is considered to be a strong implementation attack rendering many ciphers vulnerable. Unlike classical cryptanalysis, which assumes no interference to the internal operations of a cipher, the attacker in case of FA has more control over the device where the cipher is currently being

A. Baksi, *Classical and Physical Security of Symmetric Key Cryptographic Algorithms*, Computer Architecture and Design Methodologies,
https://doi.org/10.1007/978-981-16-6522-6_8

executed. As a result, among other options, Eve is able to suddenly alter an external input to the device (such as voltage level, EM radiation, heat, etc.), forcing it to run under sub-optimal condition. This type of condition can result in incorrect (faulty) output from the device. This faulty output may help the attacker to gain information on the secret key. FA gained much popularity among the researchers and has been deployed to analyse a variety of ciphers.

When it comes to analysing symmetric key cryptographic primitives, the most popular choice for FA is generally the Differential Fault Analysis or Differential Fault Attack (DFA) [BS97] (also described in Sect. 3.5.1). DFA is very powerful: Almost all (if not all) block ciphers which are considered secure with respect to classical attacks have been shown vulnerable to DFA. Note that, to the best of our knowledge, no cipher has yet been designed to have an innate DFA immunity, although there is no shortage of new cipher proposals as well as DFA countermeasures in recent years.

The root of this situation is, as we observe, a lack of theoretical results towards designing DFA-resistant primitives, akin to its classical counterpart, the Differential Attack (DA) [BS90] (see also Sect. 2.6.1). Cipher designers have been very careful to design DA resistant ciphers; but not much attention has been given to design a DFA-resistant cipher. Indeed, designing a DFA-resistant cipher looks like a very difficult task as the attacker has enormous power in this setting.

The usual DFA protection lies outside the domain of cipher design. In one end, some device/protocol-level technique is used, while in the other end, duplication-based protection is used (see Sect. 8.2.2 for more details). Duplication-based countermeasures assume the fault can alter within a predesignated boundary. Thereafter, a comparison (can be direct or by an error detection code) between the two executions is used to sense a fault. Since device/protocol-level solutions are beyond the control of the cipher designer, the best option to ensure DFA protection is duplication.[1] Given this scenario, our work analyses the problem and proposes a new type of solution, which is able to give a non-trivial search complexity against DFA completely by the cipher construction. We use the lightweight block cipher `GIFT-128` [BPP+17] as a building block for this construction, and thus manage to keep our design of relatively low-cost. Noting that the DFA protection mechanism could be costly (such as the error detecting code or the *infection* mechanism) and our design does not need duplication or any assumption related to implementation, thus opening up a new genre of low-cost DFA countermeasures.

Our Contributions

In this work, we attempt to explain DFA from a cipher designer's perspective (i.e., we interpret DFA in terms of classical attacks, which are well understood by the

[1] It may still be argued that duplication-based protections cannot be guaranteed at the cipher design level, and hence off-limit to a cipher designer.

cipher designers). We raise the question: How far a cipher designer can go to protect ciphers against DFA? In order to tackle this problem, we explore the potential offer by the SBox, one of the basic building blocks of symmetric-key algorithms. In the Substitution Permutation Network (SPN) block ciphers (and even in some f-function of Feistel network block ciphers), this is normally the only non-linear component, and it is naturally vulnerable to DFA (DFA does not work on a linear component, whereas it works very well on a non-linear one). In a nutshell, strong linearity makes it hard to attack a cipher with DFA, but too much linearity will of course render a cipher insecure or not efficient. The goal is therefore to try to find a good trade-off.

Now, since a cipher cannot be constructed only by using linear components, we naturally focus on finding a component which is somewhat the middle ground between an SBox and a linear layer. While the theoretical background is laid out in Chap. 7, we propose a cipher construction here.

The middle ground, unsurprisingly, lies in a weak class of SBoxes, which behave like a linear function in some respects. Such SBoxes have properties which are generally considered undesirable for a cipher construction. For example (non-trivial[2]), Linear Structures,[3] abbreviated as LS and defined in Definition 2.5, is something a cipher designer typically will try to avoid (see Remark 2.10). More relevant discussion on this topic can be found in Sect. 7.3. This gives rise to a paradoxical situation: An SBox which is better resistant against differential attack is weaker against DFA; and those which are better resistant against DFA are considered weak against differential attack.

To circumvent this situation, we propose to keep the main cipher (which is presumably secure against classical attacks) as it is, but add keyed permutations, L_1, L_2, as additional layers before and after it (when DFA security against both the encryption and decryption circuits is desired, only a single layer otherwise), respectively. Now, L_1, L_2 will ensure a special structure to make DFA non-trivial, and this will render the entire construction DFA resistant. At the same time, the classical security of the construction, which is governed by the main cipher, will not be hampered (refer to Sect. 8.3 for more discussion).

To validate our claims, we propose a construction of a 128-bit L $(= L_2 = L_1)$, DEFAULT-LAYER, which follows the SPN block cipher design paradigm with a 4×4 SBox. This SBox has three (non-trivial) LS. We show that this keyed permutation can provide safeguard against DFA to the non-trivial search complexity ($2^{n/2}$ for an n-bit block cipher). DEFAULT-LAYER is hardware and software friendly, and any variant of L with a multiple of 16-bit can be constructed (we recommend it to be at least 128-bit). As the DFA security scales up with the size of L (which does not happen for ciphers like AES), hence if a 256-bit variant of L is used, it will provide a DFA security of (at least) 2^{128} searches. In fact, by properly choosing the SBox, it is possible to go beyond $2^{n/2}$ security (see Table 7.4).

---

[2] Non-triviality here refers to the non-zero linear structures, since zero is always a linear structure.

[3] It is to be noted that the concept of "linearity" in the context of the classical linear cryptanalysis is different from a linear structure.

The idea of our keyed permutation DEFAULT-LAYER leads to a complete SPN-based cipher DEFAULT. It uses DEFAULT-LAYER as a component. The other component (which does have security against DFA) is named as DEFAULT-CORE which is sandwiched between two DEFAULT-LAYER instances. More discussion on these constructions can be found in Sect. 8.3 with the design rationale in Sect. 8.4. Further, detailed security analyses against classical attacks, DFA, DFA-assisted classical attacks, and side channel attacks can be found in Sect. 8.5.

The benchmarking results are given in Sect. 8.7 based on GIFT-128 [BPP+17] as a reference. We choose duplication (either spatial or temporal domain) as a reference countermeasure for benchmarking as duplication is a widely adopted fault protection method in commercial products. DEFAULT incur comparable overheads when compared with duplicated GIFT-128, both in hardware and software. As described later, DEFAULT resists a much wider class of fault models as compared to duplication, thus giving it an edge. In retrospect, our solution can be considered lightweight compared to more sophisticated duplication (such as infection or error detecting code). For instance, the recent block cipher CRAFT [BLM+19] which is proposed with fault protection as a prime target. It is a cipher designed with carefully chosen components that incur lower overhead when protected with error detection codes. As reported in [BLM+19], CRAFT incurs a 2.45× overhead when protecting against single-bit faults at the output. The cost goes higher for protecting against more faults. DEFAULT, on the other hand, explores an alternative methodology to design a cipher with natural DFA resistance and not limited to a specific number of faults. Similarly, infective countermeasures can have $\approx 3\times$ cost compared to the basic implementation (as the infection mechanism could be costly) [BSS20]. Moreover, if DEFAULT-LAYER is applied to non-lightweight ciphers like AES, the cost could be less than duplication.

As an independent contribution, we also study how to model a cipher which has an SBox with linear structures when searching for differential and linear bounds using automated tools. This is described in Sect. 8.6.

All in all, apart from being the first of its kind DFA protection, our construction achieves multiple advantages over other existing countermeasures (described in Sect. 8.8).

8.2 Background

8.2.1 DFA Models

Single Fault Adversary Versus Multiple Fault Adversary

In DFA, the *single fault model* is assumed, since this takes least assumption on the fault. Namely, the attacker can only target once during a single execution of the cipher and it is usually sufficient to attack actual ciphers. Our design protects against this model. The model we assume here, termed as the multiple fault adversary (see Sect. 3.2.2), is more relaxed (for the attacker): The attacker can inject faults as many times as she wants during the course of a single execution of a cipher, including injecting at multiple rounds. We claim the security of our construction against this model too (refer to Sect. 8.5.1).

Random Fault Model with Precision

The random fault model is usually assumed in DFA, where the attacker injects a faults without choosing/knowing the fault value which will be applied (this fault value is XORed), e.g., [PQ03] (see also Sect. 3.2.3)

Here we theorize a fault model, termed as the *precise bit flip*. In this model, one particular AND operation can be precisely targeted. This model is too strong and not yet reported as practical. Hence we do not consider this model within the scope. More relevant discussion can be found in Sect. 8.9.

8.2.2 DFA Protection

From a theoretical point of view, the state-of-the-art DFA countermeasures can be classified into three broad categories (see also Sect. 3.7):

(1) A separate, dedicated device that detects (and takes precaution) [HBB+16'2] or a shield that blocks any potential source of a fault.
(2) The underlying communication protocol between Alice and Bob ensures that a fault does not occur with a significant probability. This can be ensured, e.g., by assuming a small portion of the circuit is protected by other means [BBB+18].
(3) Duplicate the cipher execution followed by an implicit/explicit check for the equality of the executions. One may refer to [BSS20] for a recent study of such countermeasures.

One may notice that the countermeasures in the above-mentioned category (1) and (2) are basically engineering solutions, and generally outside the scope of cryptography as the cipher designer has no control (has to rely on the device manufacturer/network administrator) on the effectiveness of such countermeasures. In a slight contrast, the duplication-based countermeasures are somewhat close to what a cipher designer can specify (even though, the successful implementation of such a countermeasure still depends on someone else rather than the cipher designer).

Double Fault. One disadvantage with the generic duplication-based countermeasures is that identical faults in both computations. Hence, the result of the comparison of the two cipher outputs is zero and this case is treated as if no fault is injected. This tricks the countermeasure to release the faulty output, and works against the state-of-the-art countermeasures like infective [BSS20]. Although relatively hard to achieve in practice, it is shown to be practical in [SHS16]. We refer to it as double fault (see also Sect. 3.7). This fault can be protected by using different encodings for the two executions of the cipher (as done in CRAFT [BLM+19]), or by using triplication. However, any such method would add additional cost.

We also mention that sophisticated duplication countermeasure may require additional components as well as an external source of randomness [BSS20].

Impeccable Circuits/CRAFT. The authors of [AMR+18] propose an efficient DFA protection mechanism based on error detection codes. This idea is later extended to a block cipher, named CRAFT [BLM+19]. Although both DEFAULT and CRAFT protect against similar faults, we note that our idea works completely at algorithmic level. CRAFT employs a form of duplication and uses error detection codes. CRAFT will have different performance figures and fault coverage depending on the underlying code. In comparison, there are no unprotected or protected variants for DEFAULT. The cipher will naturally resist DFA and similar model by virtue of its construction, regardless of the implementation used. Besides, we would like to point out that for CRAFT the protection against DFA is much restricted. Any fault injection that is successful to alter the output beyond the detectable bound will make the DFA protection of CRAFT useless. Our construction is free from such limitations.

This shows the inability of current cryptographic designs to thwart DFA, although it closely resembles the classical differential attack (which is a well-understood topic). As already mentioned, DFA poses a serious threat to almost all (if not all) primitives, which cryptography has virtually no means to mitigate.

8.2.3 Feasibility of Cipher-Level Protection Against Faults

As mentioned earlier, we attempt to model DFA in terms of the classical attacks. Few fault attack models are reported in the literature which can be thought of as an extension of DFA, as those models utilize classical attacks in some form. More precisely, the algebraic, integral, impossible differential and linear attacks are shown in the literature to work in the similar setting as DFA (more can be found in Sect. 8.5.2).

It may be noted that there are other fault models which do not use any form of classical attack. While a detailed description of such attacks is out of the scope of our work, we make a short note on the fault models against which we do not claim any security. We recommend using existing (non-cipher level) countermeasures if these fault models need to be taken into account. Such fault models can be classified into the following generic classes:

- One/more bit(s) are permanently stuck at a particular value (say, zero) during the cipher execution.
- Faults that alter the operation flow in a cipher, such as an instruction skip on a microcontroller.
- Faults that do not alter the output (the so-called ineffective faults, see Sect. 3.5.2).
- Analysis methods that use statistical bias caused due to the effect of a fault.

The recently proposed fault model which has gained a serious attention in the community, Statistical Ineffective Fault Attack (SIFA) [DEK+18]/Sect. 3.5.4 relies on both ineffective fault and statistical information and therefore we do not claim security against it. To protect against SIFA, specialized countermeasures like [BKK+20'2] can be used.

One may note that it is generally hard (if not impossible) to design a cipher-level countermeasure against such fault models since they effectively change the cipher description. They assume a lot of control for the attacker and hence are more suitable in the fields of hardware/network security, etc. than cryptography (as an example, consider the extreme case of skipping the entire call to the `encrypt()` routine). These models are also hard to achieve in practice.

In some sense, the fault models we protect against can be compared to the ones that `CRAFT` [BLM+19] provides protection against. At the same time, our construction (which does not need any form of duplication/error detection) offers an advantage over `CRAFT` as our construction can protect any number of faults (not limited by the coverage of the fault detection code).

Our Fault Model

We summarize the fault model here (i.e., what we presume the attacker is able to):

- Inject fault(s) as many times as she wants during the course of a single cipher execution (which is a relaxed model in favour of the attacker).
- Target either the underlying cipher or the protection layer, or both.
- Target one or more particular word/words (respectively, bit/bits) of the protection layer (respectively, underlying cipher) at one round of her choosing.
- Choose one or more precise (non-zero) word/words (respectively, bit/bits) of her choosing as the fault value/values for the layer (respectively, cipher), which is/are effectively XORed with the target word/words of the layer (respectively, bit/bits of the cipher).
- Inject faults that are transient, i.e., not stuck at (see Sect. 3.3.1).
- Target only operands (cannot disturb any operation flow as DFA can only alter the operands).
- Target either the encryption or the decryption, but not both.
- Repeat the cipher execution as many times she wants with fixed (unknown) key and (known) inputs to the cipher and inject faults at any invocation she wishes.
- Fault analysis that follows the DFA paradigm (or one of its variants). In other words, the fault analysis model that resembles a classical attack model (and probably works as a generalized version of that classical attack) is considered within the scope.

Notice that the precise bit flip model (Sect. 3.2.1/Sect. 8.9) is out of scope. A particular multiple fault model is out of scope too, where the attacker can inject a (non-zero) difference at a round, only to nullify its effect at a later round with another (non-zero) fault. The former model suits with the statistical fault analysis

(Sect. 3.5.3) and the latter with the ineffective fault analysis (Sect. 3.5.2). Also, none of the models have been shown to be practical yet, to the best of our knowledge.

8.2.4 Working Principle for DFA

As already mentioned, DFA is closely related with the differential attack. In DA, the input difference is inserted in plaintexts (or ciphertexts in case of decryption)—hence at the beginning of cipher execution. In comparison, input difference is inserted in the form of a fault in DFA and can be applied anytime during encryption/decryption. In practice, faults are injected near the end of the cipher execution. For example, the DFA which targets the earliest rounds of AES works after the 5th round [DFL11], to the best of our knowledge. The input difference propagates through the non-linear components (generally the SBoxes). Given a pair of input–output differences for non-linear components, the attacker is able to narrow down the search space for guessing the actual value manipulated.

For better understanding, here we present a simple case study of DFA on an SBox. Let x be the unknown input to the SBox S. The attacker, Eve, knows $S(x) \oplus k$, where k is the key unknown to her. She inserts one (known) fault value δ and observes the corresponding faulty output, $S(x \oplus \delta) \oplus k$. Then she takes the difference of the faulty and the non-faulty outputs: $\Delta = S(x) \oplus S(x \oplus \delta)$. Notice that Δ is free from the unknown key. With the pair (δ, Δ), she can now eliminate a number of candidates for x since only a few x values will validate this differential transition. By inserting multiple δs, she can uniquely identify x. Once x is identified, she recovers the unknown key k by XORing $S(x)$ with the output.

The analysis procedure in DFA consists of two orthogonal terms, namely fault complexity (how many faulty encryption calls) and search complexity (computational/memory complexity required to deduce x. The general trend is to reduce the fault complexity while keeping the search complexity within an acceptable limit. In the case of last round attack, Eve has to solve each SBox independently at the last round in order to recover the entire sub-key. Therefore, it requires a higher number of faults. As we explain here, some strategies can reduce the required number of faults:

- Solve only a part of the last round key and do the exhaustive search for the rest of the key.
- Inject fault at an earlier round. Thanks to the linear layer, one fault (differential) will propagate to multiple SBoxes in parallel. This works as if multiple faults were injected at different SBoxes. Normally, 2 or 3 rounds of linear layer is sufficient to spread one fault (difference) to many (if not all) SBoxes. Hence, generally, somewhere in the 3rd last round is an optimal location, as can be seen in [PQ03, JKP12'1].

Notice that both ways increase the search complexity. In the second case, the attacker has to consider the round key of the intermediate rounds through which the differential passes.

8.3 Construction of DFA-Resistant Layer/Cipher

With the background given in Sect. 8.2.4, we first look at the problem of maximizing fault complexity. Note that fault complexity is the highest if the fault is injected at the last round. From Theorem 7.4, we see that it is possible to uniquely recover the input/output value of each SBox with not more than $\mathrm{DU}_S/2 + 1$ faults (unless there is a linear structure) when attacking the last round. This gives a provable upper bound for the attacker that she needs at most (if faults values are judiciously chosen) $\frac{\mathrm{DU}_S}{2} + 1$ faults per SBox in order the find out its input uniquely, given the SBox does not have a linear structure. Usually, for an SPN block cipher, three faults per SBox are sufficient as most block ciphers use an SBox with DU $= 4$ (except `GIFT` SBox [BPP+17], where DU is 6). In fact, in many cases, only two faults are needed to solve for any input. For example, the SBoxes are chosen in `AES` (see Sect. 2.4.1 for its description), `PRESENT` [BKL+07], `MIDORI` [BBI+15], `SKINNY-64` [BJK+16] and `GIFT` [BPP+17]; the fault values {`1`, `6`} are sufficient to retrieve all inputs uniquely. Thus, it is possible to solve the last round of the above-mentioned ciphers in just twice the number of SBoxes faults by DFA. It seems hard to force fault complexity to increase.

8.3.1 Ad-hoc DFA Protection to Any Cipher (`DEFAULT-LAYER`)

Our approach is to tackle the problem of increasing the search complexity. This means that we give the attacker the power to apply as many faults as she wants, but the search space for the analysis should be substantially larger. We note that the search complexity is minimum when injecting fault at the last round (check Sect. 8.2.4). As we already pointed out (Theorem 7.1), if an SBox S has non-zero linear structure(s), then the attacker will not be able to uniquely identify the input. Thus, she has to enumerate the remaining key candidates from the input difference–output difference relation.

Now, using an SBox with linear structure is generally considered undesirable for block cipher design as it makes classical differential attack easier (as explained in Sect. 7.2). Hence, we arrive at a paradoxical situation. If we want to design a cipher with better resistance against DFA, then it becomes weak against classical attacks and vice-versa. To find a middle ground, where the cipher is strong against both DFA and classical attacks, we propose the concept of prepending/appending an extra layer (that uses SBoxes with linear structures) to the underlying cipher (henceforth referred to as "main cipher"); refer to Fig. 8.1 for an illustration.[4] The layer L_2, as shown in Fig. 8.1a, is used to protect against DFA during the encryption process; similarly, L_1 is used for the decryption process, see Fig. 8.1b. Hence, in the

[4] The layers are invertible. Together with the underlying cipher, it technically constitutes a new (augmented) cipher. This concept is generalized to a full-fledged cipher.

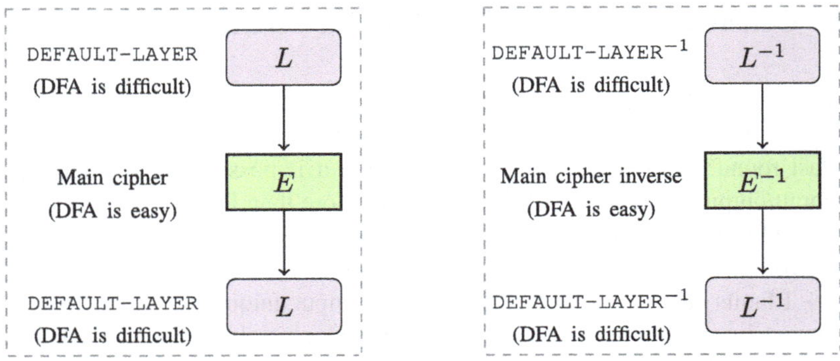

(a) Encryption **(b)** Decryption

Fig. 8.1 Main cipher augmented by DEFAULT-LAYER to resist DFA

case of encryption, the output of the main cipher (which is deemed to be protected against DFA) is not directly available to the attacker, rather she gets the output from the augmented cipher (main cipher prepended and appended with the protection layers); similar is the case for decryption. Figure 8.1 visually represents the idea. The layer L, which we name as DEFAULT-LAYER and describe it in Sect. 8.3.3, is prepended and appended to the main cipher (encryption), as in Fig. 8.1a. For decryption, L^{-1} is both prepended and appended to the main cipher inverse (shown in Fig. 8.1b), since the ciphertext $C = L \circ E \circ L(P)$, and the decryption $L^{-1} \circ E^{-1} \circ L^{-1}(C) = (L^{-1} \circ E^{-1} \circ L^{-1}) \circ (L \circ E \circ L)(P) = P$. The underlying cipher will have desirable protection against classical attacks while the additional layer will be used to thwart DFA. Since for DFA the attacker has to slowly peel out the outer rounds of the cipher, we only have to protect these rounds against DFA, while the inner cipher will provide all the security we expect from a block cipher in the black box model (as adding layers will not weaken its security).

As we assume the attacker can target both the encryption and decryption processes, the model described here can thwart DFA on both. If we assume a constrained model for the attacker, say when the decryption is done at a server which is physically protected such that Eve cannot access (as in [BBB+18, Sect. III]), then the prepended layer L in Fig. 8.1a and the appended layer L^{-1} in Fig. 8.1b can be removed, which will result in better performance.

8.3.2 *Extension to a Full-Fledged Cipher (DEFAULT)*

Aside from an ad hoc layer which is able to protect any cipher from DFA, it is also possible to construct a full-fledged block cipher. This is done by sandwiching the so-called DEFAULT-CORE (this is another keyed permutation described in Sect. 8.3.4) with DEFAULT-LAYER. The DEFAULT-CORE contains an SBox that is especially

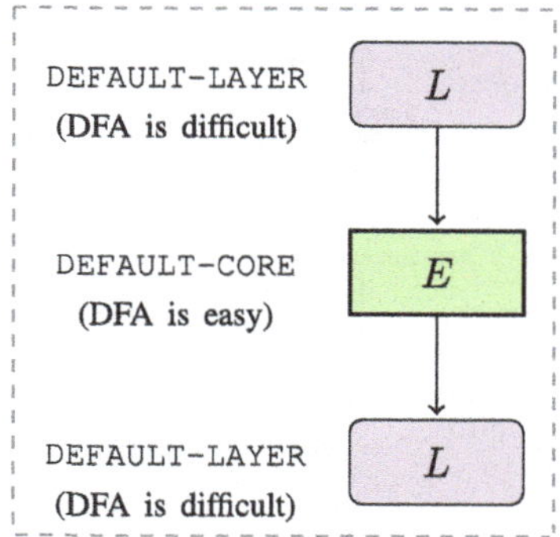

Fig. 8.2 Sandwiched construction for DEFAULT

resistant to classical linear attacks, and DEFAULT-LAYER uses an SBox that contains linear structures to resist DFA and its variants. Hence DEFAULT consists of two components (for both the encryption and decryption). DEFAULT-CORE is also following a construction similar to GIFT-128, and does not use LS SBox. Figure 8.2 shows the construction of DEFAULT. We do not reuse GIFT-128 permutation as we want to maximize the security against linear attack, even if that results in relatively low security against differential attack.

8.3.3 *Construction of* DEFAULT-LAYER

We detail the 128-bit version of our proposed DFA protecting layer (DEFAULT-LAYER) to protect 128-bit block ciphers, but we emphasize that it is adoptable to any state size.[5]

DEFAULT-LAYER is a 28-round keyed permutation[6] that receives a 128-bit message as the state $X = b_{127} \parallel b_{126} \parallel \cdots \parallel b_0$, where b_0 is the least significant bit. The state can also be expressed as $X = w_{31} \parallel w_{30} \parallel \cdots \parallel w_0$, where w_i is a 4-bit (nibble) word. We do not describe the inverse layer here for the sake of brevity, but it can be trivially derived. The round function (denoted by $\mathcal{R}$ henceforth) of DEFAULT-LAYER consists of four steps (in order): SubCells—applying a 4-bit SBox to the state, PermBits—permute the bits of the state (same as GIFT-128 [BPP+17]),[7] AddRoundConstants—XORing a 6-bit constant as well as another bit to the state (same as GIFT-128), and AddRoundKey—XORing the round key to the state (an initial AddRoundKey is also performed before the first SubCells). A visual representation is given in Fig. 8.3a.

[5] For effective DFA protection, we recommend state size of at least 128 bits.

[6] We avoid calling it a "cipher" as it is a DFA protecting layer used on top of an actual cipher.

[7] There is a slight change (the relative placement of PermBits) from the design of GIFT-128, see Sect. 8.9 for more details.

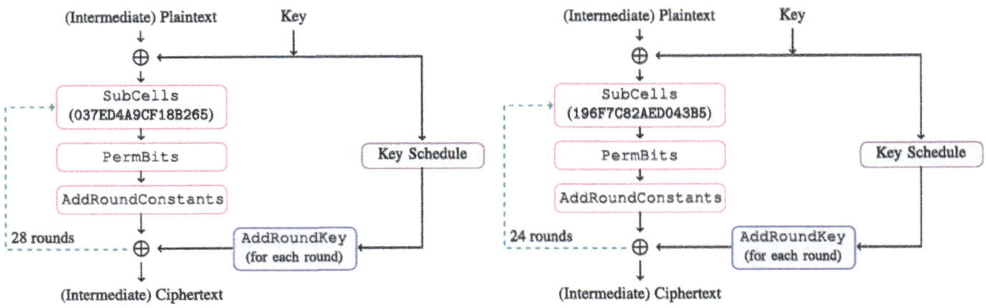

(a) DEFAULT-LAYER **(b) DEFAULT-CORE**

Fig. 8.3 Top level view of DEFAULT-LAYER (with a key schedule) and DEFAULT-CORE

SubCells

It uses the 4-bit LS SBox $S = \texttt{037ED4A9CF18B265}$. This SBox is applied to every nibble of the state: $w_i \leftarrow S(w_i)$, $\forall i \in \{0, \ldots, 31\}$. The coordinate functions (see Definition 2.3) of this SBox are given respectively (in ANF) by:

$$\begin{aligned} y_0 &= x_0 \oplus x_1 \oplus x_2, \\ y_1 &= x_0 \oplus x_1 \oplus x_0x_1 \oplus x_0x_2 \oplus x_1x_3 \oplus x_2x_3, \\ y_2 &= x_1 \oplus x_2 \oplus x_3, \\ y_3 &= x_0x_1 \oplus x_2 \oplus x_0x_2 \oplus x_3 \oplus x_1x_3 \oplus x_2x_3. \end{aligned}$$

PermBits

The bit permutation is the same as permutation P_{128} in GIFT-128 [BPP+17], it maps bits from bit location i of the internal state to bit location $P_{128}(i)$, $i = 0, 1, \ldots, 127$: $b_{P_{128}(i)} \leftarrow b_i$, $\forall i \in \{0, \ldots, 127\}$; and is given by: (0, 33, 66, 99, 96, 1, 34, 67, 64, 97, 2, 35, 32, 65, 98, 3, 4, 37, 70, 103, 100, 5, 38, 71, 68, 101, 6, 39, 36, 69, 102, 7, 8, 41, 74, 107, 104, 9, 42, 75, 72, 105, 10, 43, 40, 73, 106, 11, 12, 45, 78, 111, 108, 13, 46, 79, 76, 109, 14, 47, 44, 77, 110, 15, 16, 49, 82, 115, 112, 17, 50, 83, 80, 113, 18, 51, 48, 81, 114, 19, 20, 53, 86, 119, 116, 21, 54, 87, 84, 117, 22, 55, 52, 85, 118, 23, 24, 57, 90, 123, 120, 25, 58, 91, 88, 121, 26, 59, 56, 89, 122, 27, 28, 61, 94, 127, 124, 29, 62, 95, 92, 125, 30, 63, 60, 93, 126, 31).

Key Schedule and AddRoundKey

DEFAULT-LAYER does not have a key schedule of its own. Instead, the round keys can be extracted from key schedule of the main cipher for ease of implementation. While most of the 128-bit block ciphers use 128-bit round keys, there are block ciphers like SIMON-128/128, SKINNY-128 , and GIFT-128 that use 64-bit round keys for area savings in hardware implementation. For the latter case, we therefore suggest concatenating two 64-bit round keys to form one 128-bit round key for DEFAULT-LAYER, while keeping the same round key generating key schedule. Also, we are not aware of any security issue that may arise if the master key is used in all rounds of DEFAULT-LAYER. This indicates identical round keys (possibly the same as the master key) can also be used instead. Therefore, we keep the key schedule flexible—either the extrapolated key schedule from the main cipher can be used or the master key can be used as the identical round keys, depending on the preference of the user.

Instead of XORing half-state round keys as in GIFT-128, we XOR the full round keys (128-bits) to the state (this is essential for DFA security). The round key for the jth round, denoted by k_i^j, is bit-wise XORed to the state: $b_i \leftarrow b_i \oplus k_i^j, \forall i \in \{0, \dots, 127\}$. There is an initial round key addition before the application of the first SubCells.

8.3.4 Construction of DEFAULT-CORE (and DEFAULT)

In order to design the full-fledged cipher, we need to design the middle part of the cipher (DEFAULT-CORE), where the SBox does not have any (non-zero) linear structure. The design of the core is much alike to the DEFAULT-LAYER (hence omitted here for the sake of brevity), except for the SBox, and it has 24 rounds. The SBox of choice here is 196F7C82AED043B5, based on its very desirable cryptographic properties against linear attack (see Sect. 8.6 for details). The coordinate functions of the SBox (in ANF) are given by

$$\begin{aligned}
y_0 &= 1 \oplus x_1 \oplus x_0x_1 \oplus x_0x_2 \oplus x_3, \\
y_1 &= x_1 \oplus x_2 \oplus x_0x_2 \oplus x_3, \\
y_2 &= x_1 \oplus x_2 \oplus x_0x_3, \\
y_3 &= x_0 \oplus x_1x_2 \oplus x_3 \oplus x_0x_3 \oplus x_0x_1x_3 \oplus x_2x_3.
\end{aligned}$$

In a nutshell, the overall design of DEFAULT consists of DEFAULT-LAYER (28 rounds), followed by DEFAULT-CORE (24 rounds), followed by another DEFAULT-LAYER (28 rounds). DEFAULT-CORE is pictorially shown in Fig. 8.3b. Hence DEFAULT is an SPN block cipher with heterogeneous round structure, consisting of 80 rounds. Therefore, in comparison with duplicated GIFT-128[8] (which runs for 80 rounds in total), DEFAULT has the same number of rounds. As for the round counter, we use the same from GIFT-128, which is refreshed at the beginning of DEFAULT-LAYER/DEFAULT-CORE.

[8] We consider duplicated GIFT-128 as the closest alternative to DEFAULT (instead of a single execution of GIFT-128), since one of the most common ways for DFA resistance is to duplicate the cipher.

Table 8.1 Round constants for DEFAULT

	Round Constants	♯
DEFAULT-CORE	1, 3, 7, 15, 31, 62, 61, 59, 55, 47, 30, 60, 57, 51, 39, 14, 29, 58, 53, 43, 22, 44, 24, 48, 33, 2, 5, 11	28
DEFAULT-LAYER	1, 3, 7, 15, 31, 62, 61, 59, 55, 47, 30, 60, 57, 51, 39, 14, 29, 58, 53, 43, 22, 44, 24, 48	24

Table 8.2 DEFAULT test vectors (full cipher with 80 rounds)

1	Key	00000000000000000000000000000000
	Plaintext	00000000000000000000000000000000
	Ciphertext	02ec558a8f65dc7e53b326a1de9ade51
2	Key	ffffffffffffffffffffffffffffffff
	Plaintext	ffffffffffffffffffffffffffffffff
	Ciphertext	c823fe57c2e8b7b91db62aed3ad05e32
3	Key	98989898989898989898989898989898
	Plaintext	405b405b405b405b405b405b405b405b
	Ciphertext	cf18f09cd2d8f68ce64c3380272be64c
4	Key	329abdeb01339ab04465021f85417c3d
	Plaintext	0cebfdcbbcb01434d7111525143b94d1
	Ciphertext	7d1a340975be69a10a7d11d168b25b63

Key Schedule

Unlike DEFAULT-LAYER, DEFAULT-CORE has its own well-defined key schedule. The 128-bit key (which is the same as the master key for the underlying cipher in case of the ad hoc layer) is divided into 16-bit words: $k_7 \parallel k_6 \parallel \cdots \parallel k_0$. Each round key is extracted first and then the next round key is generated by the update: $k_7 \parallel k_6 \parallel \cdots \parallel k_0 \leftarrow (k_7 \parallel k_6 \parallel \cdots \parallel k_0) \ggg 20$ followed by $k_7 \leftarrow k_7 \ggg 1$, where $\ggg$ is an i-bit right rotation within a 16-bit word.

Round Constants and Test Vectors

Table 8.1 shows the round constants (6-bit) for DEFAULT (which are actually adopted from that of GIFT). In Table 8.2, few test vectors for the 80-round cipher DEFAULT (i.e., with the DEFAULT-LAYER—DEFAULT-CORE—DEFAULT-LAYER construction) are given[9]. As for the choice of the key schedule

[9] Visit https://bitbucket.org/anubhab001/default-cipher-public-code for our reference code.

of DEFAULT-LAYER, we use that of DEFAULT-CORE. In the DEFAULT construction, to avoid two consecutive round key additions (for efficiency), we skip that at the final round of the initial DEFAULT-LAYER and the initial round of the final DEFAULT-LAYER.

The structure of DEFAULT-LAYER for two rounds is shown (out of 28 rounds in DEFAULT-LAYER) in Fig. 8.4. The structure for the DEFAULT-CORE is also similar. The SBoxes are numbered (from 0 to 31), so are the bit positions (from 0 to 127), for better clarity. The coloured lines indicate the permutation (linear) layer. By $\bigoplus$, it is indicated that the corresponding key bits are XORed. The round constant additions are shown by $\bigoplus$ (bits at 3, 7, 11, 15, 19, 23), and the bit at 127 is flipped in each round.

8.4 Design Rationale

The fundamental goals of DEFAULT can be summarized as:

(i) protect against DFA,
(ii) applicable to wide variety of symmetric key ciphers, and
(iii) simple and lightweight.

During its design, various choices have been made and we discuss those next.

8.4.1 Design Philosophy

SPN Versus Feistel network

The first choice is to choose between SPN and Feistel network. Although implementing the inverse of Feistel construction is simple and does not require the inverse of its f-function, the non-linearity is introduced to only half of its state in each round and hence usually requires more rounds to achieve the desired security margin. On the other hand, SPN introduces non-linearity to the entire state and thus requires lesser rounds in general. The corresponding study is also simpler, so we choose to start with SPN.

Bit Permutation Versus Rotational-XOR Diffusion Versus Word-Mixing Diffusion

For SPN construction, the linear diffusion layer is usually either a bit permutation (like in PRESENT and GIFT), a rotational-XOR layer (like in SMS4 [Dt08]), or a word-mixing diffusion (like AES and SKINNY). Although the latter two provide a stronger diffusion, they can be costly in hardware and non-trivial to adopt to different

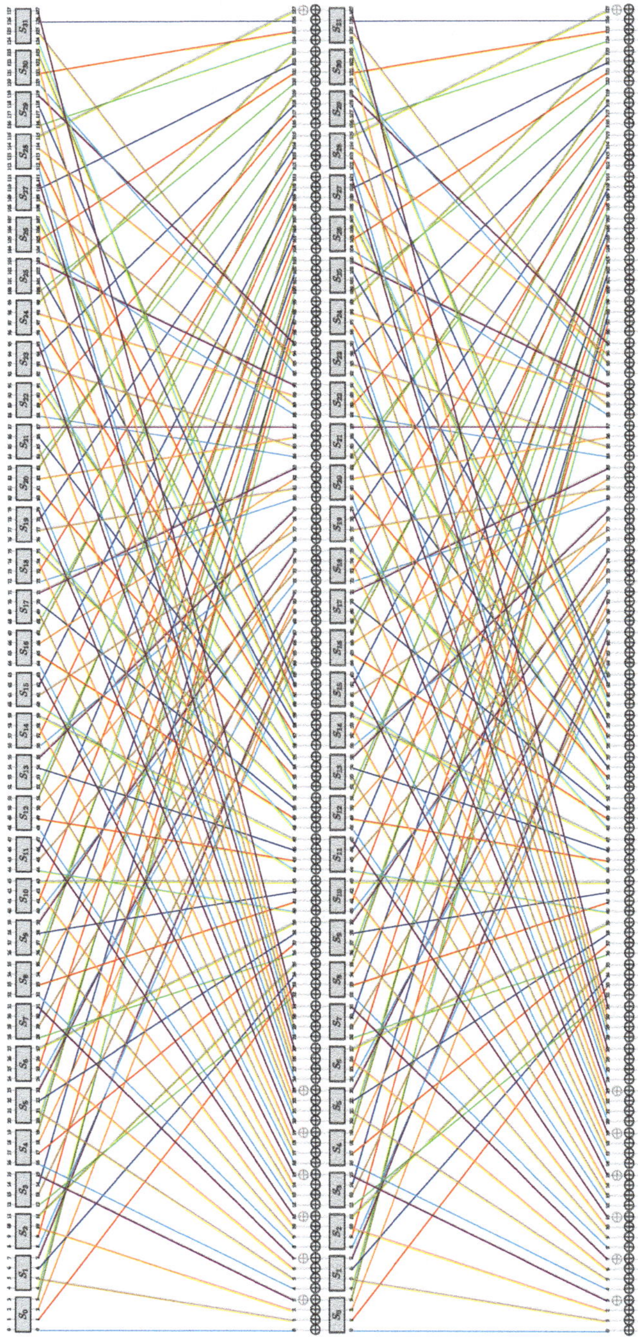

Fig. 8.4 Structure of DEFAULT-LAYER (two rounds only)

block sizes as in most cases the description is completely different for different block sizes. In hardware, the bit permutation is basically free to implement as it consists simply of circuit wiring. Moreover, from the design strategy of GIFT, we see that a bit permutation can be adjusted to various state sizes. Therefore, we choose bit permutation over other choices of the diffusion layer.

8.4.2 *Structure of the* DEFAULT PermBits

We recall here the structure of the PRESENT and GIFT bit permutations as this will be useful later to understand our security guarantees. There are essentially two levels of permutation within the PRESENT or GIFT bit permutation: the group mapping and the SBox grouping.

Group Mapping

The mapping of the output bits from a group of 4 SBoxes to another group of 4 SBoxes in the next round. This is the main difference between the PRESENT and GIFT permutation. For 4-bit SBoxes, we denote the 4 bits as bit 0, 1, 2, and 3, where bit 0 is the least significant bit. Within a group, the PRESENT permutation sends the 4 output bits from the ith SBox (indexed from 0) to bit i of the 4 SBoxes in the next round, forming a symmetrical structure. Due to this symmetrical structure, PRESENT has many symmetrical differential characteristics for a given fixed input and output differences, which results in a higher differential probability (the situation is similar for the linear cryptanalysis case). On the other hand, the GIFT permutation sends bit i from the output of the jth SBox (indexed from 0) to the bit i of the lth SBox in the next round, where $l = i - j \mod 4$. Since bit i of an SBox output is always mapped to bit i of another SBox, it makes the analysis on the propagation of the differences easier and breaks the symmetry. Therefore, we choose GIFT group mapping.

SBox Grouping

The partitioning of the SBoxes into the groups of 4 SBoxes. The SBox grouping for the 64-bit block ciphers PRESENT and GIFT-64 are the same, and the designers of GIFT extended the idea to construct SBox grouping for 128-bit block size. Similar to [BPP+17], we denote the SBoxes in round i as $S_0^i, S_1^i, \ldots, S_{g-1}^i$, where $g = n/4$ for block size n. These SBoxes can be grouped in two different ways—the Quotient Q and Remainder R groups, defined as $Qx = \{S_{4x}, S_{4x+1}, S_{4x+2}, S_{4x+3}\}$ and $Rx = \{S_x, S_{q+x}, S_{2q+x}, S_{3q+x}\}$, where $q = g/4, 0 \leq x \leq q - 1$. The SBox grouping simply maps SBoxes from Qx^i to Rx^{i+1}, where within this group, the 16-bit mapping is defined by the group mapping described above. This is the adoptable

Table 8.3 Properties of the DEFAULT (LS, Non-LS), PRESENT, SKINNY-64 and GIFT SBoxes

		DBN	LBN	LS	DU	AD max	AD min	Nl
DEFAULT LS	037ED4A9CF18B265	3	3	0, 6, 9, f	16	2	1	0
DEFAULT Non-LS	196F7C82AED043B5	2	2	0	8	3	2	4
PRESENT [BKL+07]	C56B90AD3EF84712	3	2	0	4	3	2	4
SKINNY-64 [BJK+16]	C6901A2B385D4E7F	2	2	0	4	3	2	4
GIFT [BPP+17]	1A4C6F392DB7508E	2	2	0	6	3	2	4

component of the bit permutation, as one can see that the SBox grouping is well-defined when n is a multiple of 16.

8.4.3 Selection of the DEFAULT SBoxes

Here we describe the selection process of the LS SBox (used in DEFAULT-LAYER) and the non-LS SBox providing high resistance against linear attack (used in DEFAULT-CORE). A summary of various properties, namely differential branch number (DBN, Definition 2.6), linear branch number (LBN, Definition 2.7), the linear structures with DU, maximum and minimum of algebraic degree of the coordinate functions, and non-linearity of our chosen SBoxes together with SBoxes from other lightweight ciphers (PRESENT, SKINNY-64 and GIFT) are shown in Table 8.3.

As for the size of the SBox, we decide to choose 4-bit. Although there are better (in terms of DFA security) 8-bit SBoxes (e.g., we are able to find 8-bit SBoxes with 64 LS which could ascertain up to 2^{96} DFA security for 128-bit SPN, see Table 7.1), we choose the 4-bit SBoxes for the following main reasons:

(i) to lower the cost (similar to GIFT [BPP+17]),
(ii) making the MILP modelling (described in Sect. 5.4) more efficient as generating the same for 8-bit SBoxes could be costly.

LS SBox

From the list of 302 affine equivalence (see Sect. 2.1.2) classes of SBoxes by De Cannière [DeC07], there are 10 AE classes with non-zero linear structures. Among these 10 AE classes, 8 (#293–#300) have only one non-zero linear structure, AE class #301 has three non-zero linear structures and the last AE class #302 is linear (contains the identity permutation). To maximize the number of linear structures and yet to use a non-linear permutation, we chose AE class #301 (the representative for this AE class in [DeC07] is `1032456789ABCDEF`).

Within this class, we chose an SBox with the following criteria:

1. Both the differential and linear branch numbers are 3.
2. Zero diagonal in the DDT and LAT (except the trivial case of [0, 0]).
3. In the DDT, $\forall \delta_i \in \mathbb{F}_2^4 \setminus \{0\}$, if $(\delta_i, \delta_o) = 16$, then $HW(\delta_i) \geq 2$ and $HW(\delta_o) \geq 2$.
4. In the LAT, $\forall \alpha_i \in \mathbb{F}_2^4 \setminus \{0\}$, if $(\alpha_i, \alpha_o) = 8$, then $HW(\alpha_i) + HW(\alpha_o) \geq 4$.

First, we try to optimize the differential and linear diffusion with branch number 3. Next, we avoid enabling 1-round iterative differential or linear pattern, hence we looked for empty diagonals. Lastly, for any linear transition, we select an SBox that has maximal Hamming weight for the input and output values, in order to maximize the number of active SBoxes before and after the linear transition. We end up with SBox `037ED4A9CF18B265`.

From the representative SBox, we generate all the affine equivalence SBoxes and filter it based on our SBox criteria in the following ($\forall \delta, \Delta \in \mathbb{F}_2^4 \setminus \{0\}$):

(C1) if $\mathrm{DDT}_S[\delta, \Delta] > 0$, then $HW(\delta) + HW(\Delta) \geq 3$,
(C2) if $\mathrm{DDT}_S[\delta, \Delta] = 16$, then $HW(\delta) + HW(\Delta) \geq 4$,
(C3) if $\mathrm{DDT}_S[\delta, \Delta] = 16$, then $HW(\delta) \geq 2$ and $HW(\Delta) \geq 2$, and
(C4) $\sharp\{(\delta, \Delta) \mid \mathrm{DDT}_S[\delta, \Delta] = 16 \text{ and } HW(\delta) + HW(\Delta) = 4\} \leq 1$.

In the end, we find 3840 SBoxes that satisfy our SBox selection criteria. Our LS SBox `037ED4A9CF18B265` is chosen from these 3840 SBoxes. Any of these SBoxes, combined with our `DEFAULT-LAYER` bit permutation, ensures the following properties: For any 5-round differential characteristics,

(P1) there are at least 10 active SBoxes,
(P2) if there are exactly 10 active SBoxes, then each of these active SBoxes has differential probability 2^{-1}, which totals to 2^{-10},
(P3) if there exists active SBox with differential probability 1, then there are at least 12 other active SBoxes with differential probability 2^{-1} each, which totals to 2^{-12}.

In Fig. 8.5, we show two 5-round differential characteristics; here solid lines are active bits, white boxes are active SBoxes, and red box is SBox with differential probability 1. We give an intuition on how these properties are derived from the criteria. From (C1), one can see that there are 10 active SBoxes (P1) (in Fig. 8.5a). By (C2), none of these SBoxes (in Fig. 8.5a) have differential probability 1 (P2). If there exists SBox with differential probability 1, again by (C1) and (C2), there are at least 13 active SBoxes (see Fig. 8.5b). Criterion (C3) enforces that only 1 in 13 active SBoxes can potentially have differential probability 1 (P3). The final criterion (C4) is imposed to further reduce the occurrence of differential probability 1 in differential characteristics with low number of active SBoxes.

From these properties, we can (conservatively) estimate that the probability of any differential characteristic drops by at least a factor of 2^2 for every additional

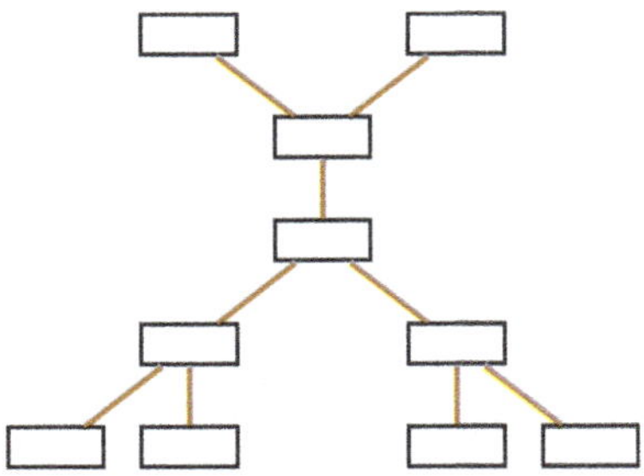

(a) 10 active SBoxes with differential probability $< 2^{-10}$

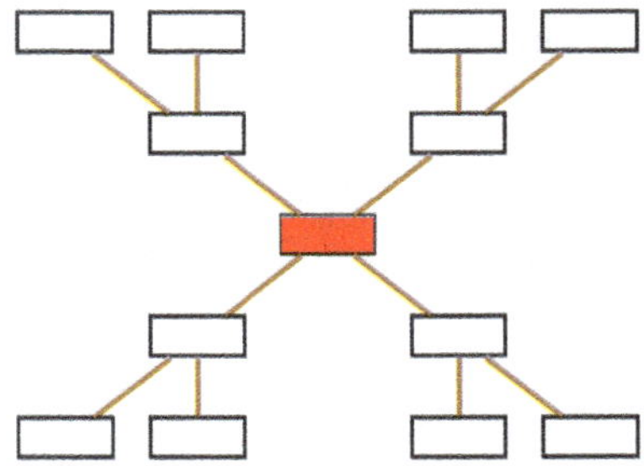

(b) 13 active SBoxes with differential probability $< 2^{-12}$

Fig. 8.5 5-round differential characteristics

round. Thus, any 32-round differential characteristic has differential probability at most 2^{-64}.

Having said that, a simple extrapolation of these results shows that with 32 rounds, any differential characteristic has differential probability at most 2^{-64}. Thus, we are confident that 40 rounds is sufficient to prevent any DFA key-recovery attack with complexity lower than 2^{64}. Later, in Sect. 8.6, we use MILP-based approach to improve the bound, thereby reducing the number of rounds to 28.

Non-LS SBox

For this SBox candidate, we focus on the linearity of the SBox as linear attacks will be the most difficult part to protect. Among the 33 AE classes with the lowest maximum linear bias 2^{-2}, AE class #32 (represented by `C0A23547691B8DEF`) and #33 (represented by `D0A23547691BC8EF`) have the least number of non-zero entries in the LAT. Statistically speaking, this gives us a higher chance of finding linear branch number 3 SBoxes. However, every 4×4 SBox with linear branch number 3 has at least one non-trivial linear structure (belonging to the AE classes #294, #297, #298, #300, #301, #302 of [DeC07]). Hence, we choose several of those SBoxes and find the linear bound using the automated technique given in Sect. 5.4. However, the bounds we obtain are not good enough. Thus, the next strategy is to select an SBox with the following linear properties:

1. $\sharp\{((\alpha_i, \alpha_o)) \mid HW(\alpha_i) = HW(\alpha_o) = 1, (\alpha_i, \alpha_o) \neq 0\} = 1$.
2. Zero diagonal in the LAT (except [0, 0]).
3. $\sharp\{((\alpha_i, \alpha_o)) \mid HW(\alpha_i) + HW(\alpha_o) = 3, (\alpha_i, \alpha_o) = \pm 4\} = 13$.
4. $\sharp\{((\alpha_i, \alpha_o)) \mid HW(\alpha_i) + HW(\alpha_o) = 3, (\alpha_i, \alpha_o) = \pm 2\} = 6$.

First, we limit the number of Hamming weight $1 \rightarrow 1$ transition to 1. Next, we avoid having a 1-round iterative linear pattern. Lastly, we minimize the number of possible Hamming weight $1 \rightarrow 2$ and $2 \rightarrow 1$ transitions. This is to encourage faster and wider propagation of the linear trail. We finally choose the SBox

196F7C82AED043B5 from the AE class #32. Similar to DEFAULT-LAYER, later we apply MILP-assisted analysis in Sect. 8.6 to finally fix the number of rounds to 24.

8.4.4 Unbiased Linear Structures

We need an extra security criterion: Each bit of the linear structures of S as well as S^{-1} must be unbiased. This is to avoid certain undesirable property of the linear layer. If we assume that the linear structures for S are {0, 1, 2, 3}, the two Most Significant Bits (MSBs) are always 0. One such SBox is 1032456789ABCDEF (the representative for AE class #301 in [DeC07, Chap. 5.4.2]). It has the property that if the first two bits of its input are known uniquely, then the first two bits of its output are also known uniquely. The attacker may be able to leverage this property by attacking the penultimate round of the cipher/protection layer, with attacking the last round. For example, consider the linear layer of PRESENT, shown in Fig. 8.6. Let us assume that the attacker is able to identify the first output bit of the SBoxes S_0, S_1, S_2, and S_3 of the penultimate round uniquely, which means she now knows all the four inputs of S_0 at the last round—thereby solving it uniquely. This way, half of the SBoxes in the last round (S_0 to S_7) can be solved uniquely, while the attacker will get $2^{n/2}$ solutions for the remaining SBoxes. This means that direct key material could be obtained for less than $2^{n/2}$ search complexity overall. This issue does not arise when each bit of the linear structures is unbiased (in which case Eve is not able to find any bit uniquely). In our chosen SBox, the linear structures being {0, 6, 9, f}, and that of for the inverse SBox being {0, 5, a, f}, this criterion is satisfied.

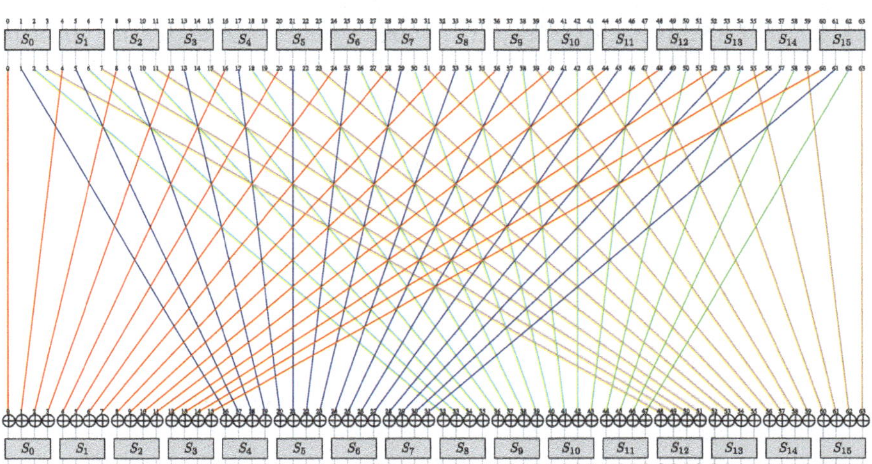

Fig. 8.6 One round (with another incomplete round) of PRESENT

Another issue which can be seen from the above discussion is that all bits from each SBox output are to be masked (by XORing the round key). Otherwise, an attacker may leverage the situation if a few bits from an SBox output are known uniquely. This is the motivation to have a full key addition, instead of half key addition as in GIFT-128.

8.5 Security Analysis

Conducting security analysis on DEFAULT is quite different from conducting security analysis on block ciphers, despite having a similar structure. This is because DEFAULT-LAYER is built on top of an existing (and presumably secure against classical attacks) cipher which only assets the desired security against DFA, and DEFAULT-CORE is used in conjunction with two instances of DEFAULT-LAYER. Although classical attacks do not pose any threat against DEFAULT-LAYER, some cryptanalytic techniques could still be applied to DEFAULT through DFA. For instance, suppose an attacker injects faults to the output of the main cipher, this difference will only propagate through the DEFAULT-LAYER and not the entire cipher, creating some form of differential attack on the DEFAULT-LAYER itself. Thus, we need to ensure that DEFAULT-LAYER is not vulnerable to classical attacks that could bypass the main cipher using DFA and target DEFAULT-LAYER directly. The desired security for the classical attacks are summarized in Table 8.4 and security evaluation against such attacks are done subsequently in Sect. 8.5. Detailed discussion on the classical attacks are omitted here (for brevity); interested readers may find it in [BPP+17, Sect. 4], for example. It may be noted that more precise differential and linear bounds are presented in Sect. 8.6. The security against DFA and side channel attacks are evaluated subsequently (Sects. 8.5.1 and 8.5.3, respectively). One may note that the LS SBox has a fixed point (0), we are confident that this will not affect the overall security as the DEFAULT-LAYER is to be used with DEFAULT-CORE/other cipher.

As DEFAULT comprises DEFAULT-CORE and (two layers of) DEFAULT-LAYER, we specify which component we are analysing and for which cryptanalysis technique is applied. This analysis is summarized in Table 8.5.

Table 8.4 Minimum security requirement of DEFAULT against classical attacks

	DEFAULT-LAYER	DEFAULT-CORE
Differential, Algebraic	2^{64} Search complexity	–
Integral, Impossible Differential	–	No distinguisher
Linear	2^{32} Search complexity	2^{64} Search complexity
Invariant Subspace	–	2^{128} Search complexity

Table 8.5 Summary of security claims for DEFAULT-LAYER, DEFAULT-CORE and DEFAULT

		DEFAULT-LAYER (28 rounds)	DEFAULT-CORE (24 rounds)	DEFAULT (80 rounds)	
DFA	DEFAULT-LAYER	2^{64}	Bypassed	2^{64}	Sect. 8.5.1
	DEFAULT-CORE	$\geq 2^{64}$	Negligible	$\geq 2^{64}$	
	Double Fault	Not applicable			
	Multiple Fault	Same security as single fault model			
Classical	Differential	$\geq 2^{64}$	$\geq 2^{114}$	2^{128}	Sect. 8.5.2
	Linear	$\geq 2^{40}$	2^{128}	2^{128}	
	Impossible Differential	Main cipher	≤ 6 rounds	Not vulnerable	
	Invariant Subspace	Main cipher	Not vulnerable	Not vulnerable	
	Algebraic	Main cipher	Not vulnerable	Not vulnerable	
	Integral	Main cipher	≤ 12 rounds	Not vulnerable	

8.5.1 Protection Against Differential Fault Attack

The following analysis applies to DEFAULT-LAYER against DFA, when it is used as a protection layer for other block ciphers, and the same analysis holds.

Single Fault Adversary

DFA on DEFAULT-LAYER.With regard to the single-fault adversary model (Sect. 3.2.2/Sect. 8.2.1) Our chosen SBox has 3 non-trivial linear structures: 6, 9, f. Hence, for any input $\alpha \in \{0, \ldots, \mathtt{f}\}$, the attacker cannot uniquely identify which among $\{\alpha, \alpha \oplus 6, \alpha \oplus 9, \alpha \oplus \mathtt{f}\}$ is the actual input to the SBox. In other words, the attacker will be able to identify one partition of the input: $\{\{0, 6, 9, \mathtt{f}\}, \{1, 7, 8, \mathtt{e}\}, \{2, 4, \mathtt{b}, \mathtt{d}\}, \{3, 5, \mathtt{a}, \mathtt{c}\}\}$, but will not be able to identify which particular input is correct.

In the last round attack of DEFAULT-LAYER, Eve has to inject faults and analyse each of the 32 SBoxes independently. That means, for each SBox, she has to do a brute-force search of 4, leading to a total search complexity of $4^{32} = 2^{64}$. If she chooses any earlier round of DEFAULT-LAYER, say the lth last round, as the target for fault injection, then she can employ different analysis strategies. First, she can observe the output differences at the last round, and attempt to solve the last round inputs. This attack procedure involves a lesser number of faults than the last round attack (as faults are spread, thanks to the linear layer), but increases the search complexity (as more unknowns are now involved). Hence, in this case, the associated search complexity will be $>2^{64}$. On the other hand, she can try to find the output differences after the lth last round from the output differences observed at the last round (doing an analysis similar to DA), and attempt to solve inputs for this round (employing DFA). In this case, two search complexities are involved; one is the DA search complexity, another being the DFA search complexity. Notice that the DA search complexity will increase as l increases. Once the output differences at the lth last round are obtained, the situation is similar to the last round attack; and the DFA search requires a brute-force search of 2^{64} anyway. Finally, she can attempt to get output/input differences at the l'th ($<l$) last round (by injecting faults at lth round, and obtaining the output differences from the last round by an analysis similar to DA). In this case, as the DFA search complexity will be $>2^{64}$, more intermediate variables will now be involved. In summary, it can be said that injecting faults at an earlier round would also have a similar effect, following Remark 7.2. Basically, attacking any number of faults anywhere in DEFAULT-LAYER would require at least 2^{64} searches.

DFA on DEFAULT-CORE.Alternatively, the adversary can still try to launch DFA on the main cipher by injecting fault(s) to the last round of the main cipher and hope that it will propagate through DEFAULT-LAYER. If so, it boils down to whether the adversary can distinguish the output difference from the main cipher with less than 2^{64} effort, otherwise it is better off attacking DEFAULT-LAYER (2^{64}). Using MILP, we found that the maximum differential probability of DEFAULT-LAYER is upper bounded by 2^{-64} (details in Sect. 8.6). Thus, the attack complexity is too high and not worthwhile.

Double-Fault Adversary

As mentioned in Sect. 8.2.2/Sect. 3.7, two identical faults can bypass several duplication-based protections. CRAFT has the ability to protect against such faults by means of different encodings in the two executions. In our context, there is only one execution of the cipher, hence the concept of double fault does not apply.

Multiple Fault Adversary

Multiple fault adversary is a more relaxed setting (see Sect. 3.2.2/Sect. 8.2.1) where an adversary is able to target multiple rounds and/or words of the cipher under operation, some protections may be bypassed. The 128-bit round function of DEFAULT-LAYER $\mathcal{R}$ (described in Sect. 8.3.3) has 2^{64} linear structures (refer to Remark 7.2). Suppose the adversary has applied multiple faults and is trying to gain information about the key from the faulty outputs of the last round. Due to the presence of 2^{64} linear structures, each faulty output will always produce a solution set of size 2^{64}, which will be the same for all the faults. Thus having applied multiple faults adversary will not be able to reduce the solution space for key down from 2^{64}. Hence, our DEFAULT-LAYER is able to protect against such an adversarial model.

Note that the attacker can inject fault at one round, only to cancel it so that the combined effect of the faults is nullified. Then using the inter-relation of the two faults, the attacker may deduce some secret information. However, this model does not follow the DFA paradigm, rather it follows the ineffective paradigm (i.e., the faults do not alter the usual computation of the cipher). Therefore, such models are not included in the scope.

8.5.2 Protection Against Classical Cryptanalysis

In the following, we apply classical cryptanalysis techniques on DEFAULT. Recall that DEFAULT has a sandwich structure of two DEFAULT-LAYER and a DEFAULT-CORE in the middle. For most on the cryptanalysis, it is sufficient to show that DEFAULT-CORE is resistant against the attack.

Differential Cryptanalysis

Using MILP, we found that the maximum differential probability of DEFAULT-LAYER is upper bounded by 2^{-64} (details can be found in Sect. 8.6). Since there are two layers of DEFAULT-LAYER, we already show that there is no meaningful differential characteristic tracing across two layers of DEFAULT with differential probability more than 2^{-128}. In addition, DEFAULT-CORE has 24 rounds and any differential characteristic will involve at least 1 active SBox per round. Thus, it trivially adds an additional factor of 2^{-24} to any differential characteristic. In summary, DEFAULT is not susceptible to differential cryptanalysis.

Linear Cryptanalysis

For the linear cryptanalysis, one may refer to [LGL+12]. Here, Eve linearly approximates the (reduced round) cipher twice, one approximation being independent to the other. Hence, the complexity for such attack is $c \times \epsilon^{-4}$, where ϵ is the linear bias and c is a small constant (typically 4), as opposed to $c \times \epsilon^{-2}$ for classical linear attack [Sti06, Chap. 3.3].

Using MILP, we found that the absolute linear bias of 11-round DEFAULT-CORE is upper bounded by 2^{-33} (details in Sect. 8.6). Thus, by simple concatenation of two 11-round linear characteristics, we can show that there is no meaningful 22-round linear characteristic in DEFAULT-CORE. In addition, there are two layers of DEFAULT-LAYER, which will only make the linear cryptanalysis even harder to realize. In summary, DEFAULT is not susceptible to linear cryptanalysis.

Even though the LS SBox has zero non-linearity, we claim this does not make the linear attack easier. We further point out that our DEFAULT SBox has linear branch number 3, which is non-existent for "optimal" SBoxes. Optimality is defined by Leander and Poschmann [LP07, Definition 1], 4-bit optimal SBoxes are bijective and have optimal values with respect to the differential cryptanalysis and linear cryptanalysis.[10] Further, it is stated in [SMS19, Theorem 1], there is no 4×4 SBox with 3 linear branch number and non-zero non-linearity.

Impossible Differential Attack

We consider the effect of impossible differential as the attack against DEFAULT-CORE. The number of rounds that an impossible differential characteristic could reach is usually inversely proportional to the diffusion speed of the permutation: the faster the diffusion, the fewer rounds that can be attacked. For instance, GIFT-64 achieves full diffusion in 3 rounds, and the designers of GIFT could, at best, find 6-round impossible differential characteristic. As proposed in [ST17'2], we generate MILP instances for all $\binom{128}{1} \times \binom{128}{1} = 16384$ differentials with both the input and output differences of Hamming weight 1 on DEFAULT-CORE. To do this, we use the methodology described for the differential case in Sect. 5.4.1 (with no redundancy) and check if those instances are satisfiable. For the 7th round, we observe that none of those is satisfiable (i.e., no impossible differential exists). Therefore, following the philosophy of [ST17'2], we are confident the full-round DEFAULT-CORE is secure against impossible differential attack.

Invariant Subspace Attack

To simplify this analysis, we assume that any (affine) subspaces are preserved over the entire DEFAULT-LAYER, the PermBits and AddRoundConstants step. Thus, we focus on subspace transition over the SubCells step in DEFAULT-CORE, namely the non-LS SBoxes layer.

There is no dimension 3 (affine) subspace transition; and among the dimension 2 transitions, most transitions can only propagate up to 3 rounds, except one: $5 \oplus \{0, 2, c, e\} \rightarrow 0 \oplus \{0, 2, c, e\}$. Notice that this affine subspace will be preserved over the AddRoundKey step if each nibble of the round key belong to $\{5, 7, 9, b\}$.

Moreover, observe that every value in $\{5, 7, 9, b\}$ has the least significant bit as 1. After the 1-bit rotation within the 16-bit word, the most significant bit is 1. For these round key nibbles to still preserve the subspace transition, it has to be either 9 or b. But after 7 key schedule updates, some of these nibbles have another 1-bit right rotation, which will break the subspace structure. Thus, we are confident that no (affine) subspace can be preserved for more than 8 rounds and DEFAULT is not vulnerable to invariant subspace attack.

[10] For a quick reference, differential and linear branch numbers of the PRESENT SBox are 3 and 2, respectively; that of GIFT are both 2; whereas the same for DEFAULT-LAYER SBox are both 3 (see Table 8.3).

Algebraic Attack

In order to check the security of DEFAULT-CORE against algebraic attacks, we check its algebraic properties using Sage (see Sect. 1.4 for more information). We are able to represent DEFAULT-CORE in terms of Boolean expression up to 4 rounds. We have observed the minimum number of monomials is 11101, at least 97 variables (out of 128) are involved and the minimum algebraic degree is 8. Further, the algebraic degree for a particular location grows by a multiple of 2. Hence, we are confident DEFAULT-LAYER will reach an algebraic degree of 64 at most in 7 rounds. Thereafter following the justification on algebraic degree as an indicator in [GJK+19, Sect. 4.2], we are confident 28 rounds of DEFAULT-LAYER would reach the maximum algebraic degree of 127.

Integral Attack

Suppose an attacker repeats the encryption multiple times and injects all possible differential fault values to a specific word in the output of the main cipher. This is similar to collecting a set of inputs (more precisely the output from the main cipher) with a specific structure to launch an integral attack. Such model is reported in [PY06] and [SSL15, Chap. 6.3]. This model is a special case of DFA where all possible faults are considered. Since the attacker does not get any extra information by using all possible faults, DEFAULT-LAYER (and hence DEFAULT) automatically protects against it.

As for DEFAULT-CORE, we could reuse some of the security analysis of GIFT-128 on our design. In particular, the designers of GIFT evaluate the longest integral distinguisher for GIFT-128 using the (bit-based) division property [Tod15] (Sect. 2.6.4) to be 11 rounds, and conclude that GIFT-128 is secure against integral attacks. Since our DEFAULT-CORE has 28 rounds, we are confident that DEFAULT-CORE is secure against integral attacks. The SOLVATORE tool [EKK+18] does not return any distinguisher in a reasonable time for the 11th and 12th rounds, and the tool reports non-existence of any distinguisher at the 13th round.

8.5.3 Protection Against Side Channels Attacks

In essence, DEFAULT-LAYER/DEFAULT is simply a bit permutation-based SPN block cipher and, as such, usual side channels attacks might apply on it. Usual countermeasures such as masking can be applied on DEFAULT.

We point out that protecting DEFAULT against side channels attacks should not make DFA easier. An additional feature of the DEFAULT-LAYER SBox is that it has lower AND operations compared to the usual SBoxes used in cipher design, hence adopting masking countermeasure [MOP07] could be more efficient.

Moreover, it is probably not needed to incorporate side channel protection for the entire 80 rounds of DEFAULT, since most of side channel attack works near the beginning or the end of the cipher execution (typically either the initial or the find round, Sect. 4.3). It may be possible to gain side channel security of DEFAULT only by making the two instances of DEFAULT-LAYER, which should be efficient due to the presence of LS.

8.6 Automated Bounds for Differential and Linear Attacks

In [MWG+11], the authors present how to find optimal differential and linear characteristics based on Mixed Integer Linear Programming (MILP, see Chap. 5 for more details), which is then tuned to work with bit permutation-based block ciphers in [SHW+13]. In general, the bound for differential (respectively linear) attack complexity is computed by the following steps. First, the minimum number of active SBoxes (say, a) for a small number of rounds (say, η) is computed. Then, a bound on the maximum differential probability (respectively linear bias) is computed by raising the maximum differential probability (respectively, maximum absolute linear bias) for an SBox to the power a, let us call this value a'. Finally, following the so-called *wide trail* strategy, the maximum expected differential probability or MEDP for short (respectively, maximum expected linear probability, MELP for short; refer to Sects. 2.6.1 and 2.6.2 for more discussion) is computed for η' by raising a' to the power $\frac{\eta'}{\eta}$. However, it may not be possible to apply this methodology in our case (because of LS), as we explain next. Hence we propose a new type of MILP modelling that is compatible with LS SBoxes.

Indeed, our special SBox with linear structures has probability 1 differential transitions (respectively, $\pm\frac{1}{2}$ linear bias). For the differential case, the above-mentioned approach will always yield an bound of $\epsilon_d = 1$ (1 is raised to the power of an integer), which naturally signifies the least possible protection against differential attack (the attack succeeds with only one chosen input difference or two chosen inputs). In case of linear cryptanalysis, it can be shown that the overall bias ϵ_l, considering only $\pm\frac{1}{2}$ biases (and assuming mutual independence of the biases), is $\frac{1}{2}$. This is obtained by substituting $\epsilon_i = \frac{1}{2}\ \forall i$ in [Sti06, Lemma 3.1]. Similar to the differential case, this also leads to the least protection against linear attack (the attack succeeds with roughly $\frac{1}{\epsilon_l^2} = 4$ known inputs). Naturally, we need to devise a way to count precisely the number of probability $\frac{1}{2}$ differential transitions and $\pm\frac{1}{4}$ linear biases.

The basic idea here is to minimize the number of active SBoxes with differential probability $\frac{1}{2}$ (respectively, $\pm\frac{1}{4}$ linear bias) while keeping the differential probability 1 (respectively, $\pm\frac{1}{2}$ linear bias) unrestricted for a particular number of rounds. The problem can be formulated as an MILP problem whose solution can be obtained by a standard solver. Upon getting the solution, let us denote the number of active SBoxes with differential probability 1 by o and that of differential probability $\frac{1}{2}$ by h. Then, the MEDP can be computed as $1^o \times (\frac{1}{2})^h = 2^{-h}$ and hence the attacker needs at least 2^h chosen differences [Sti06, Chap. 3.4], which translates to 2^{1+h} chosen inputs for the layer. To compute the MELP, let us assume the number of active SBoxes with $\pm\frac{1}{2}$ bias is o and that with $\pm\frac{1}{4}$ is h. Then, we substitute $\epsilon_i = \frac{1}{2}$ for $i = 1(1)o$, and $\epsilon_i = \frac{1}{4}$ for $i = 1(1)h$; in the piling-up lemma (see [Sti06, Lemma 3.1]) to get $\epsilon_l = 2^{o+h-1} \times (\frac{1}{2})^o \times (\frac{1}{4})^h = 2^{-h-1}$. Hence the attacker needs roughly $\frac{1}{\epsilon_l^2} = 2^{2h+2}$ known inputs [Sti06, Chap. 3.3]. Notice that the probability 1 differential transitions as well as the $\pm\frac{1}{2}$ linear biases do not play any role in the search complexity.

Notice that the MILP modelling and optimizations proposed in [Bak20'2]/Sect. 5.4 can work with LS SBoxes. The MILP modelling separately takes each point into account, while allowing probability 1 differential or $\pm\frac{1}{2}$ linear trail in the computation, but the objective function specifically ignores those points. Also, in our context with LS SBoxes, the problem associated with convex hull (explained in Sect. 5.3) appears with a lot of SBoxes. Therefore, the MILP modelling from [Bak20'2]/Sect. 5.4 is used here; together with the all-CH heuristic (see Sect. 5.4.2 for more on the heuristics) for differential and the chosen-CH heuristic for linear for all the bounds reported.

Results

For the LS SBox (used in `DEFAULT-LAYER`) `037ED4A9CF18B265`, the bounds obtained from the corresponding MILP instances are: 4.000 at the 5th round for linear, and 20.000 at the 7th round for differential. This translates to around 2^8 computations for 5 rounds against classical linear attacks and around 2^{20} computations against differential attack. Hence, we conclude 28 rounds of `DEFAULT-LAYER` would be enough to provide security of 2^{64} computations against differential attack and security of 2^{32} computations against classical linear attack, following [Sti06].

As explained in Sect. 8.5.2, we are mainly interested in the security against the classical linear attack against `DEFAULT-CORE`. Our results are obtained from the same workstation as Sect. 5.4.3, i.e., with 16× Intel Xeon E7-8880 physical cores (shared among multiple users) running Gurobi 8.1 on 64-bit Ubuntu 18.04. For the non-LS SBox (used in `DEFAULT-CORE`) `196F7C82AED043B5`, the bound obtained from the MILP instance for the linear case is 33.00 at the 11th round. Hence, the linear security at 11 rounds of `DEFAULT-CORE` is around 2^{66} computations. Hence, we conclude `DEFAULT` ensures the required DFA security (of 2^{64} computations) for `DEFAULT-LAYER`, and also the required classical security (of 2^{128} computations) for `DEFAULT-CORE`. Also, the result from the MILP instance corresponding to the differential bound of the non-LS SBox for 8 rounds is 38.00, which ensures the complexity of the differential attack on the `DEFAULT-CORE` is at least 2^{114} (as shown in Table 8.5).

More results regarding this can be found in Table 8.6 (Table 8.6a for differential and linear bounds for the LS SBox `037ED4A9CF18B265`, and Table 8.6b linear

Table 8.6 Optimal differential and linear bounds for LS SBox and non-LS SBox

(a) LS SBox: 037ED4A9CF18B265

Round(s)	1	2	3	4	5	6	7
Differential	0.000	0.000	2.000	6.000	10.000	14.000	20.000
Linear	0.000	0.000	0.000	1.000	4.000	N/A	N/A

(b) Non-LS SBox: 196F7C82AED043B5

Round(s)	1	2	3	4	5	6	7	8	9	10	11
Linear	1.000	2.000	4.000	6.000	8.000	12.000	16.000	20.000	25.000	30.000	33.000

bounds for the non-LS SBox 196F7C82AED043B5), as obtained from the MILP instances. Due to the time taken by the solver, it is hard to compute the bounds beyond that are given in Table 8.6, at least with the current modelling (with our computing resource).

8.7 Performance

We state the benchmarks for hardware and software implementations of DEFAULT. Comparison is done with GIFT-128 and a duplication-protected GIFT-128 which runs the same computation twice and compares the output. The output is released only if both the redundant computations produce the same ciphertext, otherwise it is suppressed. This is the so-called detective countermeasure [BSS20]/Sect. 3.7.1. This double redundancy is a common countermeasure adopted against fault attacks, and often found in commercial smart cards.

As a side note, it can be mentioned that the current academic researches have drifted away from the simple detective countermeasure towards more sophisticated error detection code-based or infection-based countermeasures. If such a sophisticated countermeasure is taken into account, DEFAULT can provide much better performance as the DFA protection in DEFAULT does not rely on any such costly method.

8.7.1 Hardware Benchmark

For the (approximated) hardware benchmark (round keys are computed on-the-fly), we show the area and throughput on Xilinx 7-series Kintex FPGA for GIFT-128, Duplicated GIFT-128 , and DEFAULT in Table 8.7. The underlying GIFT-128 core is implemented to execute one round per clock cycle, with 32 parallel SBoxes. DEFAULT implementation is derived from GIFT-128 core, where a multiplexer chooses SBox based on round counter. For duplication countermeasure we

Table 8.7 Hardware benchmarking for DEFAULT and GIFT-128 with/without duplication

Kintex-7 FPGA	LUTs	F/Fs	Slices	Throughput (Mbps)
GIFT-128	227	232	59	416.1
GIFT-128 Duplicated	582	456	146	415.6
DEFAULT	453	233	114	220.9

Table 8.8 Software benchmarking for DEFAULT and GIFT-128 with/without duplication

		Intel Xeon Silver 4215	ARM Cortex A-53
Speed (Cycles/Bytes)	GIFT-128	9.7 (1.000×)	61.3 (1.000×)
	GIFT-128 Duplicated	21.9 (2.258×)	124.4 (2.029×)
	DEFAULT	19.2 (1.979×)	121.9 (1.989×)
Code Size (Bytes)	GIFT-128	6624 (1.000×)	5593 (1.000×)
	GIFT-128 Duplicated	6859 (1.035×)	5818 (1.040×)
	DEFAULT	8024 (1.211×)	7085 (1.267×)

choose space redundancy, where two parallel cores of GIFT-128 are instantiated. GIFT-128 Duplicated depicts a higher area footprint as compared to DEFAULT at the advantage of throughput. As GIFT-128 Duplicated still computes 40 rounds (in parallel for redundant parts) and DEFAULT computes 80 rounds, the difference in throughput is apparent. Any optimization in GIFT-128 Duplicated to reduce the area by serialization would have a negative impact on throughput.

8.7.2 *Software Benchmark*

The (approximated) software benchmarks for GIFT-128, duplicated GIFT-128 and DEFAULT are given in Table 8.8 for the Intel Xeon Silver 4215 microprocessor and ARM Cortex A-53 microcontroller, where round keys are precomputed. The relative overheads compared to GIFT-128 are shown within parenthesis. It can be seen that the code size for DEFAULT is slightly more compared to duplicated GIFT-128; at the same time, DEFAULT is faster. For fairness, the implementation of GIFT-128 is done with the same relative placement of PermBits (see Sect. 8.3.3/Sect. 8.9).[11]

[11] Software benchmarks with the original description of GIFT-128 are noticeably worse.

The clock cycles are measured by utilizing `time()` function from `time.h` library in C, by averaging over multiple executions. The process is run on a single core with a randomly generated key and plaintext, and finally, the average is taken. Compiler optimizations are disabled to produce a consistent result. Note that the main purpose of this benchmark is to show the relative performance compared to GIFT-128 in the same setting. We would also like to note that a new efficient software representation of GIFT, named *fix-slicing*, is published recently [ANP20], drastically reducing the cycles needed for encryption on ARM Cortex-M family of microcontrollers. The fix-sliced implementation of DEFAULT would have a very similar per-round performances as GIFT-128, as the bit permutation is the same (which is mainly what fix-slicing is trying to optimize), while the SBoxes have similar cost. Overall, we expect the overheads to be similar as it scales accordingly to the number of rounds. Generally, this scaling would apply to other optimizations as well.

8.8 Conclusion

The first theoretical study on SBox with respect to its properties against differential fault attack (DFA) is detailed in Chap. 7. We observe that DFA works as a simplified model of differential attack (DA), yet the properties of an SBox which make DFA harder, make DA easier, and vice-versa. Our findings enable us to propose a cipher-level countermeasure against DFA, which has never been done earlier, to the best of our knowledge. This construction follows an SPN block cipher design paradigm and does not incur too much overhead. The core idea is based on an SBox, which is such that upon trying all possible fault values, the attacker is not able to narrow down the search space from square root bound. This work opens up a new paradigm of symmetric key cipher design by studying SBoxes with LS, which is not so well studied in the context of cryptography. Thus we conclude that this work opens up a new research direction for fault-resistant ciphers.

Our cipher DEFAULT (comprised DEFAULT-LAYER and DEFAULT-CORE) can be used as a stand-alone cipher with security claims against classical attacks on top of being resistant to DFA, or DEFAULT-LAYER can be used in conjunction with other ciphers (it supports only 128-bit state ciphers at the moment, but can be adjusted to fit other state sizes) on an ad hoc basis. Either way, we offer DFA security of $2^{n/2}$ for a n-bit layer (i.e., a non-trivial security 2^{64} with the chosen design parameters) against DFA as well as the variants of DFA that (loosely) follow classical attack paradigms, apart from the full security of 2^{n} (i.e., 2^{128} with the chosen design parameters) for the classical attacks.

In brief, the DEFAULT permutation uses the following SBoxes , together with the bit permutation of GIFT-128 [BPP+17]:

SBox with non-trivial linear structure for DEFAULT-LAYER	:	037ED4A9CF18B265
SBox with non-trivial linear structure for DEFAULT-LAYER^{-1}	:	0AD15FE2B76C8439
SBox without non-trivial linear structure for DEFAULT-CORE	:	196F7C82AED043B5
SBox without non-trivial linear structure for DEFAULT-CORE^{-1}	:	B07DCF24618E5A93

Next we summarize the advantages and limitation of our proposal:

- ✓ **First cipher design-level protection.** This solves the concern raised against existing DFA countermeasures (Sect. 8.2.2). In particular, we free the DFA protection from the hand of the cipher implementer and take it to the cipher designer entirely.
- ✓ **Applicable to (practically) all of symmetric key primitives as an ad hoc layer.** The basic concept presented here can be adopted to ensure a non-trivial DFA security on any symmetric key primitive.
- ✓ **Possible to get non-trivial DFA security.** The particular instantiation we propose offers up to $2^{n/2}$ DFA security where $n \geq 128$ is the state size of a block cipher (without jeopardizing its classical security). However, this is not the theoretical bound as can be seen from Table 7.4. Another point to note here is although a search complexity of $2^{n/2}$ may not be considered secure in the classical attack paradigm, it is not the case for practical attack paradigms like fault.
- ✓ **Lightweight.** The design being inspired from GIFT-128 can be considered as lightweight. When used in conjunction with non-lightweight ciphers, this can lead to less cost than duplicating that cipher.
- ✓ **Protected against double fault, multiple fault adversary.** A generic problem with the typical duplication-based countermeasure, termed as the double fault (see Sect. 8.2.2), is inherently protected in DEFAULT. Also, DEFAULT can protect against any number of faults (unlike some error detection-based protection where faults are not detected beyond a coverage). Moreover, it can protect against the multiple fault adversary model.
- ✓ **No need for external randomness or a protected device.** The commonly referred infective countermeasure [BSS20]/Sect. 3.7.2/Chap. 9 uses an external source of randomness. For the protocol-level countermeasures, such as [BBB+18], a part of the device is assumed to be off-limit to the attacker due to some device-level protection. In our case, there is neither a need for an external source of entropy nor a specially protected device.
- ✗ **Not full DFA security.** It is technically possible to achieve almost full DFA security (such as 2^{112} for a 128-bit state, see Table 7.4). However, it is not possible to achieve a full DFA security by this methodology.

8.9 Supplementary Discussion

Role of Multiple Fault Adversary Model

Generally, the fault attacks in symmetric key follow the single-fault adversary model, where the attacker is allowed to alter the normal flow of operation once during a single execution of the cipher. As the proponents of the fault attacks want to show the generic nature of their attacks, they tend to rely on as few assumptions as possible. To the best of our knowledge, no DFA so far works by injecting faults in more than one round of a cipher in the symmetric key setting. In the public key setting, however, such a model has been used, as in [TK10]. The Internal Differential Fault Analysis (InDFA) [SC17]/Sect. 3.5.1 (which works by altering round counters, hence altering the operation flow) uses multiple fault model. Such model essentially changes the cipher definition, and consequently is outside the scope. Other types of countermeasures are recommended for this purpose.

DEFAULT-LAYER: Comparison with Device/Protocol-Based Approaches

In a type of fault countermeasure, either a specialized device (that detects or shields) any potential source of fault, or the communication protocol (that reduces the conditions required for a successful fault attack). Both of them are off-limit to the cipher designer, as it is finally dependent on the device designer/network administrator. Both rely on the assumption of a secure component that cannot be attacked by fault. However, with a more specialized attack, it may be possible to bypass the protection. In contrast, our solution here does not depend on any such assumption and is completely definable at the cipher design level.

DEFAULT-LAYER: Usage of a Bad SBox

We use an SBox with linear structures which has 0 non-linearity and a fixed point (0) to design the DEFAULT-LAYER. As such choices are considered undesirable in cipher construction, one may doubt the security claims. We point out that, first, the main cipher is used as is (without any modification), thus not interfering with the classical security claims made by the main cipher. Second, the classical security evaluations are performed with logical argument and/or automated tools, which should provide enough confidence.

Similarities/Differences Between DEFAULT and GIFT-128

The design of DEFAULT-LAYER/DEFAULT-CORE is quite similar to (in fact, inspired from) GIFT-128. The bit permutation layers are the same in both and round constants are taken from GIFT-128. The SBoxes are different. DEFAULT-LAYER runs for 28 rounds, DEFAULT-CORE for 24 rounds, hence DEFAULT runs for $28 + 24 + 28 = 80$ rounds. In comparison, GIFT-128 runs for 40 rounds (so duplicated GIFT-128 is comparable to DEFAULT). GIFT-128 uses half key addition (key bits are XORed at the first two bits of each SBox output bits at each round), contrasting the full key addition in DEFAULT-LAYER/DEFAULT-CORE (key bits are added to all 128-bits at each round). Full key addition in DEFAULT-LAYER is required for the desired DFA security, and DEFAULT-CORE uses a similar structure for simplicity. Also, DEFAULT-LAYER does not have a well-defined key schedule, rather it relies on the key schedule from the main cipher or uses the master key as its round keys. DEFAULT-CORE has its key schedule, which is different from that of GIFT-128.

One may also note from the specification of GIFT-128 that the ordering of one round is slightly altered since AddRoundConstants has been put after PermBits/before AddRoundKey (AddRoundConstants is after AddRoundKey in GIFT-128 [BPP+17, Sect. 2]). This essentially does not affect the output from GIFT-128 since both AddRoundKey and AddRoundConstants are XOR operations. The reason for this change is to make basic software implementation simpler. In the basic implementation, one extra bit $\leftrightarrow$ nibble conversion is required for GIFT-128 setting (AddRoundConstants operates on bits and AddRoundKey operates on nibbles).

DEFAULT-LAYER: Comparison with Duplication-Based Approaches

For the comparison of DEFAULT with a duplication-based countermeasure (see Sect. 3.7), the following points can be noted. First, certain duplication-based protection can be subject to identical faults at both the computations (double fault), which will be regarded as non-faulty case. DEFAULT is free from that. Also, DEFAULT is comparable to the basic duplication on GIFT-128, as can be seen from the benchmarking results in Sect. 8.7. Hence, if any extra component is used (such as the infection mechanism), DEFAULT will have lower cost. Second, certain other type of duplication-based protection can safeguard against double fault by using different encoding for the two computations. Such encoding can be costly compared to DEFAULT and may be restricted by an error coverage.

DEFAULT-LAYER: Related Key Security

If the underlying cipher (used in conjunction with DEFAULT-LAYER) has related key security, then the overall construction retains the same claim. For DEFAULT (as a stand-alone) cipher, we are not aware of any relevant result, and it probably protects against related key attacks due to large security margin, though we do not claim related key security (akin to GIFT-128). Just in case DEFAULT fails to meet any related key analysis, we recommend to double the number of rounds (the round constants would reiterate), following the GIFT designers' statement [BPP+17, Sect. 2], "in case one wants to protect against related-key attacks as well, we advise to double the number of rounds".

Precise Bit Flip Model

Here we theorize a model where the attacker can target one particular bit, which results in flipping the bit. This model is too strong; it requires too much control for the attacker, and not yet reported as practical in the literature, to the best of our knowledge. This model differs from the SIFA model [DEK+18], as SIFA does not assume precision but relies on statistical information. We decide to keep this model out of scope for DEFAULT-LAYER as it is still a theoretical model, not to mention it will break any duplication-based countermeasure in existence.

That said, this model can be (theoretically) used against DEFAULT-LAYER. In order for this to work, the target SBox is required to be implemented as a combinational logic. The attack targets and flips one input of an AND gate, and recovers the other input to the gate. Consider, for example, the computation of the second coordinate of the LS SBox: $y_1 = x_0 \oplus x_1 \oplus x_0x_1 \oplus x_0x_2 \oplus x_1x_3 \oplus x_2x_3$; and the attacker is able to flip x_0 from the monomial x_0x_1 (marked in bold): $y_1' = \overline{\boldsymbol{x_0}} \oplus x_1 \oplus x_0x_1 \oplus x_0x_2 \oplus x_1x_3 \oplus x_2x_3$. Then by comparing y_1 and y_1', she will know whether $x_1 = 0$ or 1. Similarly, it is possible to recover other bits. One possible way to mitigate this fault (if required) could be to use look-up table based implementation of SBoxes.

The precise bit flip model is generic and can be applied to break any duplication-based protection. For example, the attacker will target one AND operation of an SBox at the last round of one of the two computations of the cipher. If the output of this operation does not change, then the countermeasure will treat this case as non-faulty and produce correct output. If the output of the AND operation changes, it will be observed by the attacker. Hence, simply by checking if the output of the countermeasure remains unaltered after injection of the fault, Eve is able to deduce whether the other input (which is not flipped) is 1 or 0.

It is to be emphasized that the usual duplication-based countermeasures such as detective (Sect. 3.7.1/Sect. 9.2.2) or infective (Sect. 3.7.2/Chap. 9/[BSS20]) cannot protect against such model. The attacker will be able to see whether or not the injected fault is effective, and this enables the attack procedure. As such, CRAFT [BLM+19] is also vulnerable to the attack. However, the randomized duplication-based countermeasure proposed in [BKK+20'2] (which protects against SIFA, see Sect. 3.7.6) can protect against this model. Otherwise, an error correction (i.e., at minimum triplication) is required to protect against such attack.

Fault Models Not Protected by DEFAULT-LAYER

In literature, the Safe Error Attack (SEA) [YJ00, YKL+01, JQY+02]/Sect. 3.5.2 model has been proposed which utilizes the cases where the faulty and non-faulty outputs are the same. Among the SEA models, one particular model is known as Ineffective Fault Attack (IFA) [Cla07]/Sect. 3.5.2. The Statistical Ineffective Fault Analysis (SIFA, [DEK+18]/Sect. 3.5.4) is not protected either; specialized countermeasure (such as [BKK+20'2]/Chap. 10) is recommended for this purpose. Another type of fault attack uses statistical information on the output distribution as it has become biased because of fault injection [Nah15]. In another direction, the fault attack model in [KPB+17] changes the instruction.

Such models do not adhere to the DFA principle and we do not claim any protection against them. Standard countermeasures are recommended against such a fault attack.

DEFAULT: Lightweight Cipher

Although the cipher DEFAULT is of 80 rounds (i.e., twice that of GIFT-128, on which it is based on), it is to be noted that DEFAULT protects against the differential fault attack. In contrast, GIFT-128 or any other cipher cannot defend against differential fault attack without at least duplication. Therefore, to have a protection against the differential fault attack, GIFT-128 has to be at least duplicated, and making GIFT-128 of 80 rounds. As noted in Sect. 8.7, the performance of DEFAULT (80 rounds) and that of duplicated GIFT-128 (80 rounds) are practically the same.

Impact of Machine Learning-Based Analysis

With respect to the machine learning-assisted differential distinguisher that is reported in [BBD+20]/Chap. 6, we note the following. As noted, it can significantly reduce the security margin returned by the MILP-based analysis (as in Chap. 5). Now, the design of DEFAULT follows from that of GIFT-128 , which uses MILP-based analysis to ascertain the differential bound, which is adopted here as well. Although ML is a promising, at the moment it is still at infancy and does not pose a direct threat to DEFAULT. With time, however, it is likely that ML-based attack will mature. In that scenario, it may challenge the security claim of DEFAULT (in Sect. 8.5).

In that case, it is important to notice, ML will challenge the security of practically all the ciphers including GIFT-128. Consequently, the number of rounds of GIFT-128 is to be increased to mitigate the threat (as in the case for related key security, Sect. 8.9). The design of DEFAULT is flexible, thus allowing us to increase the number of rounds if needed. We are confident, even though the number of rounds of DEFAULT-LAYER and/or DEFAULT-CORE is to be increased, it will be proportional to that of GIFT-128. Thus the (near-) double cost of DEFAULT to GIFT-128 will be retained.

References

[BS97] E. Biham, A. Shamir, Differential fault analysis of secret key cryptosystems, in *Advances in Cryptology - CRYPTO '97*, eds. by B.S. Kaliski (Springer, Berlin, 1997), pp. 513–525. ISBN: 978-3-540-69528-8

[BS90] E. Biham, A. Shamir, Differential cryptanalysis of des-like cryptosystems, in *Advances in Cryptology - CRYPTO '90, 10th Annual International Cryptology Conference, Santa Barbara, California, USA, August 11–15, 1990, Proceedings* (1990), pp. 2–21. https://doi.org/10.1007/3-540-38424-3_1

[BPP+17] S. Banik, S.K. Pandey, T. Peyrin, Y. Sasaki, S.M. Sim, Y. Todo, Gift: a small present. Cryptology ePrint Archive, Report 2017/622 (2017). https://eprint.iacr.org/2017/622

[BLM+19] C. Beierle, G. Leander, A. Moradi, S. Rasoolzadeh, CRAFT: lightweight tweakable block cipher with efficient protection against DFA attacks. IACR Trans. Symmetric Cryptol. **2019**(1), 5–45 (2019). https://doi.org/10.13154/tosc.v2019.i1.5-45

[BSS20] A. Baksi, D. Saha, S. Sarkar, To infect or not to infect: a critical analysis of infective countermeasures in fault attacks. J. Cryptogr. Eng. (2020), https://link.springer.com/article/10.1007

[PQ03] G. Piret, J. Quisquater, A differential fault attack technique against SPN structures, with application to the AES and KHAZAD, in *Cryptographic Hardware and Embedded Systems - CHES 2003, 5th International Workshop, Cologne, Germany, September 8–10, 2003, Proceedings* (2003), pp. 77–88. https://doi.org/10.1007/978-3-540-45238-6_7

[HBB+16'2] W. He, J. Breier, S. Bhasin, N. Miura, M. Nagata, Ring oscillator under laser: potential of pll-based countermeasure against laser fault injection, in *2016 Workshop on Fault Diagnosis and Tolerance in Cryptography (FDTC)* (IEEE, 2016), pp. 102–113

[BBB+18] A. Baksi, S. Bhasin, J. Breier, M. Khairallah, T. Peyrin, Protecting block ciphers against differential fault attacks without re-keying (extended version). Cryptology ePrint Archive, Report 2018/085 (2018), https://eprint.iacr.org/2018/085

[SHS16] B. Selmke, J. Heyszl, G. Sigl, Attack on a DFA protected AES by simultaneous laser fault injections, in *2016 Workshop on Fault Diagnosis and Tolerance in Cryptography, FDTC 2016, Santa Barbara, CA, USA, August 16, 2016*, pp. 36–46. https://doi.org/10.1109/FDTC.2016.16

[AMR+18] A. Aghaie, A. Moradi, S. Rasoolzadeh, A.R. Shahmirzadi, F. Schellenberg, T. Schneider, Impeccable circuits. Cryptology ePrint Archive, Report 2018/203 (2018), https://eprint.iacr.org/2018/203

[DEK+18] C. Dobraunig, M. Eichlseder, T. Korak, S. Mangard, F. Mendel, R. Primas, Sifa: exploiting ineffective fault inductions on symmetric cryptography. Cryptology ePrint Archive, Report 2018/071 (2018). https://eprint.iacr.org/2018/071

[BKK+20'2] A. Baksi, V.B. Kumar, B. Karmakar, S. Bhasin, D. Saha, A. Chattopadhyay, A novel duplication based countermeasure to statistical ineffective fault analysis, in *Australasian Conference on Information Security and Privacy* (2020), https://link.springer.com/chapter/10.1007/978-3-030-55304-3_27

[DFL11] P. Derbez, P.-A. Fouque, D. Leresteux, Meet-in-the-middle and impossible differential fault analysis on aes, in *Cryptographic Hardware and Embedded Systems - CHES 2011* eds. by B. Preneel, T. Takagi (Springer, Berlin, 2011), pp. 274–291. ISBN: 978-3-642-23951-9

[JKP12'1] P. Jovanovic, M. Kreuzer, I. Polian, A fault attack on the LED block cipher, in *Constructive Side-Channel Analysis and Secure Design - Third International Workshop, COSADE 2012, Darmstadt, Germany, May 3–4, 2012. Proceedings* (2012), pp. 120–134. https://doi.org/10.1007/978-3-642-29912-4_10

[BKL+07] A. Bogdanov, L.R. Knudsen, G. Leander, C. Paar, A. Poschmann, M.J. Robshaw, Y. Seurin, C. Vikkelsoe, *PRESENT: An Ultra-Lightweight Block Cipher, in CHES*, vol. 4727 (Springer, Berlin, 2007), pp. 450–466

[BBI+15] S. Banik, A. Bogdanov, T. Isobe, K. Shibutani, H. Hiwatari, T. Akishita, F. Regazzoni, Midori: a block cipher for low energy (extended version). Cryptology ePrint Archive, Report 2015/1142 (2015), https://eprint.iacr.org/2015/1142

[BJK+16] C. Beierle, J. Jean, S. Kölbl, G. Leander, A. Moradi, T. Peyrin, Y. Sasaki, P. Sasdrich, S.M. Sim, The SKINNY family of block ciphers and its lowlatency variant MANTIS. IACR Cryptology ePrint Archive (2016), p. 660, http://eprint.iacr.org/2016/660

[Dt08] W. Diffie, G.L. (translators), Sms4 encryption algorithm for wireless networks. Cryptology ePrint Archive, Report 2008/329 (2008), https://eprint.iacr.org/2008/329

[DeC07] C. De CannièE8re, *Analysis and Design of Symmetric Encryption Algorithms* (Katholieke Universiteit Leuven, Belgium, 2007). Ph.D. Thesis, https://www.esat.kuleuven.be/cosic/publications/thesis-139.pdf

[LGL+12] Z. Liu, D. Gu, Y. Liu, W. Li, Linear fault analysis of block ciphers, in *Applied Cryptography and Network Security - 10th International Conference, ACNS 2012, Singapore, June 26–29, 2012. Proceedings* (2012), pp. 241–256. https://doi.org/10.1007/978-3-642-31284-7_15

[Sti06] D.R. Stinson, *Cryptography - Theory and Practice* (Discrete Mathematics and Its Applications Series (CRC Press, Boca Raton, 2006). ISBN: 978-1-58488-508-5

[LP07] G. Leander, A. Poschmann, On the classification of 4 bit s-boxes, in *Arithmetic of Finite Fields, First International Workshop, WAIFI 2007, Madrid, Spain, June 21–22, 2007, Proceedings* (2007), pp. 159–176. https://doi.org/10.1007/978-3-540-73074-3_13

[SMS19] S. Sarkar, K. Mandal, D. Saha, On the relationship between resilient Boolean functions and linear branch number of s-boxes. Cryptology ePrint Archive, Report 2019/1427 (2019), https://eprint.iacr.org/2019/1427

[ST17'2] Y. Sasaki, Y. Todo, New impossible differential search tool from design and cryptanalysis aspects - revealing structural properties of several ciphers, in *Advances in Cryptology - EUROCRYPT 2017 - 36th Annual International Conference on the Theory and Applications of Cryptographic Techniques, Paris, France, April 30–May 4, 2017, Proceedings, Part III* (2017), pp. 185–215. https://doi.org/10.1007/978-3-319-56617-75C_7

[GJK+19] D. Goudarzi, J. Jean, S. Kölbl, T. Peyrin, M. Rivain, Y. Sasaki, S.M. Sim, Pyjamask v1.0 (2019), https://csrc.nist.gov/CSRC/media/Projects/lightweight-cryptography/documents/round-2/spec-doc-rnd2/pyjamask-spec-round2.pdf

[PY06] R.C. Phan, S. Yen, Amplifying side-channel attacks with techniques from block cipher cryptanalysis, in *Smart Card Research and Advanced Applications, 7th IFIP WG 8.8/11.2 International Conference, CARDIS 2006, Tarragona, Spain, April 19-21, 2006, Proceedings* (2006), pp. 135–150, https://doi.org/10.1007/11733447_10

[SSL15] K. Sakiyama, Y. Sasaki, Y. Li, *Security of Block Ciphers - From Algorithm Design to Hardware Implementation* (Wiley, New York, 2015). ISBN: 978-1-118-66001-0, http://eu.wiley.com/WileyCDA/WileyTitle/productCd-1118660013.html

[Tod15] Y. Todo, Structural evaluation by generalized integral property, in *Advances in Cryptology - EUROCRYPT 2015 - 34th Annual International Conference on the Theory and Applications of Cryptographic Techniques, Sofia, Bulgaria, April 26–30, 2015, Proceedings, Part I* (2015), pp. 287–314. https://doi.org/10.1007/978-3-662-46800-5_12

[EKK+18] Z. Eskandari, A.B. Kidmose, S. Kölbl, T. Tiessen, Finding integral distinguishers with ease. Cryptology ePrint Archive, Report 2018/688 (2018), https://eprint.iacr.org/2018/688

[MOP07] S. Mangard, E. Oswald, T. Popp, *Power Analysis Attacks - Revealing the Secrets of Smart Cards* (Springer, Berlin, 2007). ISBN: 978-0-387-30857-9

[MWG+11] N. Mouha, Q. Wang, D. Gu, B. Preneel, Differential and linear cryptanalysis using mixed-integer linear programming, in *Information Security and Cryptology - 7th International Conference, Inscrypt, Beijing, China, November 30-December 3, 2011*. Revised Selected Papers (2011), pp. 57–76. https://doi.org/10.1007/978-3-642-34704-75C_5

[SHW+13] S. Sun, L. Hu, P.Wang, K. Qiao, X. Ma, L. Song, Automatic security evaluation and (related-key) differential characteristic search: application to simon, present, lblock, des(l) and other bit-oriented block ciphers. Cryptology ePrint Archive, Report 2013/676 (2013), https://eprint.iacr.org/2013/676

[Bak20'2] A. Baksi, New insights on differential and linear bounds using mixed integer linear programming (full version) (2020), https://eprint.iacr.org/2020/1414

[ANP20] A. Adomnicai, Z. Najm, T. Peyrin, Fixslicing: a new gift representation. Cryptology ePrint Archive, Report 2020/412 (2020), https://eprint.iacr.org/2020/412

[TK10] E. Trichina, R. Korkikyan, Multi fault laser attacks on protected crt-rsa, in *Workshop on Fault Diagnosis and Tolerance in Cryptography* 2010–08, pp. 75–86

[SC17] D. Saha, D.R. Chowdhury, Internal differential fault analysis of parallelizable ciphers in the counter-mode. J. Cryptogr. Eng. (2017). https://doi.org/10.1007/s13389-012-0031-5

[YJ00] S. Yen, M. Joye, Checking before output may not be enough against faultbased cryptanalysis. IEEE Trans. Comput. **49**(9), 967–970 (2000). https://doi.org/10.1109/12.869328

[YKL+01] S. Yen, S. Kim, S. Lim, S. Moon, A countermeasure against one physical cryptanalysis may benefit another attack, in *Information Security and Cryptology - ICISC 2001, 4th International Conference Seoul, Korea, December 6–7, 2001, Proceedings* (2001), pp. 414–427. https://doi.org/10.1007/3-540-45861-15C_31

[JQY+02] M. Joye, J. Quisquater, S. Yen, M. Yung, Observability analysis - detecting when improved cryptosystems fail, in *Topics in Cryptology - CT-RSA 2002, The Cryptographer's Track at the RSA Conference, 2002, San Jose, CA, USA, February 18–22, 2002, Proceedings* (2002), pp. 17–29. https://doi.org/10.1007/3-540-45760-75C_2

[Cla07] C. Clavier, Secret external encodings do not prevent transient fault analysis, in *Cryptographic Hardware and Embedded Systems - CHES 2007, 9th International Workshop, Vienna, Austria, September 10-13, 2007, Proceedings* (2007), pp. 181–194. https://doi.org/10.1007/978-3-540-74735-2_13

[Nah15] P.S. Nahid Farhady Ghalaty Bilgiday Yuce, Analyzing the efficiency of biasedfault based attacks. Cryptology ePrint Archive, Report 2015/663 (2015), https://eprint.iacr.org/2015/663

[KPB+17] S.V.D. Kumar, S. Patranabis, J. Breier, D. Mukhopadhyay, S. Bhasin, A. Chattopadhyay, A. Baksi, A practical fault attack on arx-like ciphers with a case study on chacha **20**, 33–40 (2017). https://doi.org/10.1109/FDTC.2017.14

[BBD+20] A. Baksi, J. Breier, X. Dong, C. Yi, Machine learning assisted differential distinguishers for lightweight ciphers. IACR Cryptology ePrint Archive (2020), p. 571, https://eprint.iacr.org/2020/571

Chapter 9
To Infect or Not to Infect: A Critical Analysis of Infective Countermeasures in Fault Attacks

As fault-based cryptanalysis is becoming more and more of a practical threat, it is imperative to make efforts to devise suitable countermeasures. In this regard, the so-called "infective countermeasures" have garnered particular attention from the community due to its ability in inhibiting differential fault attacks without explicitly detecting the fault. We observe that despite being adopted over a decade ago, a systematic study of infective countermeasures is missing from the literature. Moreover, there seems to be a lack of proper security analysis of the schemes proposed, as quite a few of them have been broken promptly. Our first contribution comes in the form of a generalization of infective schemes which aids us with a better insight into the vulnerabilities, scopes for cost reduction and possible improvements. This way, we are able to propose lightweight alternatives of two existing schemes. Further, we analyse shortcomings of LatinCrypt' 12 and CHES' 14 schemes and propose a simple patch for the former.

9.1 Introduction

Fault attacks are becoming a real threat particularly to small-scale devices performing a cryptographic operation (see Chap. 3 for more information on fault attack). This type of attack forces a certain device to work under sub-optimal condition resulting in erroneous calculations, which is then exploited. Differential Fault Analysis or Differential Fault Attack (DFA) [BS97] (see Sect. 3.5.1 for more information), one type of fault attack, is predominantly used against symmetric key ciphers. Most, if not all, ciphers which are considered secure against classical attacks are shown to have severe weaknesses against DFA. This attack works by injecting a difference (fault) during the cipher execution, which normally results in flipping one or more bits of a register. Then, after analysing the output difference of the non-faulty and the

A. Baksi, *Classical and Physical Security of Symmetric Key Cryptographic Algorithms*, Computer Architecture and Design Methodologies,
https://doi.org/10.1007/978-981-16-6522-6_9

faulty outputs of the cipher computations; the attacker, Eve, is often able to deduce information on the secret key.

Success of DFA also has given rise to a series of works attempting to protect ciphers from this attack. Various types of countermeasures are proposed in the literature. All of these countermeasures rely on full or partial redundancy either in device, cipher implementation or the protocol. Broadly, the state-of-the-art schemes can be classified into three categories (more information in this regard can be found in Sect. 3.7):

(1) Using a separate, dedicated device. They can be either active which uses a sensor to detect any potential fault, such as [HBB16]; or passive, where a shield to block external interference [BCN+04] is used.
(2) Using redundancy in computation. These type of countermeasures commonly duplicate (fully/partially) the circuit, followed by a certain procedure which dictates what to do in case a fault is sensed directly or indirectly.
(3) Using protocol level technique. Here, the underlying protocol ensures that the conditions required for a successful fault happens with low probability [BBB+18].

Our interest lies in the second category of fault protection (further elaborated in Sect. 9.2). In particular, we focus on the so-called *infective countermeasures* (or, *infection-based countermeasures*). These countermeasures work by doing two computations of the same cipher (which we call, *actual* and *redundant*); thereafter computing the difference (XOR) between them (we denote it by Δ). This difference serves the purpose of implicitly detecting the fault—a non-zero difference implies a fault injection. This difference is further processed to produce a random vector in an unintelligible manner. This random vector is then used to corrupt (*infect*) output from the actual computation, which is then made available to Eve. Hence, the attacker gets the original (non-faulty) output if no fault is sensed by the countermeasure; or gets a random output otherwise—thus she has no meaningful information on Δ. This makes DFA impossible to mount as it requires the knowledge of Δ.

This idea of infection is proposed to counter the shortcomings of the previously proposed *detective countermeasures* (also referred to as *detection-based countermeasures*, see Sect. 9.2.2 for more details). Incidentally however, most of the infective countermeasures proposed in the literature have been broken quite soon (it is even stated in [BG13] that, "it is very difficult to design a secure infective countermeasure"). In fact, after more than a decade of their first introduction, we only have a handful of the schemes which are not broken—and yet, they generally have heavy implementation cost. Our observation is that almost all of the schemes proposed in this context are ad hoc solutions, instead of utilizing already established design standards. This motivates us to look further down into the topic to gain better understanding of the solution that the infective countermeasures intend to provide. To do a more systematic and complete study on the designs proposed, we first categorize them. Following this, we revisit the design choices to explore vulnerabilities and/or improvements.

Our Contributions

- We categorize the existing countermeasures into two types, so that a more systematic and comprehensive study is possible. This categorization is done from a cipher designer's point-of-view, which is missing in the existing literature. For convenience, we call these *Type I* (Sect. 9.2.4) and *Type II* (Sect. 9.2.4) countermeasures. The basic difference between them is that Type I schemes allow the full computations of the actual and redundant ciphers to run, then computes Δ; in contrast, Type II schemes compute the difference after each round. Type I countermeasures are further divided into two sub-categories based on existing literature; *multiplication-based* (Sect. 9.3.1) and *derivative-based* (Sect. 9.3.2). Type II countermeasures are also referred to as *cipher-level* countermeasures (described in Sect. 9.4).
- Among the schemes in the Type I category, we find two schemes which are not broken. Although, we do not find any attack to those schemes, we remark that, both of them incur a substantial hardware or software overhead. In this regard, we propose two efficient alternatives to the countermeasure described in [LRT12] (Sect. 9.3.3). Also, we propose software-friendly options for the scheme proposed in [GSS+17], (Sect. 9.3.3) based on existing ciphers.
- For the Type II schemes, we show (in Sect. 9.4.1) that CHES' 14 infective countermeasure proposed in [TBM14] is not an infective countermeasure at all; thereby refuting its security claim and showing its weakness against DFA. This also refutes the security claim made by the modified infective countermeasure in [PCM15]. The basic principle of this document is the same as that of [TBM14]; certain modifications are done on [TBM14] to make it better resilient against instruction skip attacks on a microcontroller.[1]
- Interestingly though, this CHES' 14 scheme is proposed as an improvement on the first cipher-level infective countermeasure presented in [GST12] (LatinCrypt' 12), which is broken in [BG13, TBM14]. We propose a simple patch (in Sect. 9.4.2) on the [GST12] scheme that resists the attacks presented in [BG13, TBM14].

In context of the security proofs of our countermeasures/patch, we would like to note that no separate security proof would be required. Since we reuse the existing cipher design paradigm, the security proof is the same as the cipher itself.

As for the practical evaluation of our schemes, we note the following. Our proposed countermeasures are designed at the algorithmic level. Hence, anyone with the access to an experimental set-up can validate. Any hidden vulnerability in the implementation which is transparent at the algorithmic level is not within the scope.

[1] An instruction skip is considered a separate (non-DFA) type of fault.

9.2 Background

9.2.1 Context of Differential Fault Analysis

As mentioned earlier, DFA works by injecting a difference (fault) during cipher computation. This fault injection can be done by various methods; clock/power glitch, LASER shot, to name a few. More information on such fault-injecting equipment can be found, for example, in [BBK+12]. This fault, in effect, results in a few bit(s) flip of the cipher at a round. Normally, attacker can choose the round, but unable to precisely target the words; as a result, she does not know the actual fault value (the value which is effectively XORed with the cipher state). This is a very commonly employed model and termed as random fault model [PQ03, SH13]/Sect. 3.2.3.

Now, flipping one/few bit(s) of a register works as a simplified version of the classical Differential Attack (DA) [BS90]. In DA, the difference is inserted through the chosen plaintexts (hence, works at the beginning of the cipher execution). In contrast, in DFA, the difference can be inserted at any point of time during execution (normally it is inserted near the end of the cipher execution). Now, since the difference passes through a small number of rounds of the cipher (can be even one round), the resistance against differential attack is not very strong. At the end of the cipher execution, when attacker gets the corresponding output difference, an analysis similar to DA may reveal secret information.

Although, DA and DFA work very similarly, one may notice that the same idea used to thwart DA cannot be potentially used to thwart DFA. The DA protection arises from many iterations of the cipher; which is meaningless in DFA, as attacker is able to attack any round near the end. Hence, the solutions proposed to protect against DFA require certain assumption on the underlying device/communication protocol, rather than completely relying on the cipher description (which is the case for DA).

We assume that Eve can inject faults temporarily (the fault values are not permanent—the device goes back to its normal situation once the source of fault is revoked). This contrasts to the hard fault model where particular bit(s) are permanently stuck with the fault value (one may refer to [Cla07] for an example). The stuck-at model assumes more control for the attacker and is outside the scope of DFA. Like this, modification of operation is also a strong attacker model (such as an instruction skip on a software implementation, as in [KPB+17]), and also not a DFA. In short, DFA assumes the transient fault model where only the operands are subject to alteration.

9.2.2 Early Countermeasures: Detection-Based

One of the earliest countermeasures proposed against DFA is known as detective countermeasure (or detection-based countermeasure). Conceptually, it checks

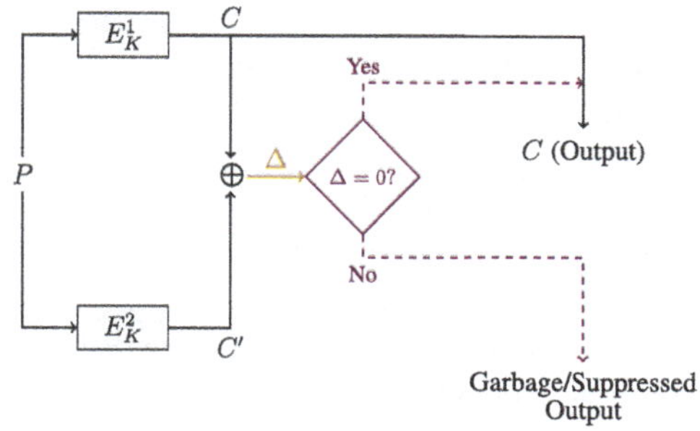

Fig. 9.1 Overview of detective countermeasures

whether any fault is injected by explicitly checking whether $\Delta = 0$. If a fault is detected (i.e., $\Delta \neq 0$), it blocks the device from producing the faulty output. In other words, either the output from the cipher is suppressed, or an invalid signal ($\perp$) is generated, or a random output is generated. This stops the attacker from getting any meaningful information regarding the faulty state, which makes DFA impossible. Ideas in this direction are commonly generated from coding & information theory, such as linear parity [KKG03], non-linear (n, k) codes [KKT05], etc. Fig. 9.1 (which is adopted from [BKK+20'2, Fig. 1]) shows the overall idea (Sect. 9.2.4 describes the notations).

9.2.3 *Evolution of Infective Countermeasures*

It is commonly argued that the concept of explicit equality checking in detection-based countermeasures is subject to bypassing that step (e.g., [KQ07] or [SSS+03, Sect. 3]). Generally, such comparisons rely only on one bit (like the zero flag in a microcontroller); any attack that is able to flip this bit renders the whole countermeasure useless.[2] Although, injecting two faults in one invocation of cipher is not a model commonly used; it is commented in [Ott04] that, 1-bit flip is an incidence which may happen "by chance". If the attacker is able to flip this particular bit, even with low probability, the entire security provided by the detection-based countermeasure would be nullified. In general, the motivation for the improved version of the detection-based countermeasures has come from removing the single point of failure. It is stated in [SSS+03, Sect. 3]: "... the usage of a decision procedure violates the guarantee of developing a highly reliable hardware fault immune cryptographic protocol".

The idea of *infection*, which avoids the dependency of security of one bit, is first proposed in the context of public key cryptography [SSS+03]. This proposal is claimed to be weak (see [YK04]), though, this idea is followed soon after in the public key cryptography such as [JC05, BO06].

[2] This case is different from the fault injection flipping only one bit.

The concept is adopted in symmetric key setting [JMR07, FRB+11, GST12, LRT12, TBM14, GSS+15, PCM15]. The infective countermeasures proposed in [JMR07, FRB+11] do not involve randomness, and are attacked in [LRT12]. The authors in [LRT12] also speculate, randomness may be required in such countermeasures; although they do not present any formal proof. Anyway, all infective countermeasures proposed thereafter adopt this idea and use randomness. Still, most schemes proposed in the literature are broken soon. In fact, in our literature survey, we observe that basically three different schemes are proposed which are not considered broken by DFA.

9.2.4 Notations and Terminologies

Before proceeding further, we define terms and notations that we use the subsequent parts.

- *Actual & redundant computations.* As mentioned earlier, infective countermeasures require two computations of the same cipher. These two are referred to as actual and redundant computations; and symbolically denoted as $C = E^1_K(P)$ and $C' = E^2_K(P)$, respectively (E is the underlying cipher parameterized by the secret key K with input P, and the superscript denotes the two separate executions). Output from the actual cipher is later infected and made available to the attacker. The notations, C and C', are used in this work to denote the output from the actual and redundant computations, respectively. When Δ is computed after the full iteration of both the actual and redundant computations are finished; i.e., in Type I countermeasures; $\Delta = C \oplus C'$.
  Also, we assume that the attacker is able to exactly repeat the same fault, in same location and during the same round on one particular device; as many times she wants (as long as it is practical). Thus, we give the attacker to exactly repeat the same fault over temporal domain, and to make Δ constant.[3]
- η. We denote by η the total number of rounds of the cipher; e.g, for `AES`, $\eta = 11$ (counting the initial `AddRoundKey` as a separate round).
- n. We use n to denote the block size of the cipher; e.g., $n = 128$ for `AES`.
- $\texttt{RoundFunction}_j(\cdot)$, $j = 1, \ldots, \eta$. We use $\texttt{RoundFunction}_j(p)$ to denote the jth round function of the underlying cipher with the corresponding input p, $j \in \{1, \ldots, \eta\}$. Note that, it does not involve the round key insertion.
  For example, in case of `AES`; $\texttt{RoundFunction}_1(p) = p$; $\texttt{RoundFunction}_j(p) = \texttt{MixColumns}(\texttt{ShiftRows}(\texttt{SubBytes}(p)))$ for $j = 2, \ldots, 10$ (Fig. 9.2 shows a pictorial description); and $\texttt{RoundFunction}_{11}(p) = \texttt{ShiftRows}(\texttt{SubBytes}(p))$. Basically, when the jth round key k_j is inserted with $\texttt{RoundFunction}_j(\cdot)$, it gives the actual

[3] We assume the most common fault attack model, where the fault value is constant but unknown to the attacker; but the same fault can be repeated over time by keeping the source of the fault unchanged. Such model is used, for example, in [BB15].

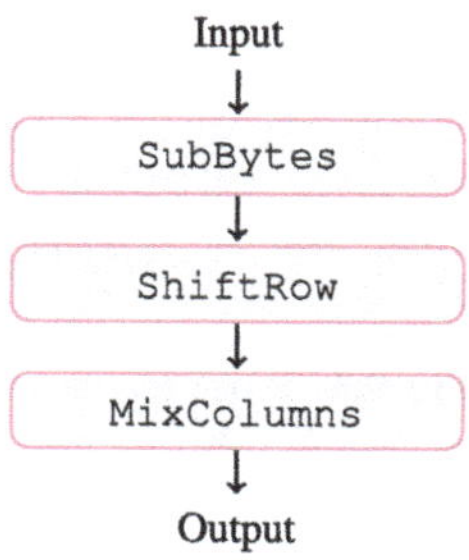

Fig. 9.2 One standard (un-keyed) round of AES-128 encryption

jth round of the cipher. In other words, $\texttt{RoundFunction}_j(\cdot)$ is the internal diffusion within the state of a cipher (not involving the round key).

- $\texttt{RoundFunction}_0(\cdot)$. By $\texttt{RoundFunction}_0(p)$, we denote the standard (most frequent) round of a cipher, with input p (without involving the round key). In case of AES, $\texttt{RoundFunction}_0(p) \equiv \texttt{RoundFunction}_2(p)$.
- ξ. Generally, cipher designers are conservative in the sense that the number of rounds (η) in a cipher is kept more than that of what would be required to reach a practical security. Often, we may not need full η rounds of iteration to ensure a practical security. Instead, with less than η iterations of $\texttt{RoundFunction}_0(\cdot)$, together with corresponding round key insertions; one can achieve good resistance against classical cryptanalysis techniques such as differential or linear attacks. We denote, by ξ, the minimum number of iterations required for a cipher to offer a practical security.
  For example, consider a reduced version of AES with 4 $\texttt{RoundFunction}_0(\cdot)$ rounds, together with corresponding AddRoundKeys. It can be proven to have no differential path with probability better than 2^{-113} [KS07]. Hence, we take $\xi = 4$ for AES.
- R. By R, we denote an n-bit random vector. When more than one random vectors are used, we denote them by $R_0, R_1, \ldots$, respectively. These vectors are generated using entropy external to the cipher computation, and hence uncontrollable & unknown to the attacker. The numeric values of R_i's change at every invocation of the countermeasure but fixed during the course of one invocation of the countermeasure.
- $\vec{0}$. We use $\vec{0}$ to denote an n-bit vector of all 0 bits.
- $\vec{1}$. Similarly, we use $\vec{1}$ to denote an n-bit vector of which most significant $n-1$ bits are 0 and the least significant bit is 1.[4]

[4] This definition of $\vec{1}$ is different from the same notation in Sect. 4.3.2.

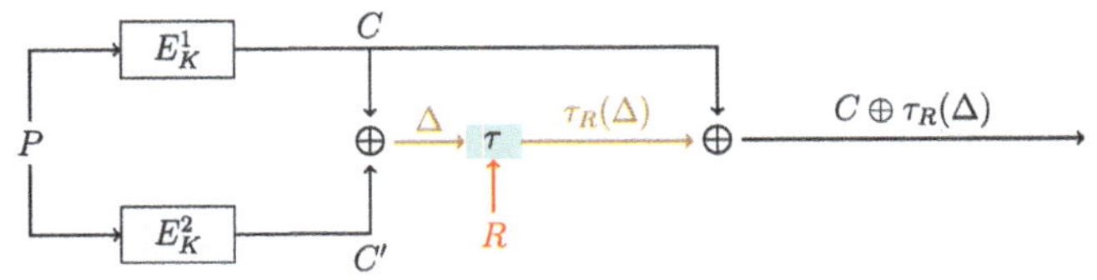

Fig. 9.3 Type I infective countermeasures

Type I

This type of countermeasures allow the two computations, $E_K^1(\cdot)$ and $E_K^2(\cdot)$ to finish their iterations full (η) iterations, the difference Δ is then obtained by XORing the outputs from the actual (C) and the redundant cipher computations (C'). Figure 9.3 shows a pictorial view.

Then, Δ is passed through a function $\tau(\cdot)$, which is parametrized by R. This $\tau_R(\Delta)$ is such that, $\tau_R(\vec{0}) = 0 \; \forall R$; and for any $\Delta \neq 0$, the distribution of $\{\tau_R(\Delta)\}$ is indistinguishable from the uniform distribution over $\mathbb{F}_2^n$.

Notice that, the attacker should not be able to deduce the input of $\tau_R(\cdot)$ except the case when output of $\tau_R(\cdot) = 0$. If, somehow, the attacker is able to do so, then she can find out Δ and hence can perform a DFA. So, one intrinsic property of $\tau_R(\cdot)$ is that, it is not invertible, except $\tau_R(\vec{0}) = 0$.

The output from $\tau_R(\cdot)$ is XORed with C (the output from the actual computation), and this is made available. Hence, the attacker gets, $C \oplus \tau_R(\Delta)$, which can be interpreted as the infected output from the actual cipher.

Examples of this type of countermeasures include the schemes in [LRT12] ($\tau_R(\Delta) = R \cdot \Delta$ over GF(2^n) multiplication) or [GSS+17] ($\tau_R(\Delta) = N(R) \oplus N(R \oplus \Delta)$, where N is defined as non-linear hybrid cellular automata).

Type II/Cipher Level

This new type of infection is introduced in [GST12], and also referred to as the cipher-level countermeasure. A schematic view is given in Fig. 9.4. The schemes belonging to this type rely on the diffusion property of the underlying cipher to spread the infection. Instead of letting both the ciphers to run full (η) rounds, these schemes try to identify the effect of a fault at an intermediate round.

Say, the registers S_0 and S_1 are computing the actual and redundant computations, respectively. In the *actual round* (respectively, *redundant round*), S_0 (respectively, S_1) is updated only. A *meaningful round* refers to either an actual or a redundant round. Then the XOR difference, δ, is computed ($\delta = S_0 \oplus S_1$)[5]; and this δ goes through a function $\sigma(\cdot)$. This function, $\sigma(\cdot)$ is such that, it outputs $\vec{0}$ only when its input is $\vec{0}$. As for the choice of $\sigma(\cdot)$, inversion in GF(2^8) is proposed per SBox

[5] In our patched version of [GST12] (Sect. 9.4.2), we do not explicitly compute δ.

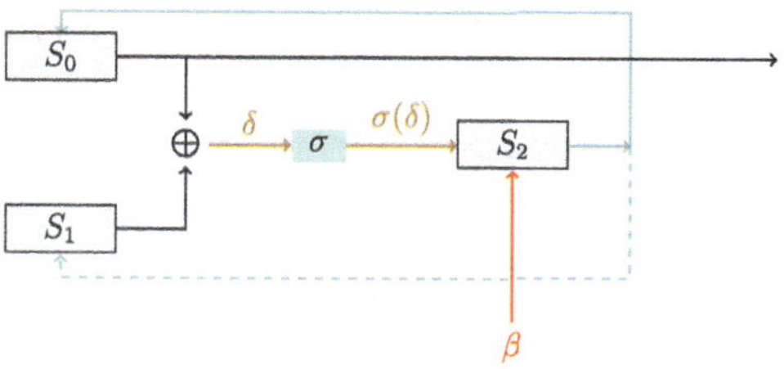

Fig. 9.4 Type II infective countermeasures (schematic)

of AES in [GST12]. The choice of $\sigma(\cdot)$ in [TBM14] is the $n:1$ OR gate, although the authors introduce/use the term BLFN instead of OR gate. The same choice for $\sigma(\cdot)$ with the same term is later used in [PCM15] as well as in its journal extension [PCM17]. The scheme also takes a random n-bit vector β such that, $\exists k_0$ for which, $k_0 \oplus \texttt{RoundFunction}_0(\beta)$ gives β. See Sect. 9.4 for more details on Type II countermeasures.

One special register, called the *dummy register*, denoted by S_2 here, is initialized with β. In the so-called *dummy rounds*, this register S_2 is updated only (so, this is not a meaningful round). This register S_2 is updated (with influence from δ) by the following rule:

$$S_2 \leftarrow \begin{cases} \beta & \text{if no fault is injected,} \\ \beta' \ (\neq \beta) & \text{if fault is injected.} \end{cases}$$

During the subsequent meaningful rounds, S_0 and S_1 are updated with influence from S_2. So, in case of a fault, the contents of S_0 and S_1 are infected, which propagates further in the subsequent rounds. This ideally makes the final output, S_0, random in case of a fault. In case of no fault, the content of S_0 and S_1 are not deviated from its actual computation—this ensures a proper execution of the cipher. In our patched version of the LatinCrypt' 12 countermeasure (proposed in [GST12]), we do not update S_0 and S_1 during meaningful rounds.

To determine whether a dummy or a meaningful round will take place, the authors in [GST12] propose to use a random bit, λ: $\lambda = 0$ means a dummy round, $\lambda = 1$ means a meaningful round. Also, one counter i is looped from 1 to 2η; within this loop, $\texttt{RoundFunction}_j(\cdot)$'s, $(1 \leq j \leq \eta)$ are performed, together with the corresponding round key insertions. Further, when this is a meaningful round (i.e., $\lambda = 1$), i is even implies an actual round takes place, and i is odd implies a redundant round takes place. We make the ordering of actual, redundant, and dummy rounds deterministic in the patched version; so we do not use λ. In both [GST12, TBM14], i is initialized by 1 (odd); which means, a redundant round always precedes the corresponding actual round. In the modified scheme in [PCM15], however, the order of execution of actual and redundant rounds are not predetermined.

9.2.5 Necessity and Sufficiency of Randomness

It is well known from [LRT12] that, randomness is required in infective countermeasures. However, they do not present any formal proof; rather, their comment is more of an informal case study with AES-128 encryption.

The attacker basically exploits the information she gains from DFA to derive the key K. If Z is the information obtained from DFA, then the amount of information available about the key can be measured by the mutual information between random variables K and Z:

$$I(K; Z) = \sum_{k \in K} \sum_{z \in Z} \Pr(k, z) \log \left(\frac{\Pr(k, z)}{\Pr(k) \Pr(z)} \right).$$

It is easy to see that $I(K; Z) = 0$ if and only if K and Z are independent. Therefore, it is sufficient that Z is uniformly distributed to make DFA fail.

For instance, in Type I infective countermeasures, the attacker gets a $Z = C \oplus \tau_R(C' \oplus C)$ as an output. Therefore, Z must be random in order to be independent of key K. Thus, for infective countermeasure, it is necessary and sufficient to have $\tau_R(C' \oplus C)$ random. On the other hand, in the detective countermeasure,, once we detect the faulty cipher, we can simply output Z as constant. Here, randomness of Z is not necessary (random Z works too).

Detection-based countermeasures set the mutual information zero by suppressing the (faulty) output. In contrast, infection-based countermeasures apply one-way functions on the output difference (which contains non-zero mutual information regarding the secret key) to reduce the mutual information to zero.

Non-zero mutual information between K and Z could imply an information leakage. For example, the mutual information between the XOR of the non-faulty and the faulty ciphertext, with the key is 32-bits in case of the diagonal fault attack on AES[SMC09]. For the design choice of Remark 9.1, the mutual information is of 1-bit in case of a fault, hence we modify it to Algorithm 9.2(b) (Sect. 9.3.3). More quantitative analysis in this direction can be found in [KSK14], where the authors use CLEFIA as the target cipher for various DFA approaches.

In all our proposed countermeasures/patch, the mutual information between the key K and the observation Z which is available to the attacker is zero (in case of a fault). Therefore, the attacker does not get any information regarding the key in case of a fault.

9.2.6 Scope and Applicability

The working procedure of infective countermeasures relies on the (non-zero) difference between the actual and redundant computations of the cipher. Hence, if both the actual and the redundant computations are infected by identical faults, which result in the same output in both cases, the corresponding difference will be zero. In this case, the countermeasure will treat this as non-faulty; and make the faulty output (from the actual cipher) available to the attacker (without infection). Hence, this type of repeated faults can be used to make the countermeasure invalid. We call this type of faults a double fault (see Sect. 3.7).

However, repeating the identical fault in spatial domain would require a very strong adversary model. Such model, although shown to be effective through experiment (the only case we know is, [SHS16]), is not common. In a more common model, the authors assume that the attacker can repeat the fault in time domain (such as, the model used in [BB15] to break an infective countermeasure). Dealing with double faults is rather tricky in infective countermeasure; and we leave this problem open for future research.

Besides double faults, infective countermeasures cannot provide safeguard against few other fault models; for example, Ineffective Fault Attack [Cla07]/Sect. 3.5.2. Another type of fault attack, known as Collision Fault Attack (CFA) [BK06]/Sect. 3.5.2, injects transient fault near the start of the cipher (E) execution. This attack works very similarly to DFA. Suppose, one computation is allowed to run as-is, where the other computation is injected with a fault (near the beginning of the cipher execution). If, it happens that, both the faulty and non-faulty outputs are equal; then the situation is similar to attacking the inverse of the cipher (E^{-1}) by DFA (as the fault can be thought to be injected near the end of execution of E^{-1}). The attacker may be able to deduce information regarding the early rounds of the cipher (or equivalently, the later rounds of E^{-1}). This may eventually help her to find information on the secret key (by an analysis similar to DFA on E^{-1}). We note that, while such model cannot be protected by Type I countermeasures (as both the actual and redundant computations produce equal output); our patched version of the LatinCrypt'12 countermeasure (Sect. 9.4.2) can indeed protect against such an attack.

Impeccable Circuits/CRAFT

The authors in [AMR+18] presented a novel idea, called *impeccable circuit*, which allows fault detection using error detecting codes. The idea is later extended to a full-fledged cipher named CRAFT [BLM+19]. Impeccable circuits/CRAFT can detect faults up to some extent (the fault coverage can be changed by changing the underlying error detection code). Also, both the constructions can detect double fault.

In comparison to that, an infective countermeasure works on a different principle. It does not rely on an error detecting/correcting code, but relies on the difference of the final/intermediate output. While the impeccable circuits/CRAFT cannot detect faults beyond the permissible coverage of the error detecting code; an infection countermeasure can take action any such fault, except double fault. However, we note that an infection scheme can be made protected against double faults by employing a similar concept (different encodings for the actual and the redundant computations).

Statistical Ineffective Fault Attack (SIFA)

SIFA [DEK+18]/Sect. 3.5.4/Sect. 10.3 is a type of IFA, which also makes use of statistical information. Infective countermeasures do not have any inherent protection against SIFA (or other IFA).

At the same time, it is worth noticing that, SIFA does not necessarily phase out infective countermeasures in general. We point out the following use-case for infective countermeasure in the presence of a SIFA model. SIFA requires a large number of faults (typically of the order of thousands), implying an extensive control over the target device. DFA, on the other hand, has been shown to work with as low as only one fault. Hence, SIFA is more complex in nature compared to DFA. So, there could be scenarios where the attacker is unable to apply SIFA may be able to mount DFA; implying the necessity of a DFA protection.

Basically, the applicability for our countermeasures can be compared to that of `CRAFT` [BLM+19]. It may be noted that, although `CRAFT` is published after SIFA, it is not protected against SIFA (this is acknowledged by the authors of `CRAFT` in [BLM+19, Sect. 1]).

For existing countermeasures against SIFA, one may refer to Sect. 3.7.6 and also to Chap. 10. It can be noted that the duplication-based countermeasure proposed in [BKK+20'2] actually relies on a detection-based approach. Instead of detection, infection can be used, if additional protection is needed.

In a nutshell, we highlight the fault attacks that are generally considered outside the scope of an infective countermeasure, together with corresponding countermeasure(s):

- **Stuck-at/hard.** Since the algorithmic description of the countermeasure can be altered in cases like stuck-at or hard, we recommend to employ some dedicated device that is able to thwart a fault injection (see Sect. 9.1).
- **Double fault.** Injecting identical faults (the so-called double fault) at both the actual and the redundant computation can skip the infection mechanism. This can be avoided by using different encodings for the two computations (e.g., similar to `CRAFT`). If the attacker cannot inject identical faults to more than two computations, then more redundant computations (such as, triplication [BBK+10]) can be used.
- **IFA (SIFA).** Any ineffective fault analysis, including SIFA, can bypass the infective mechanism as an ineffective fault does not result in a change of the target state. In such a case, one can employ some form of error correction and/or a masking countermeasure as used in the recent SIFA countermeasures [BKH+19, DDE+19, SRM19, BKK+20'2, SJR+20] (see Sect. 10.3.2 for more details).

It may be noted that, such countermeasures either are beyond the control of a cipher designer; or can be non-trivial/costly to design or implement compared to an infective countermeasure. Hence, in cases where only DFA protection is solicited, infective countermeasures can be used.

9.2.7 Connection with Side Channel Countermeasures

Side channel countermeasures are not directly considered in this work, as we focus on DFA. Here, we would like to note that SCA is already known quite powerful. If the attacker has access to side channel information then she can choose the degree to which she can use it to obtain further information—she should not be limited to recover only Δ. More precisely, she can target other potentially useful registers as well; which may allow her to recover the secret key directly, thereby a separate DFA would not be required. So, if a side channel adversarial model is considered within the scope, the infection component as well as both the actual and the redundant computations are required to be protected by some SCA countermeasure. Hence, we implicitly assume the actual and the redundant computations are already protected, and comment on protecting the infection component here.

In general, the hiding countermeasures [MOP07, Chap. 7] that aim at decreasing the SNR (by increasing noise, or decreasing signal) require minimal changes to the underlying cipher, as the countermeasures are applied at the implementation level. Hence, we are confident that adopting such a countermeasure would be relatively straightforward on top of our countermeasures.

On the other hand, the masking countermeasures [MOP07, Chap. 9]/Sect. 4.2.5 rely on randomizing the intermediate variables so that the power leakage is independent of those variables. Masking is applied on the algorithmic level, so applying it to the infective countermeasures could be of interest. In this regard, we note the following observations:

- For the multiplication-based schemes (Sect. 9.3.3), masking can be applied by following the approach similar to [LRT12]. This is done in Algorithm 9.3(a) and (b) (Sect. 9.3.3).
- For the derivative-based scheme (Sect. 9.3.3), the underlying cipher is reused as the building block. Therefore, only the difference (Δ) needs to be protected by masking. This can be done, for example, by computing $\Delta' = C \oplus C' \oplus R'$ and redefining $\tau_R(\Delta)$ as $\tau_R(\Delta', R') = E_R(R') \oplus E_R(\Delta')$, where $R' \stackrel{\$}{\leftarrow} \mathbb{F}_2^n$ with $R' \neq C \oplus C', \vec{0}$. The additional cost incurred for this protection would be that of an n-bit $\oplus$ operation and generation of an n-bit random vector R'.
- For the patched version of the LatinCrypt'12 countermeasure (Sect. 9.4.2), Eve can only target those steps where S_2 is updated (Lines 5, 6, 7, 9 of Algorithm 9.6), as S_0 and S_1 are presumably SCA protected. However, one may notice that S_2 is initialized with β (which is randomly generated) (Line 1). Hence, β inherently works as a mask to S_2, and no other masking would be necessary.

9.3 Type I Constructions

We describe the Type I schemes here with adequate details. As already mentioned, the existing schemes in Type I category can be classified in two sub-categories; we refer to them as multiplication based (Sect. 9.3.1) and derivative-based (Sect. 9.3.2). We then propose our constructions of Type I constructions (Sect. 9.3.3) followed by the corresponding benchmarking results (Sect. 9.3.4).

9.3.1 Multiplication-Based Constructions

To the best of our knowledge, the earliest infective countermeasure, which is still unbroken, is based on the GF(2^n) multiplication proposed in [LRT12] (see Algorithm 9.1). In our terminology, here $\tau_R(\Delta) = R \cdot \Delta$, where $(\cdot)$ refers to a GF(2^n) multiplication and given $R \neq \vec{0}, \vec{1}$. If $R = \vec{0}$, then C is available without infection to the attacker. On the other hand, if $R = \vec{1}$, then she gets $C \oplus \Delta$ as the output; from where she can compute Δ (as she knows C).

Hence, for AES, one has to implement a GF(2^{128}) multiplication. However, the authors acknowledge that the GF(2^{128}) multiplication is costly in hardware, although they do not provide any benchmarking result. So, they come up with the idea of substituting the GF(2^{128}) multiplication by 16 independent GF(2^8) multiplications (which replace the GF(2^{128}) multiplications in Lines 1, 2, 5 of Algorithm 9.1). The authors claim, the security of the scheme will remain unchanged, given those sixteen random multipliers are independent.

Algorithm 9.1 Multiplication based infective countermeasure: FDTC'12

Input: $C, C'; R_0, R_1, R_2$ ▷ $R_0, R_1 \neq \vec{0}$; $R_2 \neq \vec{0}, \vec{1}$
Output: C if no fault; random, otherwise
1: $a \leftarrow R_2 \cdot (C \oplus R_0)$ ▷ $\cdot$ refers to GF(2^n) multiplication
2: $b \leftarrow R_2 \cdot (C' \oplus R_1)$
3: $c \leftarrow a \oplus b$
4: $d \leftarrow R_0 \oplus R_1$
5: $e \leftarrow R_2 \cdot d$
6: $f \leftarrow (C \oplus R_0) \oplus c$
7: $g \leftarrow f \oplus e$ ▷ $g = (C \oplus R_0) \oplus R_2 \cdot (C \oplus C')$
8: **return** $g \oplus R_0$

While the original proposal remains unbroken so far; this lightweight alternative is broken soon afterwards [BG13]. We describe the attack here. Assume the attacker is able to replicate the same fault value in the temporal domain (i.e., the fault value is constant in all the injections). Under this assumption, the lightweight variant will restrict two particular values to output per GF(2^8) multiplication; corresponding to the cases when $R = \vec{0}$ or $\vec{1}$. So, per GF(2^8) multiplication, it only outputs 254 values. One of the missing values is the non-faulty output (C), whereas the other is the faulty output ($C \oplus \Delta$). As explained earlier, leaking $C \oplus \Delta$ can lead to a successful DFA. This breaks the security claim of the lightweight proposal, as attacker can exhaust

254 cases (then repeat the procedure till all the key bytes are recovered). For the original GF(2^{128}) multiplication for AES, it would require 2^{128} repeats of the same fault and storage; making it impractical.

9.3.2 *Derivative-Based Constructions*

The generic schemes of the derivative-based category follow the concept of the Boolean derivative. More precisely, they use a non-linear function $N(\cdot)$ and compute $\tau_R(\cdot)$ as the derivative of $N(\cdot)$ at Δ:

$$\tau_R(\Delta) = N(R) \oplus N(R \oplus \Delta).$$

In [GSS+15], the authors proposed such a construction by choosing a quadratic N. However, this scheme was subsequently broken in [BB15]. Recall from Sect. 9.3 that $\tau_R(\cdot)$ has to be non-invertible. This claim is shown to be incorrect in [BB15], under the assumption that attacker can keep Δ a constant. Note that derivative of a quadratic function is affine; further, the attack generates a sequence where in each bit, there is only one term of the form $\Delta_i r_i$, and rest of terms are independent of R. So for a constant Δ, it easy to find the value of r_i by seeing the distribution of these bits. Later, the authors of [GSS+15] opt for more complex and high degree function $N(\cdot)$ based on non-linear cellular automata in [GSS+17]; and this is the only yet unbroken proposal in this category, to the best of our knowledge.

9.3.3 *New Type I Schemes*

With the backdrop presented, we now introduce our proposals. The novelty of our proposals lies in the simplicity and lower implementation cost.

Multiplication-Based

As GF(2^n) multiplication has a significant hardware (where n is generally 128), we propose two lightweight alternatives. First we present a scheme that requires a significant amount of randomness in Algorithm 9.2(a). It generates n fresh random vectors (each of which is of n bits) $R_0, \ldots, R_{n-1}$ (none of which equals to $\vec{0}, \vec{1}$). They are used to generate a random $n \times n$ binary matrix M, which is then multiplied (over GF(2)) with Δ to constitute $\tau_{(\cdot)}(\cdot)$.

Since this scheme requires a total of n^2 bits of entropy; therefore, it may not appear suitable where frequent random number generations is not easy. Hence, instead of using n^2 random bits, we next propose another scheme that uses $2n$ bits of entropy, which is described in Algorithm 9.2(b). We define the ith *cyclic rotation*, $\rho^i(\cdot)$, on an n-bit vector $\vec{a} = (a_0, a_1, a_2, \ldots, a_{n-1})$, recursively as

$$\rho^0(\vec{a}) = \vec{a};$$
$$\rho^1(\vec{a}) = (a_{n-1}, a_0, a_1, \ldots, a_{n-2});$$
$$\rho^i(\vec{a}) = \rho^{i-1}(\rho^1(\vec{a})) \text{ for } i = 2, \ldots, n-1.$$

For example, with the vector $\vec{a} = (a_0, a_1, a_2, a_3)$, we have $\rho^1(\vec{a}) = (a_3, a_0, a_1, a_2)$; $\rho^2(\vec{a}) = \rho^1(\rho^1(\vec{a})) = \rho^1(a_3, a_0, a_1, a_2) = (a_2, a_3, a_0, a_1)$; and so on. Once we generate an n-bit random vector R_0 ($\neq \vec{0}, \vec{1}$), we create an $(n-1) \times n$ binary matrix M whose ith row is the ith cyclic rotation of R_0 (for $i = 0, \ldots, n-2$). Following this, we augment another randomly generated n-bit binary vector R_1 ($\neq \vec{0}, \vec{1}$) as the last row to M (to make it an $n \times n$ binary matrix). Then, we multiply M and Δ over GF(2).

In both the algorithms, if the attacker can guess M, then she will be able to deduce information about Δ. However, guessing M succeeds with probability 2^{-n^2} for Algorithm 9.2(a), and 2^{-2n} for Algorithm 9.2(b). In terms of security, either of these two can be adopted.

Remark 9.1 One may notice, the structure of M in our second alternative (Algorithm 9.2(b)) is same as a circulant matrix, except the last row (it would be a circulant matrix if last row would be equal to $n-1$th cyclic rotation of R_0). However, we observe that circulant M reveals one bit of entropy of Δ. Suppose, $\Delta = (\Delta_0, \Delta_1, \Delta_2, \ldots, \Delta_{n-1})^\top$ and $R_0 = (r_0, r_1, r_2, \ldots, r_{n-1})$. So, we have

$$\tau_R(\Delta) = M \cdot \Delta$$
$$= \begin{bmatrix} r_0 & r_1 & r_2 & \ldots & r_{n-1} \\ r_{n-1} & r_0 & r_1 & \ldots & r_{n-2} \\ \vdots & \vdots & \vdots & \ddots & \vdots \\ r_1 & r_2 & r_3 & \ldots & r_0 \end{bmatrix} \begin{bmatrix} \Delta_0 \\ \Delta_1 \\ \vdots \\ \Delta_{n-1} \end{bmatrix}$$
$$= \left[\bigoplus_{i=0}^{n-1} r_i \Delta_i, r_{n-1}\Delta_0 \oplus \left(\bigoplus_{i=0}^{n-2} r_i \Delta_{i+1} \right), \ldots, r_0 \Delta_{n-1} \oplus \left(\bigoplus_{i=1}^{n-1} r_i \Delta_{i-1} \right) \right]^\top$$

XORing all the bits of $\tau_{R_0}(\Delta)$ will give: $\bigoplus_{i=0}^{n-1} \bigoplus_{j=0}^{n-1} r_i \Delta_j = \left(\bigoplus_{j=0}^{n-1} \Delta_j \right) \left(\bigoplus_{i=0}^{n-1} r_i \right)$ = (parity of Δ) AND (parity of R_0). Now, since we assume that the attacker is able to replicate the same Δ over multiple runs; hence, Δ is constant, so is parity of Δ. Also, R_0 is random, which means; its parity is 0 with probability $\frac{1}{2}$. Hence,

$$(\text{parity of } \Delta) \text{ AND } (\text{parity of } R_0) = \begin{cases} 0, \text{ with probability } 1 & \iff \text{parity of } \Delta \text{ is } 0 \\ 0, \text{ with probability } \frac{1}{2} & \iff \text{parity of } \Delta \text{ is } 1 \end{cases}.$$

So, just by XORing all the bits, attacker is able to deduce the parity of Δ. This gives the attacker 1-bit of mutual information between the observation $C \oplus a$ and the key, in case of a fault. For this reason, this idea is modified to Algorithm 9.2(b) which is free from this shortcoming.

Algorithm 9.2(a) Multiplication based infective countermeasure (1st variant)

Input: C; C' ▷ $|C|, |C'| = n$
Output: C if no fault; random, otherwise

$\Delta \leftarrow C \oplus C'$
  ▷ Δ is represented as an n-bit vector
for $i \leftarrow 0; i < n; i \leftarrow i + 1$ **do**
  $R_i \xleftarrow{\$} \mathbb{F}_2^n$
$M \leftarrow \begin{bmatrix} R_0 \\ R_1 \\ \vdots \\ R_{n-1} \end{bmatrix}$
$a \leftarrow M \cdot \Delta$
  ▷ Multiplication is over GF(2)
return $C \oplus a$

Algorithm 9.2(b) Multiplication based infective countermeasure (2nd variant)

Input: C; C' ▷ $|C|, |C'| = n$
Output: C if no fault; random, otherwise

$\Delta \leftarrow C \oplus C'$
  ▷ Δ is represented as an n-bit vector
$R_0, R_1 \xleftarrow{\$} \mathbb{F}_2^n$
$M \leftarrow \begin{bmatrix} R_0 \\ \rho^1(R_0) \\ \vdots \\ \rho^{n-2}(R_0) \\ R_1 \end{bmatrix}$
$a \leftarrow M \cdot \Delta$
  ▷ Multiplication is over GF(2)
return $C \oplus a$

Side Channel Protection (Masking). One may note that, the authors [LRT12] do not explicitly compute Δ from C and C'. This is because they are careful to avoid what they call *combined attacks*. Such attacks work by first injecting the fault to cause a non-zero Δ, then to recover Δ by side channel attacks (such as the power leakage). In this way, attacker can solve for Δ from the side channel information, the knowledge of Δ further helps to recover the secret key by utilizing DFA. The authors first mask C and C' by XORing them with two random vectors R_0 and R_1, respectively; then applying the GF(2^n) multiplications (Lines 1, 2 in Algorithm 9.1); and later cancelling the effect of unwanted masks. As our focus is on DFA in Algorithms 9.2(a) and 9.2(b), we do not consider masking Δ. In situation, where SCA protection is needed, the following Algorithms 9.3(a) and 9.3(b) can be used (as replacement for Algorithms 9.2(a) and 9.2(b), respectively). As noted in Sect. 9.2.7, the implicit assumption to apply Algorithms 9.3(a) and 9.3(b) is that the rest (the actual and redundant computations) is already protected against SCA. As for the overhead of the masking, n-bit $\oplus$ (eight times) and $\cdot$ (two times) are to be performed, apart from generating two n-bit random vectors.

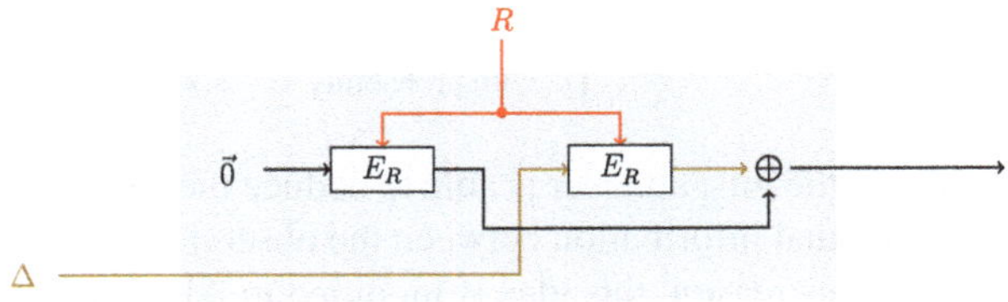

Fig. 9.5 Derivative based construction of $\tau_R(\Delta)$: Our design

Algorithm 9.3(a) Multiplication based infective countermeasure (1st variant) with masking

Input: C; C' ▷ $|C|, |C'| = n$
Output: C if no fault; random, otherwise
1: **for** $i \leftarrow 0; i < n+2; i \leftarrow i+1$ **do**
2: $R_i \xleftarrow{\$} \mathbb{F}_2^n$ ▷ $R_0, R_1 \neq \vec{0}$
3: $M \leftarrow \begin{bmatrix} R_2 \\ R_3 \\ \vdots \\ R_{n+1} \end{bmatrix}$
4: $a \leftarrow M \cdot (C \oplus R_0)$
▷ $\cdot$ refers to GF(2^n) multiplication
5: $b \leftarrow M \cdot (C' \oplus R_1)$
6: $c \leftarrow a \oplus b$
7: $d \leftarrow R_0 \oplus R_1$
8: $e \leftarrow M \cdot d$
9: $f \leftarrow (C \oplus R_0) \oplus c$
10: $g \leftarrow f \oplus e$
▷ $g = (C \oplus R_0) \oplus M \cdot (C \oplus C')$
11: **return** $g \oplus R_0$

Algorithm 9.3(b) Multiplication based infective countermeasure (2nd variant) with masking

Input: C; C' ▷ $|C|, |C'| = n$
Output: C if no fault; random, otherwise
1: $\Delta \leftarrow C \oplus C'$
2: $R_0, R_1, R_2, R_3 \xleftarrow{\$} \mathbb{F}_2^n$ ▷ $R_0, R_1 \neq \vec{0}$
3: $M \leftarrow \begin{bmatrix} R_2 \\ \rho^1(R_2) \\ \vdots \\ \rho^{n-2}(R_2) \\ R_3 \end{bmatrix}$
4: $a \leftarrow M \cdot (C \oplus R_0)$
▷ $\cdot$ refers to GF(2^n) multiplication
5: $b \leftarrow M \cdot (C' \oplus R_1)$
6: $c \leftarrow a \oplus b$
7: $d \leftarrow R_0 \oplus R_1$
8: $e \leftarrow M \cdot d$
9: $f \leftarrow (C \oplus R_0) \oplus c$
10: $g \leftarrow f \oplus e$
▷ $g = (C \oplus R_0) \oplus M \cdot (C \oplus C')$
11: **return** $g \oplus R_0$

Derivative-Based

Since the constructions in this category use the derivative of a function, our idea is to use any existing cipher, rather than the usual approach of using an ad hoc solution. With our usual notations, this construction can be described as, $\tau_R(\Delta) = E_R(\vec{0}) \oplus E_R(\Delta)$, E being any standard cipher. Figure 9.5 shows the construction.

We do not impose any restriction on E, so it can be taken as the underling cipher or can be another cipher. However, to keep overhead low, we recommend to reuse the underlying cipher or use an optimized implementation/lightweight cipher.

Advantages. This strategy gives us a few advantages over the usual ad hoc approaches such as

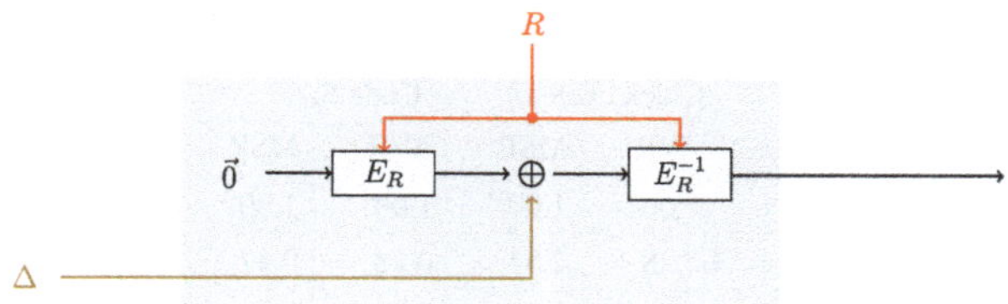

Fig. 9.6 Encrypt–XOR–decrypt-based design of $\tau_R(\Delta)$

1. The underlying cipher is already analysed thoroughly for weakness. This gives us more confidence (as breaking such a countermeasure would probably imply breaking the cipher used); particularly when compared to the new and ad hoc approaches.
2. The cipher to be protected can be reused. This avoids the necessity to build a new hardware/code afresh for a new component. In that case, we only have performance penalty in terms of throughput.
   Further, as already mentioned, instead of the full cipher; ξ rounds of `Round Function`$_0(\cdot)$, together with the corresponding round key insertion can be used with minimal compromise to security.
3. One may notice, for example, the derivative-based construction in [GSS+17] is only defined for block size, $n = 128$. Although, in theory, the concept can be generalized for other block sizes (e.g., for a 64-bit block); one has to go through the lengthy design process, and has to guarantee its security from scratch. In our construction, one can easily choose from a pool of already analysed ciphers (including the underlying cipher itself).

Remark 9.2 Since this type of construction does not require inversion, any hash function can be used in place of E. For a hash function H (which does not involve a key), we propose to use $\tau_R(\Delta) = H(R) \oplus H(R \oplus \Delta)$.

Remark 9.3 If the cipher E is invertible, then a new construction of $\tau_{(\cdot)}(\cdot)$ can be given. The cipher, E, takes a random vector R as its key and $\vec{0}$ as input. Then it XORs Δ with $E_R(\vec{0})$. This is then decrypted using the same key R. So, $\tau_R(\Delta)$ is given by $\tau_R(\Delta) = E_R^{-1}(E_R(\vec{0}) \oplus \Delta)$. We call this type of construction as encrypt–XOR–decrypt-based design. A pictorial description is given in Fig. 9.6. This may, however, cost more due to inverse key schedule, inverse SBox etc. during decryption.

9.3.4 Benchmarking Results for Type I Schemes

Here, we present benchmarking results for Type I schemes. The extra component needed to provide external randomness is not considered, following the previous papers (such as [BBB+18]). We consider `AES`encryption as the underlying cipher, so we implement it to get a perspective of overhead (so, $n = 128$, $\eta = 11$).

Table 9.1 Software benchmarking results (clock cycles, code size) for Type I schemes

		Clock cycles		Code Size		
		AVR	MSP	AVR	MSP	
AESEncryption (unprotected)		1.00	1.00	1.00	1.00	–
GF(2^{128}) Multiplication		2.28	2.57	0.28	0.47	[LRT12]/Sect. 9.3.1
Algorithm 9.2(b)		41.86	34.06	0.28	0.35	Sect. 9.3.3
Cellular Automata		≈2700	≈3400	0.82	0.88	[GSS+17]/Sect. 9.3.2
AES Encryption Derivative‡	$z = 11$	1.65	1.81	1.42	1.35	Sect. 9.3.2
	$z = 4$	0.64	0.63	1.06	1.36	

‡ z is a dummy parameter that can take any integer from $[\xi, \eta]$

For the GF(2^{128}) multiplication in [LRT12], we choose the irreducible polynomial as, $x^{128} + x^7 + x^2 + x + 1$ (the same polynomial used in AES-GCM). Also, since we do not consider the combined attacks (i.e., the fault attacks coupled with side channel attacks; see Sect. 9.3.1), we perform this field multiplication only once (instead of three times, as in Algorithm 9.1). We only implement Algorithm 9.2(b) (second hardware friendly alternative, see Sect. 9.3.3), as the other alternative (Algorithm 9.2(a)) can be considered a part of it. We also implement the non-linear hybrid cellular automata based design in [GSS+17] (Sect. 9.3.2). For the derivative-based construction that relies on a standard cipher, we take AES encryption.

All Type I schemes are implemented as stand-alone module, which means; one has to account for the additional clock cycles/circuitry needed to run the actual and the redundant cipher, as well as other subsidiary modules (such as, XORing the outputs of the actual and the redundant ciphers, etc.).

For software benchmark, we implement two versions of the derivative and encrypt–XOR–decrypt-based implementations; corresponding to $z = 4$ and $z = 11$, where we consider the cipher constituted by z-rounds of $\texttt{RoundFunction}_0(\cdot)$, with corresponding AddRoundKeys. Note that z is a parameter, whereas ξ is a constant for a given cipher (e.g., $\xi = 4$ for AES). So, z can take any integer value from $[\xi, \eta]$. Such reduced round version helps to reduce latency. In Table 9.1, we present the software performance results (both the clock cycles and code size) of our Type I proposals, along with the existing ones. These are given relative to 1.00× that of the unprotected AESencryption. The data presented here are taken as the average of multiple runs. We use open source codes available in Fair Evaluation of Lightweight Cryptographic Systems (FELICS) tool [DCK+15] for AVR and MSP architecture.

As it can be seen, the derivative scheme based on AES(with $z = 4$) outperforms other schemes in software (0.64× in AVR and 0.63× in MSP) in terms of clock cycles. Technically, these are equivalent to 8 rounds of AES, minus the key schedule. In terms of relative code size, the Algorithm 9.2(b) works similar to, if not better, than a single GF(2^{128}) multiplication [LRT12].

Table 9.2 Hardware benchmarking results (FPGA) for Type I schemes

	Spartan 3			Virtex 6		
	Slices	Slice F/Fs	4- input LUTs	Slice registers	Slice LUTs	
AESencryption (unprotected)	1931 (14)	785 (2)	3551 (13)	540†	1402 (3)	–
GF(2^{128}) multiplication	8398 (63)	–	16496 (61)	–	7264 (15)	[LRT12] Sect. 9.3.1
Cellular automata	214 (1)	136†	406 (1)	136†	210†	[GSS+17] Sect. 9.3.2
Algorithm 9.2(b)	91†	130†	88†	130†	56†	Sect. 9.3.3

(·) indicates % resource utilization † indicates negligible utilization

The data in the previous table may appear counter-intuitive; particularly noticing the GF(2^{128}) multiplication takes less clock cycles than the matrix multiplication in Algorithm 9.2(b). However, one should keep in mind that:

1. All implementations are quite basic. There are scopes to optimize the codes keeping clock cycles/code size in mind. The reason behind this is to get a fair comparison among the constructions. For example, the matrix multiplication in Algorithm 9.2(b) is done on bit-by-bit basis. In the 8-bit AVR microcontroller, this can be sped up by using byte-by-byte operations.
2. The FELICS framework takes the input and output test vectors in (arrays of) bytes. This is helpful, e.g., for AES; but causes extra overhead for bit-oriented constructions. For the bit-oriented matrix multiplication in Algorithm 9.2(b), one has to go through additional conversion from byte to bit (at the beginning) and from bit to byte (at the end), which adds unnecessary costs.

As for the hardware benchmark, we choose the Spartan 3 (3s1500fg676-4) and Virtex 6 (6vcx75tff484-1) FPGA families, and report the results in Table 9.2. Here, our hardware friendly approach, Algorithm 9.2(b) outperforms all other designs in both the families. In fact, the relative amount of resource utilization in the devices are negligibly small.

9.4 Type II/Cipher-Level Constructions

Here, we focus on the Type II constructions for infective countermeasures. At first, the n-bit random vectors β and k_0[6] are so chosen that, $\texttt{RoundFunction}_0(\beta) = \beta \oplus k_0$. Finding such pairs is easy in SPN ciphers—one can set a β randomly, then compute,

[6] It is to be observed that k_0 is not a round key.

$k_0 = \beta \oplus \texttt{RoundFunction}_0(\beta)$. Notice that, β uniquely identifies k_0; so once a k_0 is fixed (based on a β), for β' ($\neq \beta$), $\texttt{RoundFunction}_0(\beta') \oplus k_0$ will not give β. Both β and k_0 are (re-)generated at each invocation of the countermeasure, but are fixed during one invocation. These are kept secret from the attacker.

However, finding such (β, k_0) pair for Feistel constructions can be tricky in general. In a Feistel cipher, the f function can map n_1 bits to n_2 ($\neq n_1$) bits—in such a case, such k_0 does not exist (as length of β and k_0 are different). Further, the round keys can be non-linearly mixed (in contrast to the SPN ciphers, where these are commonly XORed). In this case, one has to go through a rigorous computation to get k_0 for a given β, which may turn out to be quite costly. Hence, we keep the Type II schemes for Feistel ciphers out of scope for this work. There can be Feistel ciphers, where these countermeasures are applicable at ease; but it has to be possibly checked on a case by case basis, and may not be generalized.

We use the following notations: $\neg$ for logical negation, $\wedge$ for logical AND, $\vee$ for logical OR, $+$ for arithmetic addition, $\times$ for arithmetic multiplication, $\lceil\cdot\rceil$ for the ceiling function, $x \cdot \vec{y}$ for scalar x multiplication with the vector $\vec{y}$ over GF(2), $\sharp x(\vec{y})$ for number of occurrence(s) of element x in vector $\vec{y}$.

We now briefly describe how infection in [GST12] works (see Algorithm 9.4). A variable, i, loops from 1 until it reaches 2η. At each round, a randomly generated bit, λ, determines whether this round will be a dummy round (i is not incremented) or a meaningful (actual or redundant) round (i is incremented by 1). When $\lambda = 0$, a dummy round occurs. It sets the value β to the register S_2 (S_2 is also initialized with β). When $\lambda = 1$, an actual or a redundant round occurs, depending on whether i is even or odd, respectively.

Since, $\lambda = 1$ makes i from even to odd, or vice-versa; both the actual and redundant rounds are carried out η times. The two registers, S_0 and S_1, are used to compute the actual and redundant rounds, respectively. The variables a and b are updated such that

$$a = \begin{cases} 0 & \text{if } \lambda \text{ is 1 and } i \text{ is even (actual round)}, \\ 1 & \text{if } \lambda \text{ is 1 and } i \text{ is odd (redundant round)}, \\ 2 & \text{if } \lambda \text{ is 0 (dummy round)}; \end{cases}$$

$$b = \begin{cases} 0 & \text{if } \lambda \text{ is 0 (dummy round)}, \\ \lceil i/2 \rceil & \text{otherwise (meaningful round)}. \end{cases}$$

Effectively, a determines which register (among S_0, S_1, S_2) to update; and b determines which $\texttt{RoundFunction}_{(\cdot)}(\cdot)$ and which round key/k_0 to use ($S_a \leftarrow \texttt{RoundFunction}_b(S_a) \oplus k_b$, Line 8).

Since we start i from 1 (odd), the redundant round always precedes the actual round. Three back-up registers, T_0, T_1, T_2 are also updated based on a. Basically, T_a holds a copy of S_a. Then Δ is computed by $T_0 \oplus T_1$, which is passed through a function $\sigma(\cdot)$ which has the property of returning $\vec{0}$ as output if the input is $\vec{0}$. The variable c is updated with the following rule:

$$c = \begin{cases} \sigma(T_0 \oplus T_1) & \text{if } \lambda \text{ is 1 and } i \text{ is even (actual round),} \\ \vec{0} & \text{otherwise.} \end{cases}$$

Basically, c determines whether the output of $\sigma(T_0 \oplus T_1)$ (which is $\vec{0}$ for no fault, or non-zero for a fault) can be infected to S_2 (Line 11) and S_0 (Line 12). The authors propose to use inversion in GF(2^8) per SBox of `AES`as $\sigma(\cdot)$.[7]

Finally, to infect the fault which is injected at the last round (when $i = 2\eta - 1$ or 2η); the authors propose to update S_0 one last time based on S_2 (Line 14: $S_0 \leftarrow S_0 \oplus \texttt{RoundFunction}_0(S_2) \oplus k_0 \oplus \beta$). This ensures, any infection passes through at least one $\texttt{RoundFunction}_0(\cdot)$.

If no fault is injected, then both the actual and redundant states (S_0, S_1) as well as T_0, T_1 contain the same computation of the cipher, and the dummy state S_2 as well as T_2 contain β. Injecting a fault during a dummy round ($\lambda = 0$) will not give attacker any meaningful information. If a fault is injected during an actual round (S_0), then it will infect both S_0 and S_2 (Lines 12, 13). Note that this infection will spread in S_0 in the subsequent actual rounds. Also, notice that the c is zero for a redundant round, that is due to the fact that the corresponding actual round is not computed (but the redundant round is computed). Now, the infection in S_2 will change the subsequent update of S_2 from β (Line 8); this will change T_2 from β (Line 9). Now that $T_2 \neq \beta$, in all the subsequent rounds (actual, redundant and dummy alike) T_0, T_1, and T_2 will be infected further. In turn, this will infect subsequent S_0 and S_2 further. Finally, each of $S_0, S_1, S_2, T_0, T_1, T_2$ will contain random values.

If the attacker chooses a redundant round (when S_1 is updated) to inject a fault, then it will cause infection to S_0 and S_2 in the next actual round. Then, by a similar process, all three registers will be infected.

Despite its promise, this first cipher-level protection is attacked soon afterwards in [BG13], and later in [TBM14]. The basic observation that leads to the attack is, when a fault is injected at the last round ($i = 2\eta - 1$ or 2η), infection passes through only one $\texttt{RoundFunction}_0(\cdot)$ (Line 14). One round of diffusion is not generally sufficient to resist Eve to recover information on the faulty state. Hence, the attacker can still perform DFA by injecting at the last round of the cipher computation.

---

[7] However, they keep the choice for $\sigma(\cdot)$ relaxed, so other $\sigma(\cdot)$ can also be used with their scheme.

Algorithm 9.4 Infective countermeasure: LatinCrypt'12 (for SPN)

Input: $\begin{cases} P \\ \beta; k_0 \\ \text{round keys } k_j; \\ \quad j = 1, \ldots, \eta \text{ derived from } K \end{cases}$

▷ $\texttt{RoundFunction}_0(\beta) = \beta \oplus k_0$

Output: $\begin{cases} E_K(P) & \text{if no fault} \\ \text{random} & \text{otherwise} \end{cases}$

1: $S_0 \leftarrow P$ ▷ Actual state
  $S_1 \leftarrow P$ ▷ Redundant state
  $S_2 \leftarrow \beta$ ▷ Dummy state
2: $T_0 \leftarrow 0; T_1 \leftarrow 0; T_2 \leftarrow \beta$
3: $i \leftarrow 1$
4: **while** $i \leq 2\eta$ **do**
5:   $\lambda \xleftarrow{\$} \mathbb{F}_2^1$
6:   $a \leftarrow (i \wedge \lambda) \oplus ((\neg\lambda) \times 2)$
7:   $b \leftarrow \lceil i/2 \rceil \times \lambda$
8:   $S_a \leftarrow \texttt{RoundFunction}_b(S_a) \oplus k_b$
9:   $T_a \leftarrow S_a \oplus T_2 \oplus \beta$
10:   $c \leftarrow (\lambda \wedge (\neg(i \wedge 1))) \cdot \sigma(T_0 \oplus T_1)$
11:   $S_2 \leftarrow S_2 \oplus c$
12:   $S_0 \leftarrow S_0 \oplus c$
13:   $i \leftarrow i + \lambda$
14: $S_0 \leftarrow S_0 \oplus \texttt{RoundFunction}_0(S_2) \oplus k_0 \oplus \beta$
15: **return** S_0

Algorithm 9.5 Infective countermeasure: CHES'14 (for SPN)

Input: $\begin{cases} P \\ \beta; k_0 \\ \text{security level } t \ (\geq 2\eta) \\ \text{round keys } k_j; \\ \quad j = 1, \ldots, \eta \text{ derived from } K \end{cases}$

▷ $\texttt{RoundFunction}_0(\beta) = \beta \oplus k_0$

Output: $\begin{cases} E_K(P) & \text{if no fault} \\ \text{random} & \text{otherwise} \end{cases}$

1: $S_0 \leftarrow P$ ▷ Actual state
  $S_1 \leftarrow P$ ▷ Redundant state
  $S_2 \leftarrow \beta$ ▷ Dummy state
2: $i \leftarrow 1; q \leftarrow 1$
3: $rstr \leftarrow \mathbb{F}_2^t \ni \sharp 1(rstr) = 2\eta$
4: **while** $q \leq t$ **do**
5:   $\lambda \leftarrow rstr[q]$
6:   $a \leftarrow (i \wedge \lambda) \oplus ((\neg\lambda) \times 2)$
7:   $b \leftarrow \lceil i/2 \rceil \times \lambda$
8:   $S_a \leftarrow \texttt{RoundFunction}_b(S_a) \oplus k_b$
9:   $c \leftarrow (\lambda \wedge (\neg(i \wedge 1))) \wedge \sigma(S_0 \oplus S_1)$
10:   $d \leftarrow (\neg\lambda) \wedge \sigma(S_2 \oplus \beta)$
11:   $S_0 \leftarrow (\neg(c \vee d) \cdot S_0) \oplus ((c \vee d) \cdot S_2)$
12:   $i \leftarrow i + \lambda$
13:   $q \leftarrow q + 1$
14: **return** S_0

The CHES'14 countermeasure, given in [TBM14] (see Algorithm 9.5), is similar to that of the LatinCrypt'12. The variable, c is still updated the same way, but it now returns a bit instead of an n-bit vector; as $\sigma(\cdot)$ now returns 1-bit as output (the $n:1$ OR gate, denoted by BLFN in the paper, is chosen as $\sigma(\cdot)$). Another major difference in this scheme is, it computes another 1-bit variable d, which is updated such that (Line 8):

$$d = \begin{cases} \sigma(S_2 \oplus \beta) & \text{if } \lambda \text{ is 0 (dummy round),} \\ 0 & \text{otherwise (meaningful round).} \end{cases}$$

Then, if $(c \vee d)$ is 1, then S_0 is substituted by S_2 (Line 11: $S_0 \leftarrow (\neg(c \vee d) \cdot S_0) \oplus ((c \vee d) \cdot S_2)$. This overwrites the content of S_0, and makes it random, as content of S_2 $(= \beta)$ is random. Figure 9.7 depicts the work-flow. We next show why this is a problem.

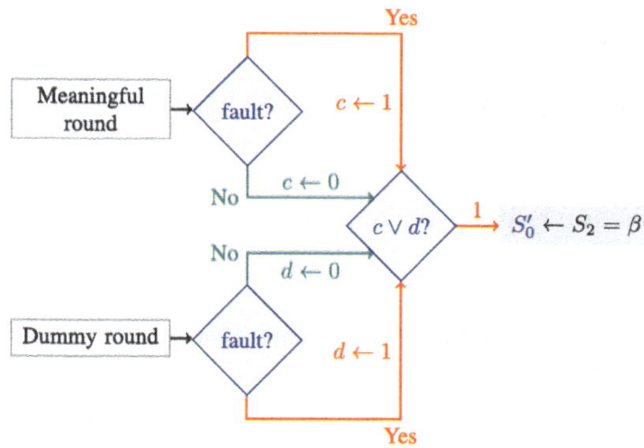

Fig. 9.7 Fault detection work-flow in CHES'14 countermeasure

9.4.1 Critical Look at CHES'14 Countermeasure

We would like to point out that the scheme proposed in CHES'14 [TBM14] does not fit in the philosophy of infective countermeasures, rather it is a fault detection technique. Being a detective countermeasure, [TBM14] does protect a cipher from DFA. However, at the same time, it is susceptible to 1-bit judgment condition, as discussed in Sect. 9.2.3. The work of [TBM14] is well cited as an infective countermeasure, such as [PCM15, BG16, PCM17, DEK+18]. For better understanding of both the infective and detective countermeasures, we feel to throw light on the proper categorization of the [TBM14] countermeasure.

In the subsequent parts, we reveal our analysis after a detailed inspection of [TBM14]. At this point, it would be helpful to recall the philosophy of infective schemes: To diffuse the fault to the entire state of the cipher in a non-deterministic way. This philosophy is already described in the existing literature, e.g., [LRT12, GSS+17]. We show that the CHES'14 countermeasure does not actually diffuse the fault in the state. Thus, it does not follow the infection philosophy.

In order to do a comparative analysis, one may recall their descriptions from Algorithms 9.4 and 9.5. We start by doing dry runs of the algorithms for two main cases (namely, for the final and the penultimate rounds). This analysis can be easily extended for earlier rounds.

Case 1: Last (Actual) Round Fault

We start by analysing a fault injection in the last actual round. Dry Run 9.1 traces the relevant steps of Algorithm 9.4. It can be seen that, once the fault is injected in the round; it is diffused. The faulty state is stored in T_0 and continues further using c before the last call to the round function, after which, the infected state is returned.

The CHES' 14 scheme (Algorithm 9.5) is traced by Dry Run 9.2. It can be noticed that the faulty state S_0' has no contribution to the final state that is returned (this is merely a random state β). The 1-bit variables c and d implicitly detect the fault in meaningful or dummy rounds, respectively. If detected, the scheme simply replaces the faulty state by a random state, thereby deviating from the basic notion of infection. We use red to indicate infection and blue to indicate substitution.

Case 2: Penultimate (Actual) Round Fault

Here, we investigate how a fault propagates through multiple rounds of Algorithms 9.4 and 9.5 by targeting the penultimate actual round for fault injection.

Like before, it can be observed from Dry Run 9.3 (tracing steps of Algorithm 9.4) and Dry Run 9.4 (tracing steps of Algorithm 9.5) that; while for the LatinCrypt' 12 countermeasure, the induced fault is diffused across the actual rounds; the CHES' 14 countermeasure ends up substituting the faulty state with a random state ($= \beta$).

It is understood from the existing literature that detection countermeasures can be rendered useless if the bit that senses the presence of fault is altered. In fact, the main motivation for adopting the infective countermeasures (instead of the detective counterpart) is to protect the first order fault plus 1-bit fault, as stated in [SSS+03, Sect. 3] (note that this paper introduced the concept of infection). If we only consider first order faults (i.e., the attacker can only target once during the cipher execution), then detective countermeasures would be sufficient (no need for infection). Hence, although the CHES' 14 countermeasure protects against first order DFA; it does not inherently imply the infective nature.

- Initial conditions: $i = 2\eta$; $\lambda = 1$
- Assume fault injection in Line 8 of Algorithm 9.4 and in Line 8 of Algorithm 9.5

Dry Run 9.1 Algorithm 9.4, Lines 6–15

$a \leftarrow 0$ ▷ i is even
$b \leftarrow \eta$
$S_0' \xleftarrow{\text{fault}} \texttt{RoundFunction}_\eta(S_0) \oplus k_\eta$
$T_0 \leftarrow S_0'$ ▷ Infection begins
$c \not\leftarrow 0$
$S_2' \leftarrow S_2 \oplus c$
$S_0' \leftarrow S_0' \oplus c$
$i \leftarrow 2\eta + 1$
$S_0' \leftarrow S_0' \oplus \texttt{RoundFunction}_{k_0}(S_2') \oplus \beta$
return S_0' ▷ S_0' is the infected state

Dry Run 9.2 Algorithm 9.5, Lines 6–14

$a \leftarrow 0$ ▷ i is even
$b \leftarrow \eta$
$S_0' \xleftarrow{\text{fault}} \texttt{RoundFunction}_\eta(S_0) \oplus k_\eta$
$c \leftarrow 1$
$d \leftarrow 0$
$S_0' \leftarrow S_2 = \beta$ ▷ No infection
$i \leftarrow 2\eta + 1$
$q \leftarrow t + 1$
return S_0' ▷ S_0' is just a random state

It may be noted that, the same weakness has been pointed out in an independent work [FCL+20]. More precisely, it is mentioned in [FCL+20, Sect. VIII.A], "Similar to the consistency check in the detection countermeasure, BLFN brings about a 1-bit judgment condition. Therefore, the malicious modification on the result of BLFN may lead to infection failure". This observation further reinforces our claim.

Hence, the inability to thwart a single-bit judgment condition invalidates the claim that this countermeasure is infective. A similar argument works for the modified countermeasure in [PCM15, PCM17], rendering it outside the domain of the infective countermeasures.

9.4.2 *Our Patch for LatinCrypt'12 Countermeasure*

One may observe that if a fault is injected in a sufficiently early round, then it passes through sufficient rounds of diffusion, thus making it random to the attacker. This is valid regardless of the nature (actual, redundant, or dummy) of the round. Hence, to make the [GST12] scheme sound, what we need is the assurance that no matter which round attacker chooses; the countermeasure goes through sufficient rounds of diffusion.

This observation leads us to propose a simple patch for the aforementioned scheme. Recall (Sect. 9.4) that attacker can choose the last meaningful round as the target for fault injection, which goes through only one round of diffusion (Line 14 of Algorithm 9.4). So, our patch for this scheme works basically by retaining the preceding part of Algorithm 9.4 (with few amendments); mainly to insert more diffusion at Line 14. In fact, we make use of the following interesting property:

$$\begin{aligned}\beta &= \texttt{RoundFunction}_0(\beta) \oplus k_0 \\ &= \texttt{RoundFunction}_0(\texttt{RoundFunction}_0(\beta) \oplus k_0) \oplus k_0.\end{aligned}$$

We now define *z-nested round function*, denoted by $\texttt{RoundFunction}^z(\cdot\,,\,\cdot)$; $z = 1, 2, \ldots, \eta$; recursively as:

$$\texttt{RoundFunction}^z(x, y) = \begin{cases}\texttt{RoundFunction}_0(x) \oplus y & \text{if } z = 1, \\ \texttt{RoundFunction}_0(\texttt{RoundFunction}^{z-1}(x, y)) \oplus y & \text{otherwise.}\end{cases}$$

It can be checked that independent of z, $\texttt{RoundFunction}^z(\beta, k_0) = \beta$. Hence, we propose to substitute Line 14 of Algorithm 9.4:

$$S_0 \leftarrow S_0 \oplus \texttt{RoundFunction}_0(S_2) \oplus k_0 \oplus \beta$$

by

$$S_0 \leftarrow S_0 \oplus \texttt{RoundFunction}^z(S_2, k_0) \oplus \beta$$

where z ($\geq \xi$) is predetermined. It may be noted that, the overhead for nested round function is on the temporal domain and not on the spatial domain.

In a nutshell, we incorporate the following amendments:

- We observe the back-up registers, T_0, T_1, T_2 are indeed redundant—the same functionality can be obtained by using actual (S_0), redundant (S_1) and dummy registers (S_2) only. So, we do not use them.
- We remove λ altogether. Instead of running the meaningful/dummy rounds in a random order, we simplify our scheme by running them in a deterministic order. Within a loop of i from 1 to η, we run the redundant round, actual round and the dummy round (in this order). After this, we compute the nested round function on a loop of j, counting from 1 to z ($\geq \xi$).
- We only infect the dummy register within loop over i, that goes from 1 to η (neither S_0, nor S_1 is infected within this loop). This makes sure that S_2 contains β (in case of no fault) or something other than β (in case of a fault). Introduction of nested round function ensures S_2 gets sufficient diffusion, no matter which round attacker targets. At then end, we infect S_0 by $S_0 \leftarrow S_0 \oplus S_2 \oplus \beta$. Since, $S_2 \oplus \beta$ works like a one-time pad (similar to the case of Type I countermeasures), in case of a fault, attacker gets no information on S_0.
- The reason we do not infect S_0 within the loop of i is, attacker can simply bypass any infection in S_0 within the loop of i by injecting fault at the last actual round.
- We do not infect the redundant state S_1 at all. This is inspired from the observation that, attacker actually gets the infected S_0 (and not S_1).
- Instead of computing δ (i.e., $\delta \leftarrow S_0 \oplus S_1$; $S_2 \leftarrow S_2 \oplus \delta$) explicitly, we do the computation implicitly ($S_2 \leftarrow S_0 \oplus S_2$; $S_2 \leftarrow S_1 \oplus S_2$). This modification helps to prevent any attack on δ (such as resetting it to $\vec{0}$ before it updates S_2; or skipping the XOR of δ with S_2).
- We choose $\sigma(\cdot)$ as the identity function for simplicity.[8] Although we do not compute δ ($= S_0 \oplus S_1$) explicitly; technically speaking, we XOR ($S_0 \oplus S_1$) directly to S_2.

One motivation for keeping random ordering of dummy/meaningful round (through λ), based on [GST12], is to have a somewhat protection against side channel analysis. The authors speculate that, since the operations in dummy rounds mimic

[8] It is recommended in [GST12] to use a non-linear function as $\sigma(\cdot)$. However, we argue this is not a necessary condition.

that of a meaningful round; attacker cannot identify a meaningful round (with probability $> \frac{1}{2}$). However, this does not amount for secure enough protection against side channel analysis. For example, say, the attacker can get λ itself by side channel analysis; thus the assumed SCA security does not hold anymore. At the same time, we emphasize that making the meaningful/dummy round computation deterministic does not incur additional vulnerability from the perspective of DFA. Another reason for keeping the random ordering would be to reduce the success rate of the fault injecting attacker [PCM15, PCM17]. In this regard, we would like to note that our infection schemes are independent on the success rate of the attacker. Even if the attacker succeeds to inject a fault with probability 1, the security claims of our schemes would still hold. Hence, we argue making the execution sequence random would not add any extra protection.

In Algorithm 9.6, we present an algorithmic description of our patched scheme. As already described, here we iterate over a loop of i, from 1 through η. Within this loop, we deterministically compute the actual, redundant, and the dummy round, in this order. After the actual and the redundant round computations are done, we update the dummy register S_2 ($S_2 \leftarrow S_0 \oplus S_1 \oplus S_2$: Lines 5, 6). S_2 is also initialized with β (Line 1). Hence, any fault in actual or redundant round will result S_2 to contain a value different from β. This change in S_2 will affect the dummy round computation, ensuring S_2 does not contain β (Line 7). In case of no fault, S_2 will continue to contain β. When this loop is iterated; the following loop over j, from 1 to z ($\geq \xi$), will ensure the diffusion is spread sufficiently on S_2 (in case of fault)/S_2 still contains β (otherwise) by additional dummy rounds. Finally, S_0 is returned after it is XORed with $S_2 \oplus \beta$ (Line 10). It can be noted that, injecting fault anywhere during the loop over j will not yield useful information to the attacker.

Now let us revisit the patched scheme for its usability against CFA (Sect. 9.2.6). When Eve injects fault at any meaningful round, it sets a β' ($\neq \beta$) at S_2; so S_2 is now infected (Lines 5, 6). The infection spreads on S_2 over subsequent rounds. Even after several rounds of iteration, if S_0 becomes equal to S_1 (which is the case CFA wants to utilize), S_2 will still be infected (i.e., S_2 will contain something $\neq \beta$). The next loop over j will also spread the infection to S_2. Hence, S_2 will be random (Line 10); this means, S_0 will be XORed by by a random vector. Thus, in case of a CFA, attacker will always get a random (infected) output; and cannot mount the attack. To the best of our knowledge, this is the first CFA countermeasure reported in the literature.

- Initial conditions: $i = 2\eta - 2$; $\lambda = 1$
- Assume fault injection in Line 8 of Algorithm 9.4 and in Line 8 of Algorithm 9.5

Dry Run 9.3 Algorithm 9.4, Lines 5–15

Iteration 1: $(i = 2\eta - 2) < 2\eta$

$\lambda = 1$ ▷ (Say)
$a \leftarrow 0$ ▷ Actual Round
$b \leftarrow \eta - 1$
$S'_0 \xleftarrow{\text{fault}} \texttt{RoundFunction}_{\eta-1}(S_0) \oplus k_{\eta-1}$
$T_0 \leftarrow S'_0$ ▷ Infection begins
$c \nleftarrow 0$
$S'_2 \leftarrow S_2 \oplus c$
$S'_0 \leftarrow S'_0 \oplus c$
$i \leftarrow 2\eta - 2 + 1$

Iteration 2: $(i = 2\eta - 1) < 2\eta$

$\lambda = 1$ ▷ (Say)
$a \leftarrow 1$ ▷ Redundant Round
$b \leftarrow \eta$
$S_1 \leftarrow \texttt{RoundFunction}_{\eta}(S_1) \oplus k_{\eta}$
$T_1 \leftarrow S_1$
$c \nleftarrow 0$
$S'_2 \leftarrow S_2 \oplus c$
$S'_0 \leftarrow S'_0 \oplus c$
$i \leftarrow 2\eta - 1 + 1$

Iteration 3: $(i = 2\eta)$

$\lambda = 1$ ▷ (Say)
$a \leftarrow 0$ ▷ Actual Round
$b \leftarrow \eta$
$S'_0 \leftarrow \texttt{RoundFunction}_{\eta}(S'_0) \oplus k_{\eta}$
$T_0 \leftarrow S'_0$ ▷ Infection continues
$c \nleftarrow 0$
$S'_2 \leftarrow S'_2 \oplus c$
$S'_0 \leftarrow S'_0 \oplus c$
$i \leftarrow 2\eta + 1$
$S'_0 \leftarrow S'_0 \oplus \texttt{RoundFunction}_0(S'_2) \oplus k_0 \oplus \beta$
return S'_0 ▷ S'_0 is the infected state

Dry Run 9.4 Algorithm 9.5, Lines 5–14

Iteration 1: $(q = t - 2) < t$

$\lambda = 1$ ▷ (Say)
$a \leftarrow 0$ ▷ Actual Round
$b \leftarrow \eta - 1$
$S'_0 \xleftarrow{\text{fault}} \texttt{RoundFunction}_{\eta-1}(S_0) \oplus k_{\eta-1}$
$c \leftarrow 1$
$d \leftarrow 0$
$S'_0 \leftarrow S_2 = \beta$ ▷ No infection
$i \leftarrow 2\eta - 1$
$q \leftarrow t - 2 + 1$

Iteration 2: $(q = t - 1) < t$

$\lambda = 1$ ▷ (Say)
$a \leftarrow 1$ ▷ Redundant Round
$b \leftarrow \eta$
$S_1 \leftarrow \texttt{RoundFunction}_{\eta}(S_1) \oplus k_{\eta}$
$c \leftarrow 1$
$d \leftarrow 0$
$S'_0 \leftarrow S_2 = \beta$ ▷ No infection
$i \leftarrow 2\eta$
$q \leftarrow t - 1 + 1$

Iteration 3: $(q = t)$

$\lambda = 1$ ▷ (Say)
$a \leftarrow 0$ ▷ Actual Round
$b \leftarrow \eta$
$S'_0 \leftarrow \texttt{RoundFunction}_{\eta}(S'_0 = \beta) \oplus k_{\eta}$
$c \leftarrow 1$
$d \leftarrow 0$
$S'_0 \leftarrow S_2 = \beta$ ▷ No infection
$i \leftarrow 2\eta + 1$
$q \leftarrow t + 1$
return S'_0 ▷ S'_0 is just a random state

In the light of [PCM15] (i.e., resilience of infective countermeasures against the instruction skip attack), one may notice the interesting observations: Under non-faulty situation, Lines 5–7 of Algorithm 9.6 are idempotent. So, it does not affect the non-faulty computation of the cipher if these lines are iterated, say, μ (> 1) times:

for $m = 1; m \leq \mu + z; m \leftarrow m + 1$ **do**
  $S_2 \leftarrow S_0 \oplus S_2$
  $S_2 \leftarrow S_1 \oplus S_2$
  $S_2 \leftarrow \texttt{RoundFunction}_0(S_2) \oplus k_0$

In case of a fault, this will help the infection to propagate further in S_2. Similarly, the Line 10 is still vulnerable to one instruction skip. As this line is idempotent under non-faulty situation, we can substitute it by

1: **for** $m = 1; m \leq \mu; m \leftarrow m + 1$ **do**
2: | $S_0 \leftarrow S_0 \oplus S_2 \oplus \beta$ ▷ $S_2 = \texttt{RoundFunction}^z(S_2, k_0)$

This ensures that the Line 10 is protected against $\mu - 1$ instruction skip attack. Altogether, it can be claimed that the overall construction can withstand up to $\mu - 1$ instruction skips. This can pave the pathway to an infective countermeasure which is guaranteed to have a certain protection against instruction skip attacks. We are confident that the arguments presented in algorithmic terms are still valid when the proposed strategy is expressed in terms of instructions instead. The underlying property that is used here still applies when the algorithm is expressed as an equivalent sequence of instructions. Hence the implementation-level idempotency condition can be ascertained from the algorithmic description.

Algorithm 9.6 Our patched version of LatinCrypt' 12 countermeasure (for SPN)

Input: $\begin{cases} P \\ \beta; k_0 \\ \text{round keys } k_j;\ j = 1, \ldots, \eta \text{ derived from } K \\ z\ (\geq \xi) \end{cases}$ ▷ $\texttt{RoundFunction}_0(\beta) = \beta \oplus k_0$

Output: $\begin{cases} E_K(P) & \text{if no fault} \\ \text{random} & \text{otherwise} \end{cases}$

1: $S_0 \leftarrow P$ ▷ Actual state
$S_1 \leftarrow P$ ▷ Redundant state
$S_2 \leftarrow \beta$ ▷ Dummy state
2: **for** $i \leftarrow 1; i \leq \eta; i \leftarrow i + 1$ **do**
3: | $S_0 \leftarrow \texttt{RoundFunction}_i(S_0) \oplus k_i$ ▷ Actual round
4: | $S_1 \leftarrow \texttt{RoundFunction}_i(S_1) \oplus k_i$ ▷ Redundant round
5: | $S_2 \leftarrow S_0 \oplus S_2$
6: | $S_2 \leftarrow S_1 \oplus S_2$
7: | $S_2 \leftarrow \texttt{RoundFunction}_0(S_2) \oplus k_0$ ▷ Dummy round
8: **for** $j \leftarrow 1; j \leq z; j \leftarrow j + 1$ **do**
9: | $S_2 \leftarrow \texttt{RoundFunction}_0(S_2) \oplus k_0$ ▷ Dummy round
10: $S_0 \leftarrow S_0 \oplus S_2 \oplus \beta$ ▷ $S_2 = \texttt{RoundFunction}^z(S_2, k_0)$
11: **return** S_0

Benchmarking

In Table 9.3, we present the software benchmarking results (clock cycles and code size) for the basic patched version of the LatinCrypt' 12 scheme (Algorithm 9.6), for AVR and MSP microcontroller instruction sets. We take `AES` encryption as the underlying cipher. As before, we use the source codes from the FELICS [DCK+15] tool. However, unlike the benchmarking in Type I schemes, which are done as stand-alone implementations, this one is done together with `AES` actual and redundant encryptions (because of the intrinsic nature of this scheme). We choose two different

Table 9.3 Software benchmarking results (clock cycles, code size) for our patch

	Countermeasure	Architecture	Clock cycles	Code size
AESEncryption (unprotected)	–	AVR	1.00	1.00
		MSP	1.00	1.00
AESEncryption (protected by Algorithm 9.6)	11-nested round	AVR	3.25	2.94
		MSP	3.69	2.31
	4-nested round	AVR	2.90	2.94
		MSP	3.13	2.58

nested round functions; corresponding to 11 and 4, respectively. As before, we do not consider the cost due to generation of external entropy. Also, we assume β and k_0 are provided to the algorithm. The figures given here are taken after averaging multiple runs with different test vectors; and are relative to unprotected AESin 1× unit (the numeric figures for this implementation can be found in Table 9.1). For the 11-nested rounds, the relative clock cycles is slightly bigger than 3×; this is due to the fact that now AES is running 3 times together with few other computations.[9] The relative code sizes are less than 3×, as part of the same code can be reused (such as the SBox). For the 4-nested round, the relative clock cycles is less than 3 for AVR as now less than 3× AES is computed; but the relative code size remains roughly the same as 11-nested rounds. However, for MSP architecture, although the relative clock cycles reduces from the 11-nested rounds; the relative code size has a slight increase (2.31× in 11-nested to 2.58× in 4-nested). This is probably due to some optimization done by the compiler, which is more efficient when almost three identical computations are running, compared to the case where one of them is slightly different.

Since the same hardware (that is used to design the underlying cipher) can be reused to build this scheme (with degraded throughput), we do not provide any separate hardware benchmarking here. Roughly, it can be estimated that 11-nested round countermeasure is comparable to 4× unprotected AES, whereas 4-nested would be around 3.12× unprotected AES. So, the throughput figures can be estimated to be downgraded roughly by this scale.

[9] The actual and the redundant computations are basically identical except near the very end, hence the compiler possibly optimizes the code that reduces the clock cycles.

9.5 Conclusion

In this work, we study the infective countermeasures, which are used to protect ciphers against certain classes of fault attacks (including the probably most common differential fault attacks in symmetric key cryptography). This is the first work studying these countermeasures in detail. Apart from providing a systematic classification, we show several new results. These results range from proposing new type of countermeasures (that relies on already established standards instead of the usual ad hoc approaches), to proposing new lightweight schemes (in both software and hardware). We also show a flaw in the scheme from [TBM14] published at CHES'14. Moreover, we fix a broken scheme with a little amendment. Our work underlies the need for more rigorous analysis of not only the standard ciphers but also the fault countermeasures.

The main differences between the two types of infective countermeasures are listed here. First, unlike Type I countermeasures, cipher-level countermeasures do not construct a separate diffusion function $\tau_R(\cdot)$, rather they rely on the round function of the underlying cipher to infect the actual computation. Second, cipher-level countermeasures rely on sensing the infection round by round, whereas Type I countermeasures let both the actual and redundant executions to finish before infection. Third, Type II countermeasures compute a so-called dummy round, which is idempotent in case of no fault, missing in Type I. Fourth, the function $\tau_{(\cdot)}(\cdot)$ in Type I is recommended to be highly non-linear; in contrast, the non-linearity in Type II countermeasures come from the non-linearity of $\texttt{RoundFunction}_0(\cdot)$, (not from $\sigma(\cdot)$) hence $\sigma(\cdot)$ can be linear. Fifth, a Type I construction can be adopted to non-SPN designs, such as a Feistel network-based block cipher or a stream cipher easily, whereas it may be rather non-trivial to do the same with a Type II construction (which is more fine-tuned for a Type II countermeasure).

Focusing only on DFA countermeasures gives us the advantage to explore the domain more extensively. For this purpose, the side channel protection is not directly considered within the scope. Our schemes can be protected by incorporating standard SCA countermeasures, refer to Sect. 9.2.7 for more details.

Multiple works can be considered in the future scope, here we list a few. First, one may look into constructing an integrated DFA and side channel countermeasure. Second, protecting against double faults (that identically affect both the actual and the redundant ciphers) is an interesting problem. Third, designing infective countermeasures which are more resilient against other types of faults, such as instruction skip or double fault or IFA (SIFA, in particular), can be an interesting direction to pursue. Finally, a cipher may be constructed which is more suitable to deploy together with an infective countermeasure (possibly equipped with an error detecting code).

References

[BS97] E. Biham, A. Shamir, Differential fault analysis of secret key cryptosystems, in *Advances in Cryptology - CRYPTO '97*, ed. by B.S. Kaliski (Springer, Berlin, 1997), pp. 513–525. ISBN: 978-3-540-69528-8

[HBB16] W. He, J. Breier, S. Bhasin, Cheap and cheerful: A low-cost digital sensor for detecting laser fault injection attacks, in *Security, Privacy, and Applied Cryptography Engineering - 6th International Conference, SPACE 2016, Hyderabad, India, December 14–18, 2016, Proceedings* (2016), pp. 27–46. https://doi.org/10.1007/978-3-319-49445-6_2

[BCN+04] H. Bar-El, H. Choukri, D. Naccache, M. Tunstall, C. Whelan, The sorcerer's apprentice guide to fault attacks. IACR Cryptology ePrint Archive, 2004:100 (2004), http://dblp.uni-trier.de/db/journals/iacr/iacr2004.html#Bar-ElCNTW04

[BBB+18] A. Baksi, S. Bhasin, J. Breier, M. Khairallah, T. Peyrin, Protecting block ciphers against differential fault attacks without re-keying (extended version). Cryptology ePrint Archive, Report 2018/085 (2018), https://eprint.iacr.org/2018/085

[BG13] A. Battistello, C. Giraud, Fault analysis of infective AES computations, in *2013 Workshop on Fault Diagnosis and Tolerance in Cryptography, Los Alamitos, CA, USA, August 20, 2013*, pp. 101–107. https://doi.org/10.1109/FDTC.2013.12

[LRT12] V. Lomné, T. Roche, A. Thillard, On the need of randomness in fault attack countermeasures - application to AES, in *2012 Workshop on Fault Diagnosis and Tolerance in Cryptography, Leuven, Belgium, September 9, 2012*, pp. 85–94. https://doi.org/10.1109/FDTC.2012.19

[GSS+17] S. Ghosh, D. Saha, A. Sengupta, D.R. Chowdhury, Preventing fault attacks using fault randomisation with a case study on AES. ACT **3**(3), 225–235 (2017). https://doi.org/10.1504/IJACT.2017.10007295

[TBM14] H. Tupsamudre, S. Bisht, D. Mukhopadhyay, Destroying fault invariant with randomization, in *International Workshop on Cryptographic Hardware and Embedded Systems* (Springer, 2014), pp. 93–111

[PCM15] S. Patranabis, A. Chakraborty, D. Mukhopadhyay, Fault tolerant infective countermeasure for AES, in *Security, Privacy, and Applied Cryptography Engineering - 5th International Conference, SPACE 2015, Jaipur, India, October 3–7, 2015, Proceedings* (2015), pp. 190–209. https://doi.org/10.1007/978-3-319-24126-5_12

[GST12] B. Gierlichs, J. Schmidt, M. Tunstall, Infective computation and dummy rounds: fault protection for block ciphers without check-before-output, in *Progress in Cryptology - LATINCRYPT 2012 - 2nd International Conference on Cryptology and Information Security in Latin America, Santiago, Chile, October 7–10, 2012. Proceedings*, pp. 305–321. https://doi.org/10.1007/978-3-642-33481-8_17

[BBK+12] A. Barenghi, L. Breveglieri, I. Koren, D. Naccache, Fault injection attacks on cryptographic devices: theory, practice, and countermeasures. Proc. IEEE **100**(11), 3056–3076 (2012). https://doi.org/10.1109/JPROC.2012.2188769

[PQ03] G. Piret, J. Quisquater, A differential fault attack technique against SPN structures, with application to the AES and KHAZAD, in *Cryptographic Hardware and Embedded Systems - CHES 2003, 5th International Workshop, Cologne, Germany, September 8–10, 2003, Proceedings* (2003), pp. 77–88. https://doi.org/10.1007/978-3-540-45238-6_7

[SH13] L. Song, L. Hu, Differential fault attack on the PRINCE block cipher, in *Lightweight Cryptography for Security and Privacy - Second International Workshop, LightSec 2013, Gebze, Turkey, May 6–7, 2013*, Revised Selected Papers (2013), pp. 43–54. https://doi.org/10.1007/978-3-642-40392-7_4

[BS90] E. Biham, A. Shamir, Differential cryptanalysis of des-like cryptosystems, in *Advances in Cryptology - CRYPTO '90, 10th Annual International Cryptology Conference, Santa Barbara, California, USA, August 11–15, 1990, Proceedings* (1990), pp. 2–21. https://doi.org/10.1007/3-540-38424-3_1

[Cla07] C. Clavier, Secret external encodings do not prevent transient fault analysis, in *Cryptographic Hardware and Embedded Systems - CHES 2007, 9th International Workshop, Vienna, Austria, September 10–13, 2007, Proceedings* (2007), pp. 181–194. https://doi.org/10.1007/978-3-540-74735-2_13

[KPB+17] S.V.D. Kumar, S. Patranabis, J. Breier, D. Mukhopadhyay, S. Bhasin, A. Chattopadhyay, A. Baksi, A practical fault attack on arx-like ciphers with a case study on chacha **20**, 33–40 (2017). https://doi.org/10.1109/FDTC.2017.14

[KKG03] R. Karri, G. Kuznetsov, M. Gössel, Parity-based concurrent error detection of substitution-permutation network block ciphers, in *Cryptographic Hardware and Embedded Systems - CHES 2003, 5th International Workshop, Cologne, Germany, September 8–10, 2003, Proceedings* (2003), pp. 113–124. https://doi.org/10.1007/978-3-540-45238-6_10

[KKT05] K. Kulikowski, M. Karpovsky, A. Taubin, Robust codes for fault attack resistant cryptographic hardware, in *Fault Diagnosis and Tolerance in Cryptography, 2nd International Workshop* (Citeseer, 2005), pp. 1–12

[BKK+20'2] A. Baksi, V.B. Kumar, B. Karmakar, S. Bhasin, D. Saha, A. Chattopadhyay, A novel duplication based countermeasure to statistical ineffective fault analysis, in *Australasian Conference on Information Security and Privacy* (2020), https://link.springer.com/chapter/10.1007/978-3-030-55304-3_27

[KQ07] C.H. Kim, J.-J. Quisquater, Fault attacks for CRT based RSA: new attacks, new results, and new countermeasures, in *Information Security Theory and Practices. Smart Cards, Mobile and Ubiquitous Computing Systems*, eds. by D. Sauveron, K. Markantonakis, A. Bilas, J.-J. Quisquater (Springer, Berlin, 2007), pp. 215–228. ISBN: 978-3-540-72354-7

[Ott04] M. Otto, Fault Attacks and Countermeasures. Ph.D. thesis, Institut für Informatik, Universität Paderborn (2004), http://citeseerx.ist.psu.edu/viewdoc/download?doi=10.1.1.133.5710&rep=rep1&type=pdf

[SSS+03] S.-M. Yen, S. Kim, S. Lim, S.-J. Moon, Rsa speedup with chinese remainder theorem immune against hardware fault cryptanalysis. IEEE Trans. Comput. **52**(4), 461–472 (2003–04). ISSN: 2326-3814

[YK04] S.-M. Yen, D. Kim, Cryptanalysis of two protocols for RSA with CRT based on fault infection, in *Fault Diagnosis and Tolerance in Cryptography, Third International Workshop, FDTC* (2004)

[JC05] M. Joye, M. Ciet, Practical fault countermeasures for chinese remaindering based Rsa, in *Fault Diagnosis and Tolerance in Cryptography, Third International Workshop, FDTC* (2005)

[BO06] J. Blömer, M. Otto, Wagner's attack on a secure CRT-RSA algorithm reconsidered, in *Fault Diagnosis and Tolerance in Cryptography, Third International Workshop, FDTC 2006, Yokohama, Japan, October 10, 2006, Proceedings* (2006), pp. 13–23. https://doi.org/10.1007/118897005C_2

[JMR07] M. Joye, P. Manet, J.-B. Rigaud, Strengthening hardware aes implementations against fault attacks. IET Inf. Secur. **1**(3), 106 (2007)

[FRB+11] J. Fournier, J.-B. Rigaud, S. Bouquet, B. Robisson, A. Tria, J.-M. Dutertre, M. Agoyan, Design and characterisation of an aes chip embedding countermeasures. Int. J. Intell. Eng. Inform. **1**(3/4), 328–347 (2011–12). ISSN: 1758-8715. http://dx.doi.org/10.1504/IJIEI.2011.044101

[GSS+15] S. Ghosh, D. Saha, A. Sengupta, D.R. Chowdhury, Preventing fault attacks using fault randomization with a case study on AES, in *Information Security and Privacy - 20th Australasian Conference, ACISP 2015, Brisbane, QLD, Australia, June 29 - July 1, 2015, Proceedings* (2015), pp. 343–355. https://doi.org/10.1007/978-3-319-19962-7_20

[BB15] S. Banik, A. Bogdanov, Cryptanalysis of two fault countermeasure schemes, in *Progress in Cryptology - INDOCRYPT 2015 - 16th International Conference on Cryptology in India, Bangalore, India, December 6–9, 2015, Proceedings* (2015), pp. 241–252. https://doi.org/10.1007/978-3-319-26617-6_13

[KS07] L. Keliher, J. Sui, Exact maximum expected differential and linear probability for two-round advanced encryption standard. IET Inf. Secur. **1**(2), 53–57 (2007)

[PCM17] S. Patranabis, A. Chakraborty, D. Mukhopadhyay, Fault tolerant infective countermeasure for AES. J. Hardware Syst. Secur. **1**(1), 3–17 (2017). https://doi.org/10.1007/s41635-017-0006-1

[SMC09] D. Saha, D. Mukhopadhyay, D.R. Chowdhury, A diagonal fault attack on the advanced encryption standard. IACR Cryptology ePrint Archive, 2009:581 (2009), http://eprint.iacr.org/2009/581

[KSK14] J. Krämer, A. Stüber, Á. Kiss, On the optimality of differential fault analyses on CLEFIA. IACR Cryptology ePrint Archive, 2014:572 (2014), http://eprint.iacr.org/2014/572

[SHS16] B. Selmke, J. Heyszl, G. Sigl, Attack on a DFA protected AES by simultaneous laser fault injections, in *2016 Workshop on Fault Diagnosis and Tolerance in Cryptography, FDTC 2016, Santa Barbara, CA, USA, August 16, 2016*, pp. 36–46. https://doi.org/10.1109/FDTC.2016.16

[BK06] J. Blömer, V. Krummel, Fault based collision attacks on AES, in *Fault Diagnosis and Tolerance in Cryptography, Third International Workshop, FDTC 2006, Yokohama, Japan, October 10, 2006, Proceedings* (2006), pp. 106–120. https://doi.org/10.1007/118897005C_11

[AMR+18] A. Aghaie, A. Moradi, S. Rasoolzadeh, A.R. Shahmirzadi, F. Schellenberg, T. Schneider, Impeccable circuits. Cryptology ePrint Archive, Report 2018/203 (2018), https://eprint.iacr.org/2018/203

[BLM+19] C. Beierle, G. Leander, A. Moradi, S. Rasoolzadeh, CRAFT: lightweight tweakable block cipher with efficient protection against DFA attacks. IACR Trans. Symmetric Cryptol **2019**(1), 5–45 (2019). https://doi.org/10.13154/tosc.v2019.i1.5-45

[DEK+18] C. Dobraunig, M. Eichlseder, T. Korak, S. Mangard, F. Mendel, R. Primas, Sifa: exploiting ineffective fault inductions on symmetric cryptography. Cryptology ePrint Archive, Report 2018/071 (2018), https://eprint.iacr.org/2018/071

[BBK+10] A. Barenghi, L. Breveglieri, I. Koren, G. Pelosi, F. Regazzoni, Countermeasures against fault attacks on software implemented AES: effectiveness and cost, in *Proceedings of the 5th Workshop on Embedded Systems Security, WESS 2010, Scottsdale, AZ, USA, October 24, 2010*, p. 7. https://doi.org/10.1145/1873548.1873555

[BKH+19] J. Breier, M. Khairallah, X. Hou, Y. Liu, A countermeasure against statistical ineffective fault analysis. Cryptology ePrint Archive, Report 2019/515 (2019), https://eprint.iacr.org/2019/515

[SJR+20] S. Saha, D. Jap, D.B. Roy, A. Chakraborty, S. Bhasin, D. Mukhopadhyay, A framework to counter statistical ineffective fault analysis of block ciphers using domain transformation and error correction. IEEE Trans. Inf. Foren. Secur. **15**, 1905–1919 (2020). https://doi.org/10.1109/TIFS.2019.2952262

[DDE+19] J. Daemen, C. Dobraunig, M. Eichlseder, H. Gross, F. Mendel, R. Primas, Protecting against statistical ineffective fault attacks. Cryptology ePrint Archive, Report 2019/536 (2019), https://eprint.iacr.org/2019/536

[SRM19] A.R. Shahmirzadi, S. Rasoolzadeh, A. Moradi, Impeccable circuits ii. Cryptology ePrint Archive, Report 2019/1369 (2019), https://eprint.iacr.org/2019/1369

[MOP07] S. Mangard, E. Oswald, T. Popp, *Power Analysis Attacks - Revealing the Secrets of Smart Cards* (Springer, Berlin, 2007). ISBN: 978-0-387-30857-9

[DCK+15] D. Dinu, Y.L. Corre, D. Khovratovich, L. Perrin, J. Großschädl, A. Biryukov, Triathlon of lightweight block ciphers for the internet of things. Cryptology ePrint Archive, Report 2015/209 (2015), https://eprint.iacr.org/2015/209

[BG16] A. Battistello, C. Giraud, A note on the security of CHES 2014 symmetric infective countermeasure, in *Constructive Side-Channel Analysis and Secure Design - 7th International Workshop, COSADE 2016, Graz, Austria, April 14–15, 2016*, Revised Selected Papers (2016), pp. 144–159. https://doi.org/10.1007/978-3-319-43283-0_9

[FCL+20] J. Feng, H. Chen, Y. Li, Z. Jiao, W. Xi, A framework for evaluation and analysis on infection countermeasures against fault attacks. IEEE Trans. Inf. Foren. Secur. **15**, 391–406 (2020). ISSN: 1556-6021

Chapter 10
A Novel Duplication-Based Countermeasure to Statistical Ineffective Fault Analysis

The Statistical Ineffective Fault Analysis, SIFA, is a recent addition to the family of fault-based cryptanalysis techniques. SIFA based attack is shown to be formidable and is able to bypass virtually all the conventional fault attack countermeasures. Reported countermeasures to SIFA incur overheads of the order of at least thrice the unprotected cipher. We propose a novel countermeasure that reduces the overhead (compared to all existing countermeasures) as we rely on a simple duplication-based technique. In essence, our countermeasure eliminates the observation that enables the attacker to perform SIFA. The core idea we use here is to choose the encoding for the state bits randomly. In this way, each bit of the state is free from statistical bias, which renders SIFA unusable. Our approach protects against stuck-at faults and also does not rely on any side channel countermeasure. We show the effectiveness of the countermeasure through an open-source gate-level fault attack simulation tool. Our approach is probably the simplest and the most cost-effective.

10.1 Introduction

Fault Attacks (FAs) have been proven to be a powerful new attack vector targeting devices performing cryptographic operations (both as software and hardware) [BBB+20]/Chap. 3. The rapid growth of low-end devices together with reducing cost and barriers to mounting advanced fault attacks is being recognized as a serious concern. This type of attack requires the attacker to force the device to perform outside its designated condition of operation, thus producing incorrect (faulty) output from the cipher operation. The attacker can do this by using a multitude of techniques such as shooting optical pulses, overheating, using hardware Trojan, etc. This faulty output, or even just the information whether the device actually produced a faulty output, can be used by an attacker, to deduce information on the secret state of the

A. Baksi, *Classical and Physical Security of Symmetric Key Cryptographic Algorithms*, Computer Architecture and Design Methodologies,
https://doi.org/10.1007/978-981-16-6522-6_10

underlying cipher and ultimately the secret key. Fault attacks are shown to be powerful enough to compromise the security of a cipher which is considered secure with respect to theoretical cipher evaluation criteria. It is also shown in the literature that a fault attack can be carried out with cheap equipment, thus making this type of attack a serious concern.

The earliest and probably the most common fault attack model in the symmetric key setting is the Differential Fault Attack (DFA) [BS97]/Sect. 3.5.1. In a DFA, the device is allowed to run normally (without a fault) once. Next, the attacker injects a fault that effectively toggles a bit (or few bits) in the cipher execution. The difference between the faulty and the non-faulty output lets the attacker learn information on the secret key.

In contrast to DFA, the Safe Error Attack (SEA) [YJ00, YKL+01, JQY+02]/Sect. 3.5.2 makes use of the cases where the fault injection does not change the output from the non-faulty case. One particular case of SEA, known as Ineffective Fault Attack (IFA) [Cla07], is of interest here. In an IFA, the attacker injects a potential disturbance, but the cases where the disturbance does not effectively change the execution of the cipher. In another direction of fault analysis, statistical information of a variable is observed (Statistical Fault Attack, SFA). The distribution which becomes biased as a result of fault injection can be used [Nah15]/Sect. 3.5.3.

The recently proposed Statistical Ineffective Fault Attack (SIFA) [DEK+18]/Sect. 3.5.4 combines IFA and SFA. Like IFA, SIFA makes use of the cases where a fault injection does not result in a change in the output. Also, similar to SFA, the statistical distribution of bias of a variable caused by the effect of fault is used to recover the variable.

Another class of attacks, known as the Side Channel Attacks (SCAs) [MOP07]/Chap. 4, is also capable of finding information on the secret key from a device running a cipher. Generally, fault attack countermeasures are not capable of inherently protecting against SCAs, hence a separate protection is commonly needed.

On top of being a direct way to mitigate the security of the ciphers as shown in [RAD19'1] and [GPT19], SIFA is also able to bypass duplication based countermeasures. Those duplication-based countermeasures have been proposed to counter DFA. Such countermeasures work by implementing two instances of the same cipher execution, which we call the actual and the redundant computations, following [BSS20]/Chap. 9. Assuming a fault can alter at most one of the executions, it is explicitly detected by the detective countermeasures, whereas infective countermeasures implicitly detect the difference [BSS20]. If a fault injection does not alter the course of non-faulty execution of the cipher, this case is considered as if no fault is injected by the duplication-based countermeasures. Since SIFA utilizes the cases where the fault injection does not alter the normal execution of the cipher, those countermeasures cannot (at least in the current form) protect against SIFA.

Quite recently, four countermeasures dedicated to protect against SIFA have been proposed in the literature (see also Sect. 3.7.6). In one work, Breier et al. [BKH+19] use a triplication of the circuit to correct up to 1-bit error. Since at most 1 bit is assumed to subject to SIFA, the majority of the three will correct the error and hence the attacker will receive the correct (non-faulty) output. In [DDE+19], Daemen et al.

suggest to use an error detection mechanism based on Toffoli gates which follow the reversible computing paradigm. Any successful fault would result in a garbage output and hence will be detected. Saha et al. in [SJR+20] present a combination of masking (which is used as a countermeasure to SCAs, [MOP07, Sect. 9]) and encoding. In the *impeccable circuit II* [SRM19], an error correction facility is introduced (extending the idea of impeccable circuits [AMR+18]) to protect against SIFA.

Given this backdrop, our approach neither does any error correction nor any other expensive technique. Instead, we basically use duplication (and comparison) in a way that removes the bias utilized by SIFA. Hence, our countermeasure is by far the simplest and least expensive. While it may be required to implement the cipher almost completely from scratch in other countermeasures (which can be non-trivial), our countermeasure can be easily implemented to protect any symmetric key cipher.

It has been argued [DEK+18] that the duplication-based countermeasures do not protect against SIFA. Indeed this argument is valid with respect to the countermeasures proposed till date, such as [LRT12]. This also extends to DFA countermeasures proposed even after SIFA is published, e.g., [BLM+19] or [BSS20]. On the other hand, the SIFA countermeasures [BKH+19, DDE+19, SJR+20, SRM19] rely on some form of triplication (including error correcting codes) of the cipher execution possibly with masking. The cost for any such countermeasure is more than thrice the cost of the unprotected cipher. Hence, the community seems to accept the norm that it is not possible to have a SIFA countermeasure with cost less than thrice the cost of the basic (unprotected) implementation of the cipher. Our detailed analysis reveals that, while triplication can protect against SIFA, a sophisticated version of duplication can also do the same. The idea of our protection stems from the existing duplication-based DFA countermeasures [BSS20].

Our Contributions

We extend the idea of duplication to accommodate randomized encoding to destroy statistical bias (which is exploited by SIFA). This is done by choosing an encoding based on a 1-bit random parameter λ. If $\lambda = 0$, both the actual and redundant computations are done as is, and this is referred to as *actual logic*. When $\lambda = 1$, we encode $\forall x$ bits of the state as $(x \oplus 1)$ for both the actual and the redundant computations, we refer to this as *inverted logic*. In other words, we invert the bits (0 is encoded as 1, and 1 is encoded as 0) when $\lambda = 1$. This removes the statistical bias (since $\lambda \xleftarrow{\$} \{0, 1\}$ and is kept hidden from the attacker), thereby forestalling SIFA. This idea can be used atop the duplication-based countermeasures, namely detective and infective ones [BSS20], depending on the security warranted. The proper output from the cipher is given only if there is no difference between the computations of the actual logic and the inverted logic (i.e., the fault is ineffective/no fault is injected). In such a situation, reverse encoding is performed on the actual execution (if $\lambda = 1$), or (if $\lambda = 0$) the output is returned as is. Otherwise (i.e., the fault is effective), necessary steps are taken, such as a random output is produced or the output is suppressed.

Hence, unlike the existing SIFA countermeasures, our solution does not rely on (any form of) error correction. Instead, it simply blocks the attacker to get information on the statistical bias. The attacker is able to see the cases of ineffective faults in our proposed solution, however, this does not help the attacker to gain any extra information as we explore later on.

The proposed simple and low-cost idea can be applied to any symmetric key cipher with minimal changes to the implementation. We do not claim any inherent SCA protection, though, SCA countermeasures can be applied easily. Our approach also has advantages when put in perspective to existing countermeasures. For example, our approach does not increase side channel leakage, which is the case for [BKH+19].

We describe further details in Sect. 10.4 with Fig. 10.2 showing a visual representation of our idea. The security evaluation of it is done by an open source tool used by the authors of [SRM19], with the `PRESENT-80` cipher (similar to [SJR+20]) in Sect. 10.4.3. We subsequently present benchmarking results in Sect. 10.4.4.

10.2 Fault Attack Preliminaries

10.2.1 Differential Fault Attack (DFA)

As mentioned already, DFA is likely the most commonly used FA technique in the symmetric key community. It has been successfully applied against most, if not all, symmetric key ciphers. First, it lets the cipher run as it is (without any fault). It then injects faults at some later round of the cipher. Then it uses the difference of the faulty and non-faulty outputs that works as a variant of the Differential Attack [BS90]/Sect. 2.6.1.

10.2.2 General Countermeasures Against Fault Attacks

In general, the countermeasures against the fault attacks can be classified into three broad categories, as we discuss here.

(1) **To use a specialized device.** This device is separate from the cipher design and dedicated for protection against such attack. Examples include a sensor that detects a potential fault [HBB16].
(2) **To use redundancy.** Commonly, this class of countermeasures duplicates (can be fully or partially) the circuit. After this, a recovery procedure (which dictates what to do in case a fault is sensed) takes place. Based on the recovery procedure, two types of countermeasures exist [BSS20, Sect. 2]/Sect. 9.2. First, in the detection-based countermeasures, the XOR of the non-faulty and faulty is explicitly computed. If this results in zero, the output from one of the computations (the so-called actual computation) is directly made available; otherwise,

the output is suppressed/a garbage or random output is given. In the second category, the infective countermeasures do not explicitly compute the difference. Instead, such countermeasures implicitly sense the presence of a fault [BSS20]/Sect. 3.7.2/Chap. 9). By using a sophisticated mechanism (termed as, infection), either the non-faulty output (in case no fault is sensed) or a random output is given (otherwise).

(3) **To use the communication protocol.** The protocol is used in such a way that the conditions required for a successful fault to happen with low probability [DKM+15, BBB+18]. For example, in order to utilize DFA, the inputs to the cipher have to be unchanged, the probability of which can be reduced by using a suitable protocol.

10.3 Statistical Ineffective Fault Attack (SIFA)

As mentioned earlier, SIFA [DEK+18] is a new type of fault attack which combines the concept of ineffective and statistical faults. In SIFA, the attacker exploits the bias (which is caused by the fault injection) of one/more state bits. Unlike DFA, SIFA does not need the non-faulty output, but at the same time, it requires more fault injections compared to DFA.

Here, we present an example of SIFA for better clarity, more information on SIFA can be found in [DEK+18]. Suppose a device is more prone to bit reset ($1 \rightarrow 0$) than bit set ($0 \rightarrow 1$) due to a fault injection. This can be achieved by, for example, by fixing a particular intensity of the optical fault injection set-up. This results in the bias of bit flip. In other words, the probability that the output will be changed depends on whether the target bit is actually 0 (high probability) or 1 (low probability). This bias where fault does not change the output (i.e., ineffective) can be observed by the attacker statistically. When such a state bit is known, the procedure can be repeated to get information on the other state bits and finally the secret key can be recovered. In one extreme, the probability of bit reset (respectively, set) can be 1, in which case the fault model is known as stuck-at 0 (respectively, stuck-at 1).

10.3.1 Duplication-Based Countermeasures and Need for Specialization

From the types of fault protection given in Sect. 10.2.2, the duplication-based countermeasures are the closest to cipher design and hence are of particular interest (see also Sect. 3.7/Sect. 9.2).

As noted earlier, such countermeasures can be classified into detective and infective [BSS20]/Sect. 3.7.2/Chap. 9. The schematic for the two types are shown in Fig. 10.1. In a detection-based or detective countermeasure; the XOR difference of

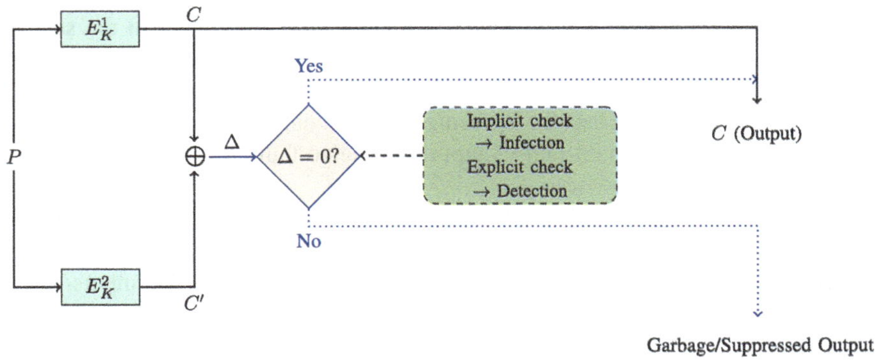

Fig. 10.1 Schematic for detective and infective countermeasures

the actual computation (C from E_K^1) and the redundant computation (C' from E_K^2), denoted by Δ, is explicitly computed (here P is the input and K is the secret key, both are common to both of the computations). If $\Delta = 0$, then no fault is detected and C is given as output. Otherwise, a garbage output (could be random or a pre-determined constant) is given or the output is suppressed. In an infection-based or infective countermeasure, the XOR difference Δ of $C \oplus C'$ is computed. However, instead of taking the 'if–else' decision, Δ is implicitly used to compute $\tau_R(\Delta)$ such that $\tau_R(0) = 0$ but $\tau_R(d) = R' \neq 0$ for $d \neq 0$, where R is randomly generated and hence R' is also random (it is possible that $R' = R$). Thereafter $C \oplus \tau_R(\Delta)$ is given as output. Hence, when fault is sensed, Δ is non-zero and the attacker gets a random output. Otherwise (in case of no fault), the actual output C is returned. The XOR difference can be computed at the end of the cipher execution (such as [LRT12]), or after each round (the basic idea is introduced in [GST12]); depending on which, a further classification of the infective countermeasures is possible, as discussed in [BSS20]/Chap. 9.

Both the detective and the infective countermeasures are suitable to protect against DFA, except for one specific type of DFA. This type is named double fault [BSS20] and is shown practical in [SHS16]. In this case, both the actual and the redundant computations are injected with the identical faults. As a result, the XOR difference is 0 and the countermeasure in place senses it as a case of no fault. Impeccable circuits [AMR+18] attempts to solve the problem by employing different encodings for the two computations and finally with an error detection mechanism. This idea is later extended to a block cipher named `CRAFT` [BLM+19]. Employing such technique can increase the cost (depending on the error-detecting code used). For example, protecting against single-bit faults at the output has 2.45× overhead for `CRAFT` [BLM+19].

However, it may be noted that none of these duplication-based countermeasures is able to protect against SIFA. SIFA only makes use of the cases where the fault injection does not alter the regular flow of the cipher. All the countermeasures, including impeccable circuits/`CRAFT`, treat such a case as no fault. This underlies the need for specialized countermeasures for SIFA, which is described next.

10.3.2 Existing SIFA Countermeasures

To the best of our knowledge, four specialized countermeasures aiming at protecting against SIFA have been proposed in the literature . We describe those here for better clarity.

Repetition Code

Breier et al. propose an error correction based on binary repetition code [BKH+19] and taking majority. Assuming the fault injection can alter at most 1 bit, the error correction will fix it back to its original content. Hence, the attacker will not get any information whether the fault has occurred or not. This actually blocks the attacker's ability to mount SIFA.

Masking and Repetition Code

Saha et al. [SJR+20] propose a two phase countermeasure. The first phase (called, *transform*) is based on masking that aims at protecting faults induced into the state of the cipher. Further, under a stronger attack model where the attacker can inject fault with high precision within the computation of individual sub-operations like SBox,, [SJR+20] proposes an *encoding* which allows error correction. The countermeasure is tested with a LASER-based fault injection experiments and shown to be sound in a practical setting. Depending on the attacker's capability, the overhead limits to just that of a masking or error correction with masking. As an additional benefit, the implemented masking protects the design against side channel attacks as well.

Error Detection through Toffoli Gate and Masking

Daemen et al. [DDE+19] propose an error detection mechanism based on Toffoli gates. This countermeasure acts as a combined SCA and a SIFA that targets at most 1 bit. The non-linear components are designed using Toffoli gates in such a way that a single-bit flip would result in a garbage output. On top, the entire circuit is masked. It may be mentioned that the concept relies on non-standard gates.

Error Correction

Shahmirzadi et al. [SRM19] extend the idea of [AMR+18] to incorporate error correction, and verifies with the open source tool VERFI presented in [AWM+19].[1] The error correction is done through an error correcting code as the authors note shortcoming of repetition with majority voting.

From the discussions, a few basic characteristics of the existing SIFA countermeasures can be noted. Except [DDE+19], the rest (namely, [BKH+19, SJR+20, SRM19]) depends on some form of error correction, thus making the cost of such countermeasure at least triple of the unprotected cipher. Error correction also suffers from the coverage of the underlying error correcting code being used. The concepts of [DDE+19] and [SJR+20] require masking, which is a costly operation. The scheme in [DDE+19] uses detection, but relies on non-standard Toffoli gates. As elaborated in Sect. 10.4, our proposed approach relies on simple duplication. Hence, no customized gate such as Toffoli or costly operation such as masking would be required. This makes our proposal the least expensive in the category.

10.4 Our Proposed Solution

Here, we describe our proposed approach in more detail. As mentioned, the basic idea of our approach is to use duplication in such a way that the attacker does not get any useful information.

As we noted in Sect. 10.3.1, the basic duplication-based countermeasures fail to protect against SIFA. Using an error detecting mechanism is said not to have protection against the same [BLM+19]. Therefore, we use a novel idea that changes the encoding of the bits to its inversion with probability $\frac{1}{2}$ so that the statistical bias is removed. A basic pictorial and algorithmic descriptions are given in Fig. 10.2 and Algorithm 10.1. Figure 10.2 shows a form of quadruplication, though both the branches for $\lambda = 0$ and $\lambda = 1$ are not taken at the same time.

Here, we choose a random bit λ ($\lambda \xleftarrow{\$} \{0, 1\}$). It is regenerated at each invocation and is kept secret from the attacker.

As a part of duplication, we run two instances of the cipher, namely the actual (denoted by E_K^1) and the redundant (denoted by E_K^2), where E denotes the cipher and K is the secret key, and the corresponding outputs are denoted by C and C'. However, depending on λ, we either choose the logic as is (if $\lambda = 0$), or the inverted logic where 0 is encoded as 1 and 1 is encoded as 0 (if $\lambda = 1$). So, the input to E, denoted by P is passed as is if $\lambda = 0$; but as its inversion $\overline{P}$ otherwise. The actual and the redundant computations for the inverted logic are denoted, respectively, by $\overline{E_K^1}$ and $\overline{E_K^2}$, and the corresponding outputs by $\overline{C}$ and $\overline{C'}$.

[1] Source is available at https://github.com/emsec/VerFI.

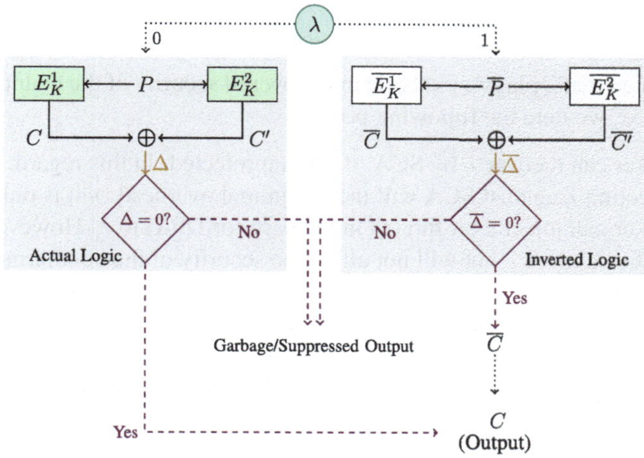

Fig. 10.2 Schematic for our SIFA protection (with detection mechanism)

After this, a recovery mechanism is applied. Here, we adopt detection for the purpose of illustration. Instead of detection, infection could be used if the attacker is powerful enough to flip the 1-bit judgement condition, together with another DFA-type fault at the cipher instance (more details in this regard can be found in [BSS20]). We choose detection as DFA protection is not the focus of this work. In some sense, we also refute the claim made in [DEK+18] that duplication-based countermeasures like detection or infection cannot protect against SIFA.

As a part of the detection mechanism, the XOR difference (Δ for the actual logic and $\overline{\Delta}$ in case of the inverted logic) is computed at the end of the cipher execution. For the actual logic, the output C is given if $\Delta = 0$, otherwise, a garbage output is given or the output is suppressed. Similarly, for the inverted logic, if $\overline{\Delta} = 0$ the actual output C is computed by inverting logic of $\overline{C}$ and given, otherwise a garbage output is given/no output is given.

Hence, the encoding of the actual and the redundant computations are same and is determined by the random bit λ. We assume the attacker can target at most one of the computations. Under our countermeasure, the encoding of that computation changes uniformly. More precisely, the probabilities of bit set and bit reset for one particular bit are both equal. As λ is kept secret, the encoding used is not known to the attacker. Suppose, the probability for bit set and that of bit reset for an (unprotected) cipher are p_0 and p_1, respectively. Because of the random encoding, both the probabilities are now $(p_0 + p_1)/2$. This destroys the statistical property utilized by SIFA, thus rendering it useless.

Security of λ

It may be noted that λ plays a vital role in the overall security of the countermeasure. In this context, we note the following points:

- The attacker can recover λ by SCA, if left unprotected. In this regard, we remark that, protecting λ against SCA will incur minimal overhead as it is only one bit.
- The attacker can inject a bit flip (as in the case for DFA) to λ. However, this will only flip the value of λ but will not affect the security of the countermeasure.
- The attacker can also inject a biased fault. This may lead to biased distribution of λ, and can compromise the security. However, we would like to note that this would result in a second-order SIFA. We do not consider this model within the scope as it is not yet proposed in the literature, to the best of our knowledge.

Stuck-at Fault Protection

For simplicity, consider stuck-at 0 fault only. Regardless of the attacker's ability, half of the time a particular bit will be encoded 0 and rest half of the time as 1. If the attacker is able to inject a stuck-at 0 fault, then half of the times it will result in changed output (the cases where that bit is encoded as 1) and hence be detected (by detective mechanism). Therefore, such cases are not useful to SIFA (as SIFA only makes use of ineffective faults). Rest half of the time, the attacker will know that the fault injection resulted in a stuck-at fault. However, since it is assumed that the attacker does not know whether it is stuck-at 0 or stuck-at 1, such information will not be useful. We thus conclude our proposal can resist against stuck-at based SIFA.

Algorithm 10.1 SIFA Countermeasure (Ours)

Input: P; K
Output: C if no fault (SIFA); suppress output, otherwise

$\lambda \xleftarrow{\$} \{0, 1\}$ ▷ λ is unknown to the attacker
if $\lambda = 0$ **then** ▷ Actual logic
  $\Delta = E_K^1(P) \oplus E_K^2(P)$
  if $\Delta = 0$ **then**
    return $C = E_K^1(P)$
else ▷ Inverted logic
  Compute $\overline{P}$ from P
  $\Delta = \overline{E_K^1}(\overline{P}) \oplus \overline{E_K^2}(\overline{P})$
  if $\overline{\Delta} = 0$ **then**
    $\overline{C} = \overline{E_K^1}(\overline{P})$
    Compute C from $\overline{C}$
    return C

Table 10.1 XOR and AND operations in inverted logic

(a) $\overline{y} = \overline{\mathrm{XOR}}(\overline{x}_0, \overline{x}_1)$

x_0	x_1	$\overline{x}_0$	$\overline{x}_1$	y	$\overline{y}$
0	0	1	1	0	1
0	1	1	0	1	0
1	0	0	1	1	0
1	1	0	0	0	1

(b) $\overline{y} = \overline{\mathrm{AND}}(\overline{x}_0, \overline{x}_1)$

x_0	x_1	$\overline{x}_0$	$\overline{x}_1$	y	$\overline{y}$
0	0	1	1	0	1
0	1	1	0	0	1
1	0	0	1	0	1
1	1	0	0	1	0

10.4.1 Adopting Inverted Logic to Symmetric Key Ciphers

In order to see how the inverted logic works, first we show the inverted XOR ($\overline{\mathrm{XOR}}$) and inverted AND ($\overline{\mathrm{AND}}$) operations in Table 10.1. With this, it is now possible to convert an SBox to its inversion.

Now, we discuss how to implement any symmetric key cipher in the inverted logic. If the circuit is already described in terms of XOR and AND gates (typical for stream ciphers), implementing it in inverted logic should be straightforward. For a typical block cipher, the circuit is described in terms of a linear layer and non-linear layer (such the SBoxes).

Adopting to Linear Layer

Overall, the linear layer (see Sect. 2.1.3) can be classified into three categories— bit permutation as in `PRESENT` [BKL+07], binary non-singular matrix as in `MIDORI` [BBI+15], and non-singular matrix over higher order finite field such as `AES`. However, at a closer look, all the three categories can be described as a multiplication by binary non-singular matrices. In particular, bit permutation corresponds to binary permutation matrices.

Since bit permutation does not have any impact on the inverted logic implementation, we only consider the case with binary matrices. Suppose, we want to implement the matrix $M = (m_{i,j})$ for $i, j = 0, 1, \ldots, n-1$. Considering the system of affine equations, $\vec{y}^{\top} = M\vec{x}^{\top}$ where $\vec{y} = (y_0, \ldots, y_{n-1})$ and $\vec{x} = (x_0, \ldots, x_{n-1})$, we can equivalently write, $y_k = m_{k,0}x_0 \oplus m_{k,1}x_1 \oplus \cdots \oplus m_{k,n-1}x_{n-1}$ for each $k = 0, \ldots, n-1$. Since in the inverted logic each variable x_i will be inverted as well as the entire result will inverted, we deduce

$$\begin{aligned}
\overline{y_k} &= \overline{m_{k,0}\overline{x_0} \oplus m_{k,1}\overline{x_1} \oplus \cdots \oplus m_{k,n-1}\overline{x_{n-1}}} \\
&= m_{k,0}\overline{x_0} \oplus m_{k,1}\overline{x_1} \oplus \cdots \oplus m_{k,n-1}\overline{x_{n-1}} \oplus 1 \\
&= m_{k,0}(x_0 \oplus 1) \oplus m_{k,1}(x_1 \oplus 1) \oplus \cdots \oplus m_{k,n-1}(x_{n-1} \oplus 1) \oplus 1 \\
&= \underbrace{m_{k,0}x_0 \oplus m_{k,1}x_1 \oplus \cdots \oplus m_{k,n-1}x_{n-1}} \oplus \underbrace{m_{k,0} \oplus m_{k,1} \oplus \cdots \oplus m_{k,n-1}} \oplus 1 \\
&= y_k \oplus (\text{parity of } k\text{th row of } M) \oplus 1
\end{aligned}$$

Hence, converting a binary matrix to its inversion is straightforward. For example, the matrix used in `AES MixColumns` (Sect. 2.4.1) can equivalently described as a 32×32 binary matrix. In this case, the parity for each row is 1. Therefore, for `AES MixColumns`, the same source code/hardware description will work for both the original logic and the inverted logic.

Adopting to Non-linear Layer

To begin with, consider the Boolean function, $y = x_0 \oplus x_0x_1$. The inverted function, $\overline{y}$ is given by $\overline{y} = \overline{\overline{x_0} \oplus \overline{x_0}\overline{x_1}} = (1 \oplus x_0) \oplus (1 \oplus x_0)(1 \oplus x_1) \oplus 1 = x_1 \oplus x_0x_1 \oplus 1$.

With this example, now consider the `PRESENT` SBox, `C56B90AD3EF84712` [BKL+07] (see Sect. 2.4.2 for a short description of `PRESENT`), which is represented by the following coordinate functions (in ANF):

$$\begin{aligned}
y_0 &= x_0 \oplus x_2 \oplus x_1x_2 \oplus x_3, \\
y_1 &= x_1 \oplus x_0x_1x_2 \oplus x_3 \oplus x_1x_3 \oplus x_0x_1x_3 \oplus x_2x_3 \oplus x_0x_2x_3, \\
y_2 &= 1 \oplus x_0x_1 \oplus x_2 \oplus x_3 \oplus x_0x_3 \oplus x_1x_3 \oplus x_0x_1x_3 \oplus x_0x_2x_3, \\
y_3 &= 1 \oplus x_0 \oplus x_1 \oplus x_1x_2 \oplus x_0x_1x_2 \oplus x_3 \oplus x_0x_1x_3 \oplus x_0x_2x_3.
\end{aligned}$$

Now the inverted SBox is given by the coordinate functions (in ANF) by (these are obtained by inverting each variable)

$$\begin{aligned}
y_0 &= 1 \oplus x_0 \oplus x_1 \oplus x_1x_2 \oplus x_3, \\
y_1 &= x_0 \oplus x_2 \oplus x_1x_2 \oplus x_0x_1x_2 \oplus x_3 \oplus x_0x_1x_3 \oplus x_0x_2x_3, \\
y_2 &= 1 \oplus x_1 \oplus x_0x_2 \oplus x_3 \oplus x_0x_3 \oplus x_0x_1x_3 \oplus x_2x_3 \oplus x_0x_2x_3, \\
y_3 &= 1 \oplus x_2 \oplus x_0x_1x_2 \oplus x_3 \oplus x_1x_3 \oplus x_0x_1x_3 \oplus x_2x_3 \oplus x_0x_2x_3.
\end{aligned}$$

This leads us to the SBox `DE8B701C25F649A3`. The Algorithm 10.2 shows the implementation of `PRESENT-80` cipher in the inverted logic (which we call $\overline{\texttt{PRESENT}-80}$).

It may be noted that the the SBox in the inverted logic is not same as its inverse, in general. For example, the inverse of the `PRESENT` SBox is `5EF8C12DB463079A`. If an SBox with this property exists, it could be beneficial to reduced cost for a

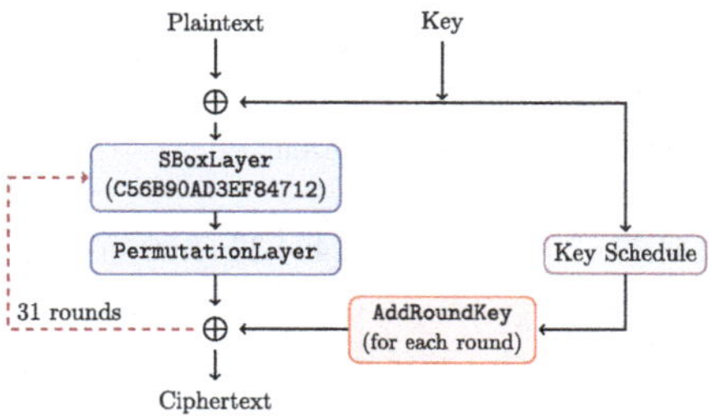

(a) Actual logic (PRESENT-80)

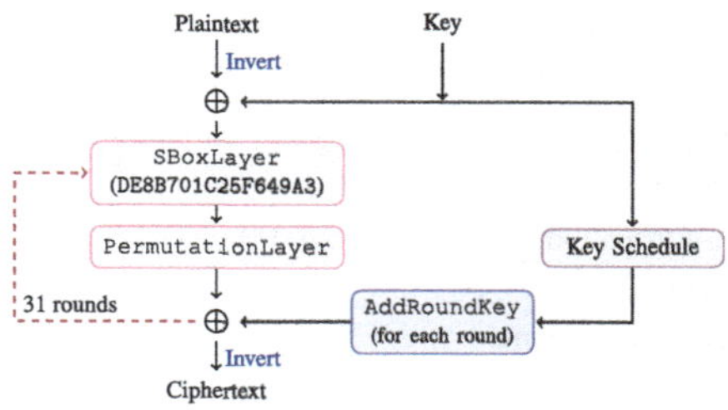

(b) Inverted logic ($\overline{\text{PRESENT-80}}$)

Fig. 10.3 Overview of PRESENT-80 in actual and inverted logic

combined encryption and decryption circuits if our countermeasure is adopted. We leave this problem open for future research.

Effect on the Key Schedule

The key schedule algorithm will not change in the inverted logic. Hence, no extra cost/protection would be necessary. The round key addition operations are done by XOR.

Adopting to PRESENT-80

Now both the linear and the non-linear layers are adopted to the inverted logic, we explain how the overall design would be adopted for PRESENT-80. The key schedule, the round key additions (including the final round key addition at the end) and the bit permutation layer would remain unchanged. The state is inverted before and after the encryption function. Also the SBox is changed in $\overline{\text{PRESENT} - 80}$ as DE8B701C25F649A3. Figure 10.3 gives visual representation for PRESENT-80 in both the actual and the inverted logic.[2]

[2] The non-linear layer, linear layer, and the key addition (respectively, sBoxlayer, pLayer, addRoundKey in Sect. 2.4.2) are described as SBoxLayer, PermutationLayer, and AddRoundKey for better readability.

Table 10.2 Area overhead of the proposed countermeasure

PRESENT-80	Gate equivalent (45nm technology)		
Cipher implementation	Combinational	Non-combinational	Total
Unprotected	1620.00	844.32	2464.32 (1.00×)
Proposed countermeasure	3570.99	2264.66	5835.65 (2.37×)

Algorithm 10.2 Convert PRESENT-80 to inverted logic ($\overline{\texttt{PRESENT-80}}$)

Input: P; K
Output: C
1: Run Key Schedule ▷ No change (from PRESENT-80)
2: Compute $\overline{P}$ ▷ Invert the bits of P
3: $S_t = \overline{P}$ ▷ State is initialized
4: **for** $i \leftarrow 1; i \leq 31; i \leftarrow i + 1$ **do**
5: $S_t \leftarrow \texttt{AddRoundKey}_i(S_t)$ ▷ Add the corresponding round key; No change
6: $S_t \leftarrow \texttt{SBoxLayer}(S_t)$ ▷ Use SBox DE8B701C25F649A3
7: $S_t \leftarrow \texttt{PermutationLayer}(S_t)$ ▷ No change
8: $S_t \leftarrow \texttt{AddRoundKey}_{32}(S_t)$ ▷ Add the last round key; No change
9: $C \leftarrow \overline{S_t}$ ▷ Ciphertext is the inverted state

10.4.2 Benchmarks

Benchmark of the proposed countermeasure is described in terms of PRESENT-80 [BKL+07] cipher (similar to [SJR+20]). Table 10.2 reports hardware benchmark the area (ASIC) overheads of the countermeasure targeting the 45 nm Nangate Open Cell Library[3] PDK[4] v13 library designed using the FreePDK45 kit[5] in terms of NAND2 equivalents. Similar to [BSS20]/Chap. 9, we do not consider the cost for generating randomness.

As for the software benchmark, we note that the inverted implementation can be done with only a little change to the actual cipher implementation. Thus, the code size will only be marginally increased. The time taken would also be basically the same as the number of rounds for the cipher is invariant over actual or inverted implementation.

[3] This is an open-source standard cell library designed for research purpose.

[4] PDK: Process Design Kit.

[5] The FreePDK45 kit is an open-source PDK (i.e., it does not correspond to any real process and cannot be fabricated), from https://github.com/mflowgen/freepdk-45nm.

10.4.3 Evaluation

The evaluation of the proposed countermeasure is done through fault attack simulation on the gate-level netlist of the protected `PRESENT-80` cipher [BKL+07], using VERFI [AWM+19] (the same tool used in [SRM19]).[6] It may be noted that the FA simulation through VERFI considers much finer level granularity, not just the state bits, thereby conforming to the SIFA-2 model used in [SJR+20].

In Fig. 10.4, we show an evaluation of our countermeasure through the simulation tool VERFI when biased faults are applied at random locations at the 30th round. The data shown in both Fig. 10.4a (unprotected, naïve duplication is used) and Fig. 10.4b (protected by our countermeasure) are collected over a simulation of 120000 runs for `PRESENT-80`, for the cases where the fault is ineffective (therefore the naïve duplication-based countermeasure would treat those cases as no fault).

It can be noted that the Fig. 10.4a and b provide similar probability, except for the target SBox : 1 in Fig. 10.4a. The naïve duplication based countermeasure cannot protect against this attack, since a statistical bias (i.e., the 16 inputs do not have equal probability) can be observed in that SBox output. This information is then utilized by the attacker.

When our countermeasure is applied, the statistical bias (which is caused by the fault) is removed, as seen in Fig. 10.4b. Thus, we conclude our countermeasure is capable of removing the statistical bias, stopping the attacker learning any information.

More specifically, it can be observed from Fig. 10.4a that these inputs to the `PRESENT` SBox): `0`, `1`, `2`, `7`, `9`, `a`, `c`, and `d` appear with 0-probability, while the rest appear with (around) $\frac{2}{16}$ probability. This is caused by SIFA through a stuck-at 1 fault at the second MSB during computation of the SBox. In other words, only the outputs from the SBox with second MSB as 1 are allowed: `c`, `5`, `6`, `d`, `e`, `f`, `4` and `7`. With our countermeasure, all the inputs have (almost) equal probability of $\frac{1}{16}$.

10.4.4 Comparison with Existing Countermeasures

The hardware cost of protecting the 3×3 SBox χ of XOODOO [DHP+18] by the SIFA countermeasures proposed in [BKH+19, DDE+19, SJR+20, SRM19] using two 65nm ASIC technologies, namely UMC (uk65lscllmvbbh_120c25_t) and Faraday (fse0k_d generic_core_ss1p08v125c) are given in Table 10.3. The authors in [SRM19] use $[7, 4, 3]_2$ and $[11, 4, 5]_2$ codes. Since χ is of three bits, we use a $[6, 3, 3]_2$ code with the generator matrix:

$$\begin{bmatrix} 0&1&1&1&0&0 \\ 1&0&1&0&1&0 \\ 1&1&0&0&0&1 \end{bmatrix}$$

[6] Our source code can be found at https://github.com/vinayby/VerFI.

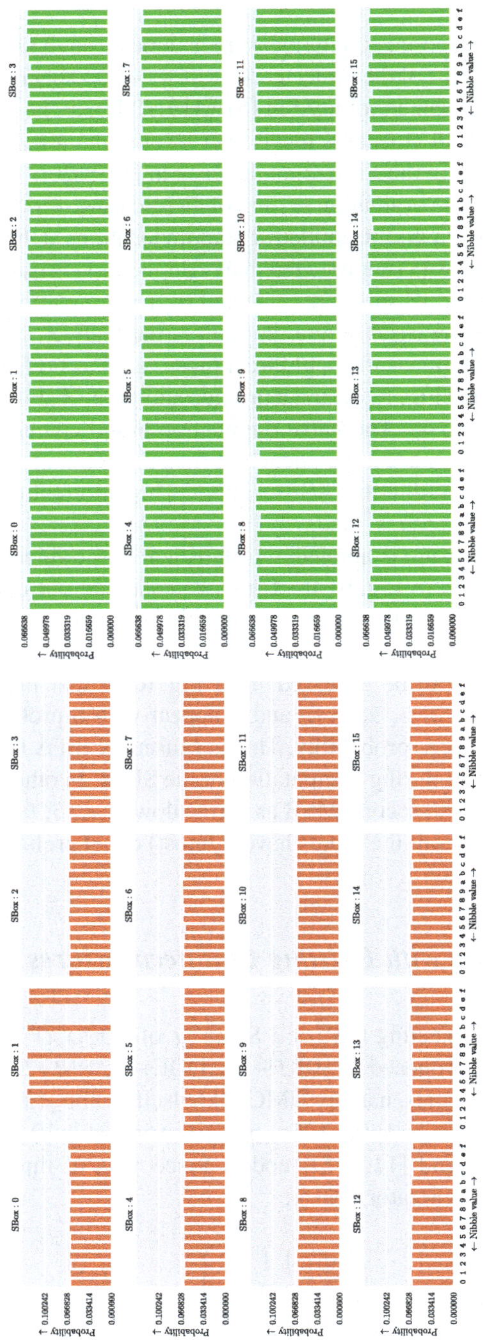

(a) Without our countermeasure (naïve duplication)

(b) With our countermeasure

Fig. 10.4 Evaluation of our countermeasure through VERFI simulation

Table 10.3 Comparison of cost of protecting χ (of XOODOO) by SIFA countermeasures

	Gate Equivalent				Method
	UMC (65nm)		Faraday (65nm)		
χ (unprotected) [DHP+18]	14.58	(1.0×)	14.06	(1.0×)	–
Breier et al. [BKH+19]	45.14	(3.1×)	42.97	(3.1×)	Error correction
Daemen et al. [DDE+19]	No standard cell library exists				Reversible computing
Saha et al. [SJR+20]	>45.14	(>3.1×)	>42.97	(>3.1×)	Masking + Error correction
Shahmirzadi et al. [SRM19]	71.53	(4.9×)	68.75	(4.9×)	Error correction
Ours	27.77	(1.9×)	25.78	(1.8×)	Remove bias by duplication

All in all, we conclude that our solution has the least overhead compared to the rest of the countermeasures.[7] Further, the evaluation is done through gate-level simulation and hence is comparable to SIFA-2 of [SJR+20] (which is a stronger SIFA model).

The proposal of [DDE+19] inherently requires masking (thus also protects against SCA), similar is the case for [SJR+20]. From a different angle, only [DDE+19] relies on error detection, while the rest rely on some form of error correction (hence the cost is at least tripled). Also, [DDE+19] needs Toffoli gates which is not available in the standard gate libraries, to the best of our knowledge. For this reason, we do not provide any benchmark for the protection proposed in [DDE+19]. The cost for applying [SJR+20] protection is at least that of [BKH+19]. In comparison, our proposal simply depends on duplication with randomized bit encoding. It does not require masking or specialized gate, nor it is restricted by the error coverage of the underlying code. In essence, the inverted logic-based cipher can be thought of as a cipher (which could also share components with the actual logic based cipher). Also, our countermeasure works at the cipher design level (thus it has an edge when adopting to any symmetric key cipher), whereas other countermeasures work at the implementation level.

[7] The error correction based techniques offer the same level of security to our error detection-based solution.

10.4.5 Connection with Side Channel Countermeasures

Our proposal does not have any inherent side channel protection. We mention that side channel countermeasures can be easily adopted as essentially the inverted logic-based implementation works like a cipher. Hence, no special technique would be necessary.

As already mentioned, the randomly generated bit λ is needed to be protected from a side channel attacker, aside from the usual protection for the actual and redundant computations. This would add minimal cost since λ is only of 1-bit.

It is to be mentioned that our proposal does not inherently increase side channel leakage (which typically happens for [BKH+19]). In fact, the basic concept used here is somewhat similar to that of the *Masked Dual-Rail Pre-charge Logic* which is proposed as a countermeasure to the side channel attacks in 2005 [PM05]. However, this method is shown not secure in [SS06, DGP+09].

Since the countermeasure uses two different SBoxes (actual and inverted logic), each SBox can depict minor differences in their side channel leakage (for example, operating in two memory locations may result in distinct time signatures). If such a model is considered within scope, we propose to combine the two SBoxes (say, $n \times n$) to one combined ($n \times 2n$) SBox. For example, the combined SBox in case of `PRESENT-80` would be the 4×8 SBox: `CD`, `5E`, `68`, `BB`, `97`, `00`, `A1`, `DC`, `32`, `E5`, `FF`, `86`, `44`, `79`, `1A`, `23`. Once this SBox has been fetched from memory, either the most significant n bits or the least significant n bits are obtained by masking the unnecessary part.

One idea to perform SCA is to check if there is any inversion of state at either the beginning or the end (since this operation is done only for the inverted logic). To prevent that, we propose to compute dummy inversion of the state both at the beginning and at the end regardless of the value of λ, but then choose the inverted state only when $\lambda = 1$.

10.5 Conclusion

In this work, we present a duplication based SIFA countermeasure. The basic idea is to use randomized encoding for the state bits. This removes the statistical bias caused by ineffective faults. Consequently, the attacker cannot mount SIFA. Our countermeasure by far is the least expensive. More precisely, ours is the only one in the category to have the overhead cost of less than thrice that of an unprotected cipher. The overhead cost (like throughput, latency, operating frequency) would be practically same as that of duplication (there will be a trade-off such parameters depending on how the duplication is performed). The verification is done through a gate-level SIFA simulation tool. Our countermeasure, being at the cipher design level, is almost readily adoptable to any symmetric key cipher (even when some other countermeasure, such as side channel, is needed).

In the future scope, one may consider a combined SIFA and side channel countermeasure atop our design. Also, our design does not protect against double fault (see Sect. 10.3.1/Sect. 3.7) since both the actual and redundant computations are using the same encoding. Hence, one may think of extending our work to protect against double fault. Another interesting problem could be to search for an SBox whose inverse (in actual logic) is same as itself in inverted logic.

References

[Nah15] P.S. Nahid Farhady Ghalaty Bilgiday Yuce, Analyzing the efficiency of biasedfault based attacks. Cryptology ePrint Archive, Report 2015/663 (2015), https://eprint.iacr.org/2015/663

[DEK+18] C. Dobraunig, M. Eichlseder, T. Korak, S. Mangard, F. Mendel, R. Primas, Sifa: exploiting ineffective fault inductions on symmetric cryptography. Cryptology ePrint Archive, Report 2018/071 (2018), https://eprint.iacr.org/2018/071

[BSS20] A. Baksi, D. Saha, S. Sarkar, To infect or not to infect: a critical analysis of infective countermeasures in fault attacks. J. Cryptograp. Eng. (2020), https://link.springer.com/article/10.1007

[AMR+18] A. Aghaie, A. Moradi, S. Rasoolzadeh, A.R. Shahmirzadi, F. Schellenberg, T. Schneider, Impeccable circuits. Cryptology ePrint Archive, Report 2018/203 (2018), https://eprint.iacr.org/2018/203

[LRT12] V. Lomné, T. Roche, A. Thillard, On the need of randomness in fault attack countermeasures - application to AES, in *2012 Workshop on Fault Diagnosis and Tolerance in Cryptography, Leuven, Belgium*, September 9, 2012, pp. 85-94. https://doi.org/10.1109/FDTC.2012.19

[BLM+19] C. Beierle, G. Leander, A. Moradi, S. Rasoolzadeh, CRAFT: lightweight tweakable block cipher with efficient protection against DFA attacks. IACR Trans. Symmetric Cryptol. **2019**(1), 5–45 (2019). https://doi.org/10.13154/tosc.v2019.i1.5-45

[HBB16] W. He, J. Breier, S. Bhasin, Cheap and cheerful: a low-cost digital sensor for detecting laser fault injection attacks, in *Security, Privacy, and Applied Cryptography Engineering - 6th International Conference, SPACE 2016, Hyderabad, India, December 14–18, 2016, Proceedings* (2016), pp. 27–46. https://doi.org/10.1007/978-3-319-49445-6_2

[DKM+15] C. Dobraunig, F. Koeune, S. Mangard, F. Mendel, F. Standaert, Towards fresh and hybrid re-keying schemes with beyond birthday security, in *Smart Card Research and Advanced Applications - 14th International Conference, CARDIS 2015, Bochum, Germany, November 4–6, 2015*. Revised Selected Papers (2015), pp. 225–241. https://doi.org/10.1007/978-3-319-31271-2_14

[BBB+18] A. Baksi, S. Bhasin, J. Breier, M. Khairallah, T. Peyrin, Protecting block ciphers against differential fault attacks without re-keying (extended version). Cryptology ePrint Archive, Report 2018/085 (2018), https://eprint.iacr.org/2018/085

[SHS16] B. Selmke, J. Heyszl, G. Sigl, Attack on a DFA protected AES by simultaneous laser fault injections, in *2016 Workshop on Fault Diagnosis and Tolerance in Cryptography, FDTC 2016, Santa Barbara, CA, USA, August 16, 2016*, pp. 36–46. https://doi.org/10.1109/FDTC.2016.16

[AWM+19] V. Arribas, F. Wegener, A. Moradi, S. Nikova, Cryptographic fault diagnosis using verfi. IACR Cryptology ePrint Archive, 2019:1312 (2019), https://eprint.iacr.org/2019/1312

[BKL+07] A. Bogdanov, L.R. Knudsen, G. Leander, C. Paar, A. Poschmann, M.J. Robshaw, Y. Seurin, C. Vikkelsoe, PRESENT: an ultra-lightweight block cipher, in *CHES*, vol. 4727 (Springer, Berlin, 2007), pp. 450–466

[DHP+18] J. Daemen, S. Hoffert, M. Peeters, G.V. Assche, R.V. Keer, Xoodoo cookbook. Cryptology ePrint Archive, Report 2018/767 (2018), https://eprint.iacr.org/2018/767

[PM05] T. Popp, S. Mangard, Masked dual-rail pre-charge logic: dpa-resistance without routing constraints, in *Cryptographic Hardware and Embedded Systems - CHES 2005* eds. by J.R. Rao, B. Sunar (Springer, Berlin, 2005), pp. 172–186. ISBN: 978-3-540-31940-5

[DGP+09] E. De Mulder, B. Gierlichs, B. Preneel, I. Verbauwhede, Practical dpa attacks on mdpl, in *2009 First IEEE International Workshop on Information Forensics and Security (WIFS)* (2009–12), pp. 191–195

[BS97] E. Biham, A. Shamir, Differential fault analysis of secret key cryptosystems, in *Advances in Cryptology - CRYPTO '97* ed. by B.S. Kaliski (Springer, Berlin, 1997), pp. 513–525. ISBN: 978-3-540-69528-8

[JQY+02] M. Joye, J. Quisquater, S. Yen, M. Yung. Observability analysis - detecting when improved cryptosystems fail, in *Topics in Cryptology - CT-RSA 2002, The Cryptographer's Track at the RSA Conference, 2002, San Jose, CA, USA, February 18–22, 2002, Proceedings* (2002), pp. 17–29. https://doi.org/10.1007/3-540-45760-75C_2

[RAD19'1] K. Ramezanpour, P. Ampadu, W. Diehl, A statistical fault analysis methodology for the ascon authenticated cipher, in *IEEE International Symposium on Hardware Oriented Security and Trust, HOST 2019, McLean, VA, USA, May 5–10, 2019*, pp. 41–50. https://doi.org/10.1109/HST.2019.8741029

[BKH+19] J. Breier, M. Khairallah, X. Hou, Y. Liu, A countermeasure against statistical ineffective fault analysis. Cryptology ePrint Archive, Report 2019/515 (2019), https://eprint.iacr.org/2019/515

[DDE+19] J. Daemen, C. Dobraunig, M. Eichlseder, H. Gross, F. Mendel, R. Primas, Protecting against statistical ineffective fault attacks. Cryptology ePrint Archive, Report 2019/536 (2019), https://eprint.iacr.org/2019/536

[SJR+20] S. Saha, D. Jap, D.B. Roy, A. Chakraborty, S. Bhasin, D. Mukhopadhyay, A framework to counter statistical ineffective fault analysis of block ciphers using domain transformation and error correction. IEEE Trans. Inf. Foren. Sec. **15**, 1905–1919 (2020). https://doi.org/10.1109/TIFS.2019.2952262

[SRM19] A.R. Shahmirzadi, S. Rasoolzadeh, A. Moradi, Impeccable circuits ii. Cryptology ePrint Archive, Report 2019/1369 (2019), https://eprint.iacr.org/2019/1369

[SS06] D. Suzuki, M. Saeki, Security evaluation of dpa countermeasures using dualrail precharge logic style, in *Cryptographic Hardware and Embedded Systems - CHES 2006* eds. by L. Goubin, M. Matsui (Springer, Berlin, 2006), pp. 255–269. ISBN: 978-3-540-46561-4

[YJ00] S. Yen, M. Joye, Checking before output may not be enough against faultbased cryptanalysis. IEEE Trans. Comput. **49**(9), 967–970 (2000). https://doi.org/10.1109/12.869328

[YKL+01] S. Yen, S. Kim, S. Lim, S. Moon, A countermeasure against one physical cryptanalysis may benefit another attack, in *Information Security and Cryptology - ICISC 2001, 4th International Conference Seoul, Korea, December 6-7, 2001, Proceedings* (2001), pp. 414–427. https://doi.org/10.1007/3-540-45861-15C_31

[Cla07] C. Clavier, Secret external encodings do not prevent transient fault analysis, in *Cryptographic Hardware and Embedded Systems - CHES 2007, 9th International Workshop, Vienna, Austria, September 10–13, 2007, Proceedings* (2007), pp. 181–194. https://doi.org/10.1007/978-3-540-74735-2_13

[MOP07] S. Mangard, E. Oswald, T. Popp, *Power Analysis Attacks - Revealing the Secrets of Smart Cards* (Springer, Berlin, 2007). ISBN: 978-0-387-30857-9

[GPT19] M. Gruber, M. Probst, M. Tempelmeier, Statistical ineffective fault analysis of GIMLI (2019). arXiv:1911.03212

[BS90] E. Biham, A. Shamir, Differential cryptanalysis of des-like cryptosystems, in *Advances in Cryptology - CRYPTO '90, 10th Annual International Cryptology Conference, Santa Barbara, California, USA, August 11–15, 1990, Proceedings* (1990), pp. 2–21. https://doi.org/10.1007/3-540-38424-3_1

[GST12] B. Gierlichs, J. Schmidt, M. Tunstall, Infective computation and dummy rounds: fault protection for block ciphers without check-before-output, in *Progress in Cryptology - LATINCRYPT 2012 - 2nd International Conference on Cryptology and Information Security in Latin America, Santiago, Chile, October 7–10, 2012. Proceedings* (2012), pp. 305–321. https://doi.org/10.1007/978-3-642-33481-8_17

[BBI+15] S. Banik, A. Bogdanov, T. Isobe, K. Shibutani, H. Hiwatari, T. Akishita, F. Regazzoni, Midori: a block cipher for low energy (extended version). Cryptology ePrint Archive, Report 2015/1142 (2015), https://eprint.iacr.org/2015/1142

[BBB+20] A. Baksi, S. Bhasin, J. Breier, D. Jap, D. Saha, Fault attacks in symmetric key cryptosystems (2020), https://eprint.iacr.org/2020/1267

Chapter 11
Concluding Remarks

11.1 Synopsis

Motivation and Background

This book covers a diverse set of topics in symmetric key cryptography. Chapter 1 gives a concise motivation of the overall topic and lays the foundation for the rest. Next, Chap. 2 glances over the basic information on cipher construction and how to analyse with classical methods with adequate clarity. Here the attacker can only access the input–output information. However, with the widespread use of low-end devices that perform cryptographic operations, the attacker can employ two different attack models. Such models take the environmental factors into consideration (this is possible as those devices are commonly deployed within the reach of the attacker). The first such model we discuss here is called fault attack. Under this model, the attacker can actively alter the regular flow of cipher execution. This is discussed in Chap. 3, thereby providing sufficient background for the subsequent chapters which deal with fault countermeasures. The second model is the side channel attack, under which the attacker passively observes the external characteristics of the device to deduce information about the secret key. This is covered in Chap. 4, but is kept relatively short (since it is not the main focus for the research contribution). Though Chaps. 2, 3 and 4 contain a few previously unreported results (and point out a few potential research problems), the primary objective is to provide the theoretical backing for the research contribution.

Research Contribution

The research contribution in this document can be conceptually split into classical cryptanalysis (Chaps. 5, 6) and fault countermeasures (Chaps. 7, 8, 9 and 10). In the

A. Baksi, *Classical and Physical Security of Symmetric Key Cryptographic Algorithms*, Computer Architecture and Design Methodologies,
https://doi.org/10.1007/978-981-16-6522-6_11

classical cryptanalysis category, we visit the concept of formulating certain classical attacks in terms of MILP instances (Chap. 5). In the process, we discover a problem in a well-known modelling that may lead to incorrect results. To patch it up, we propose our own modelling (this modelling is later used in Chap. 8), and experiment to reduce the solution time with various heuristics. Chapter 6 attempts to convert the problem of finding some non-randomness property of a cipher to a multi-class classification problem, which is then fed to a machine learning model. This methodology is the first-of-its-kind, and it typically reduces the search complexity to cube root of what would otherwise be expected. In terms of fault countermeasures, a total of three directions are explored. Our first target is the differential fault attack (DFA). Although the model of DFA is quite similar to the classical differential attack (DA)—which is easy to protect against; it remains elusive to design a countermeasure that can protect against both. It turns out to be somewhat paradoxical (Chap. 7), since something that makes DA easier makes DFA harder and vice versa. Ultimately, we look for the middle ground, but it calls for a paradigm shift in cipher design. By going against the well-established design paradigm, we propose a lightweight cipher named `DEFAULT` (Chap. 8). One of the main claim of `DEFAULT` is that, it does not rely on duplication of any form (nor does it rely on any device/protocol level assumption) to protect against DFA, which is also the first-of-its-kind. The most common countermeasures against DFA needs a duplicated execution of the cipher. The infective countermeasure is considered the most prominent in this category, which we explore next (Chap. 9). We do a systematic analysis of the infective countermeasures (that also partly helps us in Chap. 10), propose new low-overhead schemes, discover a problem with an existing scheme, and patch another scheme which was considered broken. The final contribution (Chap. 10) comes in the form of a duplication based countermeasure to the statistical ineffective fault attack (SIFA). Our countermeasure improves from the previously known countermeasures as those require triplication at minimum (our countermeasure offers same security but has lower overhead).

We are optimistic that our works have contributed non-trivially to improve the state of the art in symmetric key cryptography. We expect it will be useful to the future researchers as a reference/motivation.

Open Problems

- Characterize SBoxes for which at least 1 zero transition in DDT/LAT is included in the corresponding convex hull created with non-zero transitions (Sect. 5.3).
- Find an $n \times n$ SBox with DU $= 2^{n-1}$ or prove it does not exist (Remark 7.5).
- Disprove or find an example to $\mathrm{MinF}_S \geq 3$ (Remark 7.7).

Publicly Available Material

- (Chap. 4) `CHASKEY` side channel trace from software implementations
- (Chap. 8) `DEFAULT` reference code
- (Chap. 10) SIFA countermeasure simulation

11.2 Interesting Problems for Future Research

While potential problems for research specific to each work are already discussed in respective chapters, here we point out some of the general directions that may be of interest to future researchers.

Generalization of SCA Resistance of XOR to SBox

As observed in Sect. 4.3.2, the XOR operation creates an ambiguity with respect to CPA. The generalized case for an SBox, under the Hamming weight model with noise-free scenario, can be described as follows. For an $n \times n$ SBox S, if S^{-1} has a linear structure (Definition 2.5) at $\mathbf{1}$, the correlation coefficients corresponding to $S(\alpha)$ and $S(\alpha \oplus \Phi)$ will be of equal magnitude (but with opposite signs) $\forall\alpha$, where $\Phi = S^{-1}(x) \oplus S^{-1}(x \oplus \mathbf{1})$ $\forall x$ (assuming noise-free scenario). Put it in a different way, S has a linear structure at Φ such that the corresponding output difference is at $\mathbf{1}$ for all input differences. As a consequence, it will not be possible to uniquely identify which between α and $\alpha \oplus \Phi$ is the correct input to the SBox, for any α. For example, consider the 4×4 SBox $S =$ `126CDE39F58BA047` (see Table 2.2 for its DDT). The inverse SBox, S^{-1} has a linear structure at `f` (=$\mathbf{1}$) and $\Phi =$ `5`. Hence, it can be shown that, the correlation coefficients corresponding to and $S(\alpha)$ and $S(\texttt{5} \oplus \alpha)$ will have equal magnitude but opposite signs, $\forall\alpha \in \{\texttt{0}, \texttt{1}, \ldots, \texttt{f}\}$. This may be useful to design a (somewhat) side channel resistant cipher, at least against the Hamming weight (software) model.

Cipher with an SBox with One Non-zero LS SBox to Match `GIFT-128`

In general, the SBoxes with non-zero linear structure(s) show some interesting properties (despite being bad against the differential attack); like somewhat DFA resistance (Chap. 7), somewhat SCA resistance (Sect. 11.2), low AND count (this makes adopting a side channel countermeasure easier), and so on. In theory, it may be possible to design a cipher with one non-zero LS such that it is able to match (or even improve) the efficiency of `GIFT-128` while offering full (128-bit) classical security. Notice that, the SBox of `DEFAULT-LAYER` (Chap. 8) uses four LS and takes (roughly) same number of rounds as `GIFT-128` to offer square root (64-bit) classical security (discussed in Sect. 8.5). With two LS, it can be expected that the overall construction would behave comparable to GIFT-128 with respect to the classical attacks. The MILP modelling in Chap. 5 can be used in this case to find differential and linear bounds. The same bit permutation as GIFT-128 can be used to take advantage of an improved software implementation [ANP20].

Extending Coverage for MILP Modelling

The MILP modelling in Chap. 5 can be further extended to cover other ciphers like SKINNY [BJK+16] or AES which use a binary non-singular matrix (without loss of generality) as the linear layers (currently this modelling only supports bit permutation-based linear layer). Further, the division property [Tod15] can be searched using MILP modelling, that can be considered in the scope as well.

Extension of Machine Learning-Based Distinguisher

The machine learning-based distinguisher (Chap. 5) can be improved, for example, using an automated tool that can efficiently find a suitable architecture. Extension of the current differential model is also a possibility; such as using another differential on decryption (implementing the so-called *boomerang attack* [Wag99]). Further, such models can be integrated in the cipher design phase, so that a newly designed cipher can have security claim against such attacks.

Others

- One may be interested in a different modelling of the impossible differential attack than that of [ST17'2] so that input or output differences with Hamming weight >1 can be found.
- It would be interesting to see if full DFA protection can be achieved, which is currently not possible with DEFAULT like construction (noted in Sect. 8.8).
- SCA on stream ciphers is not that common. One recent work [SJB20] proposes an SPA, though, there are scopes to improve this model. Most notably, it suffers from the following two problems. First, the attack procedure is not automated (one has to manually sift through leakage information). Second, it has no resilience against wrong prediction (if one prediction about Hamming weight is wrong, the entire system will likely become inconsistent).

References

[ANP20] A. Adomnicai, Z. Najm, T. Peyrin, Fixslicing: a new gift representation. Cryptology ePrint Archive, Report 2020/412 (2020), https://eprint.iacr.org/2020/412

[Tod15] Y. Todo, Structural evaluation by generalized integral property, in *Advances in Cryptology - EUROCRYPT 2015 - 34th Annual International Conference on the Theory and Applications of Cryptographic Techniques, Sofia, Bulgaria, April 26-30, 2015, Proceedings, Part I* (2015), pp. 287–314. https://doi.org/10.1007/978-3-662-46800-5_12

[Wag99] D. Wagner, The boomerang attack, in *Fast Software Encryption* ed. by L. Knudsen (Springer, Berlin, 1999), pp. 156–170. ISBN: 978-3-540-48519-3

[ST17'2] Y. Sasaki, Y. Todo, New impossible differential search tool from design and cryptanalysis aspects - revealing structural properties of several ciphers, in *Advances in Cryptology - EUROCRYPT 2017 - 36th Annual International Conference on the Theory and Applications of Cryptographic Techniques, Paris, France, April 30–May 4, 2017, Proceedings, Part III* (2017), pp. 185–215. https://doi.org/10.1007/978-3-319-56617-75C_7

[SJB20] S.M. Sim, D. Jap, S. Bhasin, Dapa: differential analysis aided power attack on (non-)linear feedback shift registers (extended version). Cryptology ePrint Archive, Report 2020/1241 (2020), https://eprint.iacr.org/2020/1241

[BJK+16] C. Beierle, J. Jean, S. Kölbl, G. Leander, A. Moradi, T. Peyrin, Y. Sasaki, P. Sasdrich, S.M. Sim, The SKINNY family of block ciphers and its lowlatency variant MANTIS. IACR Cryptology ePrint Archive (2016), p. 660, http://eprint.iacr.org/2016/660

Index

A

Add–rotation–xor. *see* construction/arx. *see* construction/arx

Advanced encryption standard. *see* cipher/aes

Ae. *see* sbox/ae

Aead. *see* cipher/aead

Aes. *see* cipher/aes

Afa. *see* attack/fault/afa

Affine equivalence. *see* sbox/ae

Algebraic attack. *see* attack/algebraic

Algebraic fault attack. *see* attack/fault/afa

Algebraic normal form. *see* boolean function/anf

Almost perfect non-linear. *see* boolean function/apn

Anf. *see* boolean function/anf

Api, 10, 49, 50, 121, 142

Apn. *see* boolean function/apn

Application programming interface. *see* api

Application specific integrated circuit. *see* device/asic

Arx. *see* construction/arx

Ascon. *see* cipher/ascon. *see* cipher/ascon

Asic. *see* device/asic

Associated encryption with assoicated data. *see* cipher/aead

Attack, 7, 8, 29, 38, 40, 42, 82, 142, 173, 177, 179, 180, 182, 185, 186, 198, 201, 206, 208–210, 277, 280

- algebraic, 44, 46, 74, 182, 203
- boomerang, 280
- cpa, 100, 103
- cube tester, 46
- dfa, 249
- differential, 29, 38, 42, 46, 72, 110–112, 115, 117, 118, 121, 122, 126, 133, 142–144, 146, 154, 157, 158, 160, 164, 170, 173–175, 178, 184, 187, 193, 195, 196, 198–201, 204, 205, 208, 213, 220, 223, 258, 278–280
  - medp, 43, 204
- dpa, 103
- fault, 5, 7–9, 49, 59–62, 66–68, 71, 73, 74, 82, 163, 164, 167, 169, 171–174, 177, 178, 182, 184, 185, 199, 200, 203, 209, 210, 213, 217, 220, 222, 226–228, 230, 241, 243, 249, 255, 256, 258, 259, 261, 263, 264, 269, 277
  - afa, 63, 72, 74, 75, 81, 82
  - cfa, 63, 76, 77, 227, 245
  - dfa, 60–63, 69–77, 80–82, 85–87, 163, 164, 167, 170, 171, 173–175, 178–187, 191, 194, 196, 198–200, 203, 205, 206, 208, 209, 213, 217, 218, 220–222, 224, 226–230, 233, 239, 242, 245, 249, 256–260, 264, 278–280
  - dfia, 78–80
  - double. *see* countermeasure/double fault
  - fima, 68, 80, 87
  - fsa, 78–80
  - hard, 62, 220, 228, 259
  - idfa, 63, 75
  - ifa, 77, 80, 213, 227, 228, 249, 256
  - impossible differential, 175
  - indfa, 61–63, 76, 210
  - ineffective, 62, 69, 80, 86, 182, 183, 227, 228, 256, 258, 259, 264, 269, 272

A. Baksi, *Classical and Physical Security of Symmetric Key Cryptographic Algorithms*, Computer Architecture and Design Methodologies, https://doi.org/10.1007/978-981-16-6522-6

instruction skip, 61, 62, 86, 182, 219, 220, 246, 247, 249
integral, 63, 75, 203
lfa, 75
linear, 63
multiple, 61, 81, 84, 200, 209
nueva, 78–80
nufva, 78
pfa, 62, 68, 80
precise bit, 60, 181, 183, 212
sea, 77, 213, 256
sfa, 79, 80, 256
sifa, 68, 80, 87, 88, 183, 212, 213, 228, 249, 256, 257, 259–264, 269, 271, 272
single, 60, 84, 180–182, 210
statistical, 183
stuck-at, 78, 87, 182, 183, 220, 228, 259, 264, 269
transient, 62, 183
impossible differential, 110, 182, 202, 280
integral, 46, 182, 203
division, 46, 203
superpoly, 46
linear, 38, 42, 44, 110, 111, 113, 117, 121, 122, 126, 128, 133, 182, 187, 193, 198, 201, 204, 205, 223, 280
melp, 44, 204, 205
piling-up, 44, 205
side channel, 5, 7, 49, 78, 79, 82, 86, 99–101, 142, 180, 203, 204, 229, 233, 245, 249, 256, 257, 261, 264, 271, 272, 277, 279, 280
cpa, 101, 104, 107, 279
dpa, 102
em, 99–101, 104–106
power, 99–101, 104–106
spa, 102, 280
template, 103
time, 29, 99, 272

B

Benchmark. *see* device/benchmark
Bit permutation, 208
Black box, 49, 59
Block cipher. *see* cipher/block
Boolean function, 13, 19, 203, 266
affine, 13, 15, 46, 169, 170
anf, 13, 188, 189, 266
apn, 16, 164
cnf, 13, 82
component function, 13, 15
coordinate function, 13, 15, 170, 189, 194, 266
degree, 13, 18, 44, 46, 194, 203
derivative, 43, 46, 167, 231, 234
non-linearity, 14, 15, 18, 163, 171, 194, 201
satisfiability. *see* sat
vectorial, 13

C

Cfa. *see* attack/fault/cfa
Chaskey. *see* cipher/chaskey
Cipher
3-des, 76
acorn, 25, 62, 74, 75
aead, 19, 25, 26, 44, 48, 62, 74, 80
aegis, 25, 48, 74, 75
aes, 14, 19, 25, 26, 30, 48, 61, 62, 70, 73, 75, 79, 179, 180, 184, 185, 191, 222, 223, 225, 226, 230, 231, 236, 237, 247, 248, 265, 266, 280
ascon, 80, 143, 145, 149, 156, 157, 160
block, 19, 20, 24–28, 33, 35, 36, 62, 64, 71–75, 77, 82, 163, 164, 178–180, 182, 185–187, 189, 203, 204, 208, 209, 249, 260, 265
mode, 20, 21, 62, 73
padding, 20
tweakable, 20, 85, 87
chacha, 62, 73
chaskey, 30, 36, 100, 104, 143, 157, 160, 279
clefia, 72, 226
cloc, 74
clx, 44
craft, 65, 84, 85, 180, 182, 183, 200, 227, 228, 260–262
default, 179, 180, 182, 186–191, 194, 197–203, 205–213, 279, 280
des, 72
gift, 14, 26, 30, 35, 36, 110, 111, 113–115, 123, 126, 128, 133, 145, 161, 178, 180, 185, 187–191, 193, 194, 198, 202, 203, 206–211, 213, 279
gimli, 80, 143–145, 147, 148, 154, 155, 158–160
gorstl, 73
grain, 22, 74
hash, 19, 22, 24, 26, 28, 73, 74
hash function, 73, 74
hc-128, 72

idea, 76
keeloq, 6
knot, 143, 149, 156, 157, 160
kuznyechik, 77
led, 20, 74, 79
lizard, 74
mac, 19, 24, 36, 73, 100, 104, 105, 143
markov, 144, 145, 147
mibs, 72
midori, 70, 71, 185, 265
minalpher, 74
mode, 19, 24, 36, 73, 77, 236
non-markov, 145, 154, 160
paeq, 76
permutation. *see* permutation
piccolo, 72, 74
present, 14, 19, 26, 30, 33–35, 70, 71, 79, 111, 115, 126, 128, 185, 191, 193, 194, 197, 258, 265–269, 272
prince, 20, 70
pyjamask, 203
rc4, 6, 7, 73
salsa, 73
sha-1, 7, 73
sha-3, 73, 74
shacal1, 73
silc, 74
simon, 30, 188
siphash, 36
skinny, 20, 185, 188, 191, 194, 280
sms4, 191
speck, 30, 146, 154
sprout, 74
stream, 19, 21, 22, 44, 46, 72–75, 147, 249, 280
fsr, 21, 22, 25, 44, 74, 150
streebog, 74
tiaoxin, 74, 75
trivium, 61, 73, 74
twine, 72
twofish, 70, 72
xoodoo, 269
Classical attack. *see* attack/classical
Cmos, 99, 100
Cnf. *see* boolean function/cnf
Collision fault attack. *see* attack/fault/cfa
Competition, 54
aes, 54
caesar, 54, 74–76, 80, 149
dpa, 54, 102
estream, 54, 73
lwc, 48, 54, 143, 147, 149
sha-3, 54, 73
Complementary metal oxide semiconductor. *see* cmos
Component function. *see* boolean function/component function
Conjunctive normal form. *see* boolean function/cnf
Construction, 71, 277
arx, 29, 36
feistel, 27–29, 72, 179, 191, 238, 249
monkey duplex, 148, 149
spn, 26, 29, 30, 35, 64, 70–72, 121, 149, 163, 169, 170, 179, 185, 189, 191, 194, 203, 208, 237
sponge, 147
Convolutional neural network. *see* ml/ann/cnn
Coordinate function. *see* boolean function/coordinate function
Correlation power analysis. *see* attack/side channel/cpa
Countermeasure, 7, 9, 60, 73, 82, 85, 87, 178, 180–182, 203, 204, 209–211, 213, 228, 242, 243, 245, 249, 257, 258, 260–264, 268, 269, 271–273, 277–279
detection, 60, 65, 83–85, 87, 206, 212, 220, 221, 226, 228, 241, 242, 258–260, 263, 264
double fault, 84, 182, 200, 209, 211, 227, 228, 249, 260, 273
duplication, 278
fault, 259
hiding, 103, 229
impeccable circuit, 227, 260, 261
impeccable circuit ii, 257
infection, 60, 83–87, 178, 180–182, 200, 206, 209, 212, 218, 219, 221–230, 240–243, 246, 247, 249, 256, 259, 260, 263, 278
cipher level. *see* countermeasure/infection/type ii
type i, 219, 224, 226, 227, 230, 235, 236, 238, 244, 247, 249
type ii, 219, 224, 225, 237, 239, 249
masking, 85, 86, 103, 203, 228, 229, 233, 261, 271, 272
prevention, 83, 84, 86
rekeying, 83, 86, 87
sifa, 88, 183, 228, 256–258, 261, 262, 269, 272, 273, 279
tweak, 83
tweak-in-plaintext, 83, 87
Cpa. *see* attack/side channel/cpa

Cryptanalysis. *see* attack
Cube attack. *see* attack/integral
Cube tester. *see* attack/cube tester

D

Da. *see* attack/differential
Dbn. *see* sbox/dbn
Dc. *see* attack/differential
Ddt. *see* sbox/ddt
Default. *see* cipher/default
Detection. *see* countermeasure/detection
Detective countermeaure. *see* countermeasure/detection
Device, 268
- asic, 5, 47, 48, 268, 269
- benchmark, 48, 111, 126, 158, 180, 206, 211, 230, 235, 247, 258, 268, 271
  - hardware, 206, 237, 248
  - software, 207, 236, 247, 268
- fpga, 5, 206, 237
- hardware, 29, 82, 102, 189, 266
- lightweight, 33, 47, 111, 180, 191, 209, 230, 234, 249
- microcontroller, 5, 24, 36, 48, 62, 82, 85, 86, 100, 104–106, 182, 207, 208, 219, 221, 236, 237, 247
- microprocessor, 5, 126, 158, 205, 207
- software, 29, 82, 102, 104, 208, 266, 279, 280

Dfa. *see* attack/fault/dfa
Dfia. *see* attack/fault/dfia
Difference distribution table. *see* sbox/ddt
Differential attack. *see* attack/differential
Differential branch number. *see* sbox/dbn
Differential fault attack. *see* attack/fault/dfa
Differential fault intensity analysis. *see* attack/fault/dfia
Differential power analysis. *see* attack/side channel/dpa
Differential uniformity. *see* sbox/du
Double fault. *see* countermeasure/double fault
Dpa. *see* attack/side channel/dpa
Du. *see* sbox/du

E

Electromagnetic leakage. *see* attack/side channel/em
Expandanded ddt. *see* sbox/expanded ddt

F

Fa. *see* attack/fault
Fair evaluation of lightweight cryptographic systems. *see* tool/felics
Fault. *see* attack/fault
Fault intensity map analysis. *see* attack/fault/fima
Fault sensitivity analysis. *see* attack/fault/fsa
Feedback shift register. *see* cipher/stream/fsr
Feistel. *see* construction/feistel
Field programmable gate array. *see* device/fpga
Fima. *see* attack/fault/fima
Fsa. *see* attack/fault/fsa

G

Gift. *see* cipher/gift
Gimli. *see* cipher/gimli
Grey box, 49
Gurobi. *see* tool/gurobi

H

Hamming distance, 14, 100, 102
Hamming weight, 17, 79, 100, 102, 195, 196, 202, 279, 280
Hard fault. *see* attack/fault/hard
Hardware benchmark. *see* device/benchmark/hardw
Hash. *see* cipher/hash

I

Ifa. *see* attack/fault/ifa
Impeccable circuit. *see* countermeasure/impeccable circuit
Impeccable circuit ii. *see* countermeasure/impeccable circuit ii
Implementation. *see* device
Impossible differential fault attack. *see* attack/fault/idfa
Indfa. *see* attack/fault/indfa
Ineffective fault analysis. *see* attack/fault/ifa
Infection. *see* countermeasure/infection
Infective countermeaure. *see* countermeasure/infection
Instruction skip. *see* attack/fault/instruction skip
Integral attack. *see* attack/integral
Internal differential fault attack. *see* attack/fault/indfa

K

Key schedule, 20, 27, 31, 34, 36, 62, 72, 188–190, 202, 211, 267
Knot. *see* cipher/knot

L

La. *see* attack/linear
Lat. *see* sbox/lat
Lc. *see* attack/linear
Lfa. *see* attack/fault/lfa
Lighter. *see* tool/lighter
Lightweight cryptography. *see* device/lightweight
Lightweight cryptography competition. *see* competition/lwc
Linear approximation table. *see* sbox/lat
Linear attack. *see* attack/linear
Linear fault attack. *see* attack/fault/lfa
Linear layer, 19, 27, 31, 35, 70, 71, 112, 123, 149, 179, 184, 191, 197, 265, 267, 280
  bit permutation, 34, 35, 70–72, 123, 187, 188, 191, 193, 202, 204, 207, 209, 210, 265, 267, 280
  matrix (binary), 33, 48, 265, 266, 280
  matrix (general), 32, 70, 265, 266
Linear structure. *see* sbox/ls
Long short-term memory network. *see* ml/ann/lstm
Ls. *see* sbox/ls
Lwc. *see* competition/lwc

M

Mac. *see* cipher/mac
Machine learning. *see* ml
Maximum expected differential probability. *see* attack/differential/medp
Maximum expected linear probability. *see* attack/linear/melp
Medp. *see* attack/differential/medp
Melp. *see* attack/linear/melp
Message authentication code. *see* cipher/mac. *see* cipher/mac
Microcontroller. *see* device/microcontroller
Milp, 8, 10, 46, 50, 110–118, 122–126, 131–133, 142, 144, 194, 196, 197, 200–202, 204, 205, 213, 278, 280
Minf. *see* sbox/minf
Mixed integer linear programming. *see* milp
Ml, 8, 51, 81, 142–144, 149–155, 158, 160, 213, 278, 280
  ann, 52, 53, 146, 147, 160
    cnn, 158
    lstm, 158, 159
    mlp, 52, 143, 155, 158, 159
  svm, 160
Monkey duplex. *see* construction/monkey duplex

N

Neural network. *see* ml/ann
Non-uniform error value analysis. *see* attack/fault/nueva
Non-uniform fault value analysis. *see* attack/fault/nufva
Nueva. *see* attack/fault/nueva
Nufva. *see* attack/fault/nufva

P

Permutation, 19, 25, 29, 34, 36, 43, 44, 147, 151, 153, 169, 209
Persistent fault attack. *see* attack/fault/pfa
Piling-up. *see* attack/linear/piling-up
Power leakage. *see* side channel/power
Precise bit flip. *see* attack/fault/precise bit
Present. *see* cipher/present

S

Safe error analysis. *see* attack/fault/sea
Sage. *see* tool/sage
Sat, 73, 74, 81, 82, 154
Satisfiability. *see* sat
Satisfiability modulo theory. *see* smt
Sbox, 14, 16–19, 26, 27, 31, 34, 35, 43, 44, 48, 60, 70–72, 77, 80, 82, 88, 101, 110–119, 121, 122, 124, 128, 131–133, 145, 149, 150, 163–165, 167, 169–175, 179, 184, 185, 187–189, 191, 193–199, 201–206, 208–212, 224, 248, 261, 265–267, 269, 272, 273, 279
  ae, 18, 19, 34, 117, 167, 194, 196, 197
  algebraic degree. *see* boolean function/degree
  apn. *see* boolean function/apn
  ch, 110–121, 123, 124, 126, 131, 132, 205
  component function. *see* boolean function/component function
  coordinate function. *see* boolean function/coordinate function
  dbn, 16, 18, 113, 142, 194, 195

ddt, 15, 18, 43, 112–117, 119, 121–124, 126, 145, 146, 164, 167, 173, 175, 195, 279
du, 15, 16, 18, 43, 172, 175, 185, 194
expanded ddt, 164, 165, 174
lat, 16–18, 44, 112, 116, 117, 121, 124, 126, 128, 195, 196
lbn, 17, 113, 194–196, 201, 202
ls, 16, 163, 167, 169–175, 179, 185, 187, 188, 194–201, 204, 205, 208, 209, 212, 279
minf, 165, 167, 173
non-linearity. *see* boolean function/non-linearity
Sca. *see* attack/side channel
Sea. *see* attack/fault/sea
Side channel. *see* attack/side channel
Sifa. *see* attack/fault/sifa
Sifa countermeasure. *see* countermeasure/sifa
Signal to noise ratio. *see* snr
Simple power analysis. *see* attack/side channel/spa
Smt, 82, 154
Snr, 103, 229
Software benchmark. *see* device/benchmark/software
Solvatore. *see* tool/solvatore
Spa. *see* attack/side channel/spa
Spn. *see* construction/spn
Statistical ineffective fault attack. *see* attack/fault/sifa
Stream cipher. *see* cipher/stream
Substitution box. *see* sbox
Substitution permutation network. *see* construction/spn

T

Tool
felics, 49, 236, 237, 247
gurobi, 10, 111, 116, 123, 126, 131, 133, 205
keras, 142
lighter, 48
sage, 10, 45, 117, 121, 203
sat, 74
solvatore, 203
tensorflow, 142
verfi, 262, 269
Type i infective countermeasure. *see* countermeasure/infection/type i
Type ii infective countermeasure. *see* countermeasure/infection/type ii

W

White box, 49

The manufacturer's authorised representative in the EU is Springer Nature Customer Service Centre GmbH, Europaplatz 3, 69115 Heidelberg, Germany. If you have any concerns regarding our products, please contact ProductSafety@springernature.com

Printed and bound by CPI Group (UK) Ltd, Croydon, CR0 4YY

30/09/2026

02225291-0008